Structural Integrity

Volume 25

The *Structural Integrity* book series is a high level academic and professional series publishing research on all areas of Structural Integrity. It promotes and expedites the dissemination of new research results and tutorial views in the structural integrity field.

The Series publishes research monographs, professional books, handbooks, edited volumes and textbooks with worldwide distribution to engineers, researchers, educators, professionals and libraries.

Topics of interested include but are not limited to:

- Structural integrity
- Structural durability
- Degradation and conservation of materials and structures
- Dynamic and seismic structural analysis
- Fatigue and fracture of materials and structures
- Risk analysis and safety of materials and structural mechanics
- Fracture Mechanics
- Damage mechanics
- Analytical and numerical simulation of materials and structures
- Computational mechanics
- Structural design methodology
- Experimental methods applied to structural integrity
- Multiaxial fatigue and complex loading effects of materials and structures
- Fatigue corrosion analysis
- Scale effects in the fatigue analysis of materials and structures
- Fatigue structural integrity
- Structural integrity in railway and highway systems
- Sustainable structural design
- Structural loads characterization
- Structural health monitoring
- Adhesives connections integrity
- Rock and soil structural integrity.

**** Indexing: The books of this series are submitted to Web of Science, Scopus, Google Scholar and Springerlink ****

This series is managed by team members of the ESIS/TC12 technical committee.

Springer and the Series Editors welcome book ideas from authors. Potential authors who wish to submit a book proposal should contact Dr. Mayra Castro, Senior Editor, Springer (Heidelberg), e-mail: mayra.castro@springer.com

More information about this series at https://link.springer.com/bookseries/15775

Grzegorz Lesiuk · Mieczyslaw Szata ·
Wojciech Blazejewski · Abílio M. P. de Jesus ·
José A. F. O. Correia
Editors

Structural Integrity and Fatigue Failure Analysis

Experimental, Theoretical and Numerical Approaches

Springer

Editors
Grzegorz Lesiuk
Faculty of Mechanical Engineering
Wrocław University of Science
and Technology
Wrocław, Poland

Mieczyslaw Szata
Faculty of Mechanical Engineering
Wroclaw University of Science
and Technology
Wrocław, Poland

Wojciech Blazejewski
Faculty of Mechanical Engineering
Wroclaw University of Science
and Technology
Wrocław, Poland

Abílio M. P. de Jesus
INEGI, Department of Mechanical
Engineering, Faculty of Engineering
University of Porto
Porto, Portugal

José A. F. O. Correia
CONSTRUCT & INEGI, Department
of Civil Engineering
Faculty of Engineering
University of Porto
Porto, Portugal

ISSN 2522-560X ISSN 2522-5618 (electronic)
Structural Integrity
ISBN 978-3-030-91846-0 ISBN 978-3-030-91847-7 (eBook)
https://doi.org/10.1007/978-3-030-91847-7

This Springer imprint is published by the registered company Springer Nature Switzerland AG
The registered company address is: Gewerbestrasse 11, 6330 Cham, Switzerland

Preface

This volume contains full papers presented at the first Virtual Conference on Mechanical Fatigue (VCMF 2020), which was organised by the University of Porto (FEUP, Portugal), the Wroclaw University of Science and Technology (Poland), University of Electronic Science and Technology of China (China), Siberian Federal University (Russia), and the ESIS/TC12 Technical Committee (European Structural Integrity Society—ESIS), between 9 and 11 September 2020. This conference was intended to be a forum of discussion of new research concepts, equipment, technology, materials and structures and other scientific advances within the field of mechanical fatigue and fracture. The first edition of the VCMF 2020 event has reached more than 60 participants from more than 20 nationalities demonstrating the vitality of this new event.

This volume of the Springer Structural Integrity book series gathers 34 works presented in the first edition of the VCMF 2020 event which were distributed along five chapters covering the following topics: cyclic plasticity and internal structure; ultra-low, low-, highand giga-cycle fatigue; mechanisms of fatigue damage; fatigue thresholds; short and long crack growth; fatigue life prediction; fatigue behaviour modelling and simulation; creep-fatigue interactions; probabilistic fatigue and fracture; failure analysis and case studies; fatigue life extension; damage evaluation and fatigue design; variable amplitude loading; thermal fatigue; multiaxial fatigue; fatigue in biomaterials; high-temperature fatigue; fatigue-corrosion; environmental assisted fatigue; influence of manufacturing processes on fatigue behaviour; structural integrity assessments accounting for fatigue; and applications and design codes (e.g. pressure vessels, metallic bridges, wind towers, offshore structures, etc.).

The Organizing Committee of the VCMF 2020 deeply acknowledges all authors that contributed to the success of this event and particularly submitted their works to this book series. The members of the Advisory and Scientific Committees are also fully acknowledged for their support. Finally, a word of appreciation for the Organizing Committee members as well as students and other FEUP/INEGI/IC staff for their tireless support. Also, Springer is fully acknowledged for their support to the Structural Integrity Book Series.

Wrocław, Poland — Grzegorz Lesiuk
Wrocław, Poland — Mieczyslaw Szata
Wrocław, Poland — Wojciech Blazejewski
Porto, Portugal — Abílio M. P. de Jesus
Porto, Portugal — José A. F. O. Correia
September 2021

Contents

Part I
Experimental Characterization & Microstructural Aspects of Fatigue and Fracture

Chapter 1
Experimental Verification of the Survivability Model Under Mixed I+II Mode Fracture for Steels of Rolling Rolls

S. V. Belodedenko, V. I. Hanush, and O. M. Hrechanyi

Abstract The model of fatigue mixed fracture, which follows from the amalgamation rule of resource safety indices, has been experimentally confirmed. It does not require finding an effective SIF, but is based on the construction of survivability curves for pure modes. Survivability in a mixed process is defined as the result of combining individual fracture processes. Peculiarities of steels of rolling rolls fracture under conditions of cyclic deformation according to the scheme of four-point asymmetric bending are revealed. Although such a test scheme is designed to obtain the fracture of the II mode, in existing studies, this has not been achieved. The crack from the notch, practically, immediately went away at an angle ~45° toward the tensile zone, which indicates a mixed I+II fracture. With sufficient accuracy to predict survivability, the exponent of the Paris's region of the fatigue crack growth diagram (FCG) can be taken as $n_{\mathrm{I}} = 4$ (I mode), $n_{\mathrm{II}} = 3$ (II mode). As a parameter of schematization, it is possible to use the value of SIF at an FCG rate of 10^{-7} m/cycle. Its value for steels of rolling rolls can be 60–70 MPa$\sqrt{\mathrm{m}}$ in mode I. For mode II, this index decreases by 2.5–3 times.

Keywords Mixed fracture · Steels of rolling rolls · Survivability curve

1.1 Introduction

For a considerable of time, the rolling roll operates in a damaged state, when it is slowly destroyed. This process has a mixed character, when at the same time the defect (crack) is affected by deformations of the I, II and III fracture modes. Among them, II mode plays an important role. This type of fracture acts both during the initiation of pits and lamination (spalling) of contact fatigue in the work places of the rolls, as well as on the development of defects arising in the end (neck) zones of the rolls. The contribution from the destruction of the II mode increases with the

S. V. Belodedenko (✉) · V. I. Hanush · O. M. Hrechanyi
Department "Machines and Units of Metallurgical Production", National Metallurgical Academy of Ukraine, Dnipro, Ukraine

G. Lesiuk et al. (eds.), *Structural Integrity and Fatigue Failure Analysis*, Structural Integrity 25, https://doi.org/10.1007/978-3-030-91847-7_1

reduction of the length of the rolls. This situation is typical for pipe mill rolls and continuous casting machine rollers.

When studying the rules of amalgamating reliability indicators, an algorithm was developed to determine the overall level of safety of the object when exposed to it of degradation processes system [1]. This algorithm was used to predict the survivability of rolling rolls under the action of mixed fracture. The latter is presented as the action of a system of degradation processes [2]. Based on the analysis of the resistance of materials to fracture, a calculation model of the parameters of the kinetic diagram of crack growth (CG) for I, II and III modes was proposed. Although the model is put forward at the level of a hypothesis, the results of its use are quite acceptable.

Therefore, the paper is devoted to the experimental study of cyclic fracture processes in the II mode and the mixed mode. The aim of the research was:

- verification of the calculated model of resistance to fatigue fracture, steels used for the production of rolling rolls, and
- verification of the algorithm for predicting survivability under mixed fracture.

1.2 General Regularities of Metals Behavior at the II Mode

Until about the mid-1970s of the last century, the study of fracture in the II mode was not actively conducted [3]. It was considered that the critical SIF for this mode of fracture $K_{\mathrm{II}c}$ is much larger than the value $K_{\mathrm{I}c}$, so the destruction of the II mode will not have time to develop before the violation of the integrity of the structure due to the I mode. A significant impetus in the study of mode II is associated with the conclusion that this type of deformation is responsible for the initiation of fracture on contact (pitting, fretting) [4, 5]. It was found that the defects develop at much lower contact pressures than those considered acceptable. This is a consequence of the fact that the threshold SIFs in the II mode for steels K_{IIth} are 1.7–3.5 times lower than the same SIFs in the I mode K_{Ith} [5]. At the same time, there are experimental results where the inequality $K_{\mathrm{IIth}} > K_{\mathrm{Ith}}$ is stably observed [6].

Regarding critical SIFs, it was found that in some cases the value of $K_{\mathrm{II}C}$ is 30% lower than $K_{\mathrm{I}C}$, and the critical crack in the II mode can be twice as short as in the I mode [7]. The extension of research in the field of shear resistance to the class of non-metals (including elastic-brittle materials) expanded the understanding of the relationship between the parameters of fracture models in modes I and II. Such relations follow from the diagrams of limit conditions for mixed fracture. According to them, the effective SIF K_{eff} is determined by the individual acting K_{I}, K_{II} and K_{III}. The simplest circular form of the mixed fracture diagram, which was proposed by Irwin in 1957, follows the equality $K_{\mathrm{I}C} = K_{\mathrm{II}C}$. According to one of the most commonly used criteria of maximum tangential stresses (MTS), it turns out that $K_{\mathrm{I}C}/K_{\mathrm{II}C} = \alpha_1 = 1.155$. This relationship appears in the famous 3D-Richard's criteria [8]:

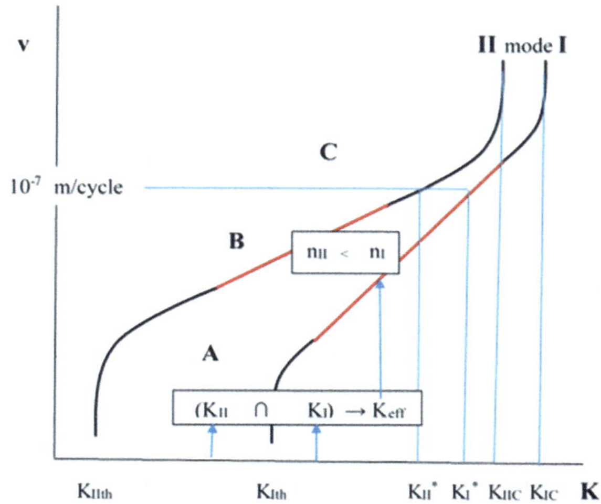

Fig. 1.1 Scheme of fatigue crack growth diagrams for I and II fracture modes (it is conditionally shown that the critical SIF under static load K_{ic} coincides with the cyclic critical SIF K_{ifc})

$$K_{\mathrm{eff}} = \frac{K_{\mathrm{I}}}{2} + \frac{1}{2}\sqrt{K_{\mathrm{I}}^2 + 4(\alpha_1 K_{\mathrm{II}})^2 + 4K_{\mathrm{III}}^2}. \tag{1.1}$$

Despite the fact that there are several different forms of criteria for mixed fracture, most of them correspond well with each other, especially if the I mode dominates [9, 10]. In general, the ratio of fracture toughness $K_{\mathrm{II}C}/K_{\mathrm{I}C}$ can be both greater and less than one and is in the range: $(\sqrt{3}/2) < (K_{\mathrm{II}C}/K_{\mathrm{I}C}) < \sqrt{3}$.

To predict the survivability or period of crack growth to its critical level, it is necessary to have a model (diagram) of fatigue crack growth (FCG), which relates the rate of FCG v with the current SIF. One of its important parameters is the indicator of the degree or slope of the diagram n in the middle section (B, Fig. 1.1).

If we take into account that the ratio K_{IIth} / K_{Ith} varies in the range, the size of which is practically an order of magnitude larger than the size of the range of change of the ratio $K_{\mathrm{II}C}/K_{\mathrm{I}C}$ (ie, $(K_{\mathrm{IIth}}/K_{\mathrm{Ith}}) \gg (K_{\mathrm{II}C}/K_{\mathrm{I}C})$), then the inequality follows from this: $n_{\mathrm{I}} > n_{\mathrm{II}}$ (Fig. 1.1). In general, this position is confirmed by experimental results [5, 7].

Mixed fracture criteria are usually uncontested for static loading. At the same time, the authors of the 3D-criterion indicate the scope of its application as $\Delta K_{\mathrm{Ith}} < \Delta K_{\mathrm{eff}} < \Delta K_{\mathrm{I}C}$ [8]. The equivalent rate of the FCG v is chosen according to the effective K_{eff}, determined from the current K_{I} and K_{II}, for the FCG diagram, obviously, of the I mode (Fig. 1.1). That is, it is supposed to expand the criterion for a case of cyclic loading. However, this technique should be used with caution. In this aspect, a more productive technique is to construct the crack growth path [11]. This algorithm also requires the presence of FCG diagram parameters.

It is known that mixed fracture significantly changes the rate of FCG v. It is only the presence of crack deformation according to the II mode during storage of the main deformation according to the I mode ($K_{\mathrm{II}}/K_{\mathrm{I}} = 0.1$) that increases the FCG rate v almost twice [12].

1.2.1 Approaches to an Estimation of Survivability at the Mixed Fracture

The choice of characteristics to be investigated depends on the survivability prediction algorithm. In contrast to the above-mentioned known approaches, a completely different concept of assessing the consequences of a mixed process is proposed here. If in traditional approaches to forecasting the amalgamating the action of processes is due to the definition of effective SIF ($K_{\rm I} \cap K_{\rm II} \rightarrow K_{\rm eff}$, Fig. 1.1), then in this case the amalgamation is carried out by summing the damage in the form of relative survivability periods $1/N_g$. This procedure is preceded by the step of determining the survivability curves for pure fracture modes. To do this, use the parameters of FCG, for example, $n_{\rm I}$ and $n_{\rm II}$, $\Delta K_{\rm II}$* and $\Delta K_{\rm I}$* (Fig. 1.1).

Proactive maintenance strategies use a complex indicator to assess the technical condition—the resource safety index (RSI) β_{Pi}, calculated for the reliability P and for the degradation process i [1]. For their complex (system), the general safety index will be:

$$\beta_{\Sigma P} = \lg\left(\sum \frac{U_i}{10^{\beta_{Pi}}}\right)^{-1}, \tag{1.2}$$

where U_i is the relative importance of failure (criticality) with probability Q_i.

For cyclic loading of a mechanical system element at reliability $P = 0.5$ Eq. (1.2) is transformed as:

$$\beta_{\Sigma} = \lg N_{\Sigma} = \lg\left(\Sigma \frac{U_i}{N_i}\right)^{-1}. \tag{1.3}$$

Lifetime at the non-stationary process N_Σ is determined through the lifetime of *Ni* in stationary processes as follows:

$$N_{\Sigma} = \frac{1}{\Sigma\left(\frac{U_i}{N_i}\right)}. \tag{1.4}$$

For cyclic processes, the criticality U_i is determined by the relative duration of the process c_i, which corresponds to the level of the block load. Also, the criticality depends on the accumulated damage to destruction a_0. Its value is usually in the range $d_0 = 0.2\ldots 2.0$ and determines the danger of the process. The smaller the value of d_0, the more intense the degradation process, the more dangerous it is. Therefore: $U_i = c_i/d_0$. With this in mind, we have:

$$N_{\Sigma} = \frac{d_0}{\Sigma\left(\frac{c_i}{N_i}\right)}. \tag{1.5}$$

In this form, a formula is obtained that coincides with the known formula for summing the damage taking into account the factor of nonstationarity. If there are no data on the influence of the shape of the block on the accumulation of damage, take $d_0 = 1$. For several degradation processes acting on the element simultaneously, their relative duration is the same and $c_i = 1$. This situation corresponds to a combined load or mixed fracture. Then, it is possible to specify Eq. (1.5) for survivability at the mixed I+II fracture: $N_{g\Sigma} = \left(\Sigma\left(1/N_{g\mathrm{I(II)}}\right)\right)^{-1}$.

The survivability of pure $N_{gI(\mathrm{II})}$ modes is the period of crack growth from the size a_0 to ac, which is determined from the integrated equation of the constant section of the FCG diagram (B, Fig. 1.1). The following form of this equation is used:

$$v = 10^{-7}\left(\frac{\Delta K_{\mathrm{I(II)}}}{\Delta K^*_{\mathrm{I(II)}}}\right)^{n_{\mathrm{I(II)}}}. \tag{1.6}$$

1.3 Materials and Test Methods

The scheme of four-point asymmetrical bending (4PAB) was chosen to determine the fracture parameters at the II mode (Fig. 1.2). This scheme is well combined with the scheme of three-point or transverse bending (3PB), which obtains the parameters of fatigue fracture in the I mode. Prismatic specimens are made of steels 9HS (analogy 90CrSi: ultimate strength $\sigma_b = 803$ MPa, yield point $\sigma_y = 658$ MPa, unit elongation $\delta = 0.16$) and H12F1 (analogy X155CrVMo12-1: $\sigma_b = 885$ MPa, $\sigma_y = 723$ MPa, $\delta = 0.15$). The task was to obtain the twice amplitude of SIF ΔK_i*, which corresponds to the rate CG $v = 10^{-7}$ m/cycle. This characteristic is a kind of marker of changes in the rate of CG. A decrease in the value of ΔK_i* corresponds to an increase in the rate v.

Depending on the total span $L = L_1 + L_2$ and the relative values of the shoulders $L_{1r} = L_1/L$, $L_{2r} = L_2/L$ is determined by the shear force Q, as well as the bending

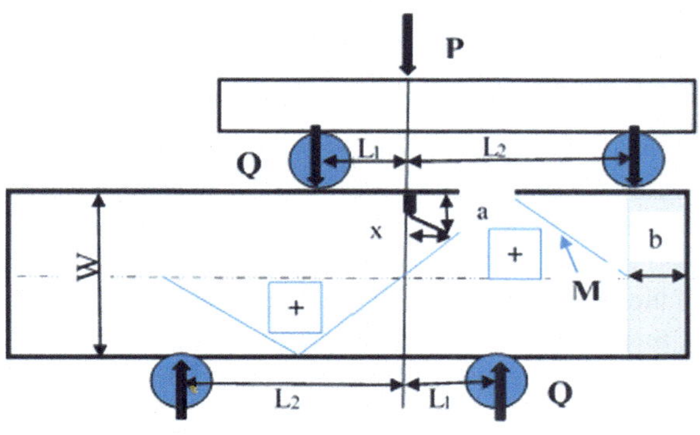

Fig. 1.2 Test scheme of the specimen with the cross-sectional dimensions **b** · W on a four-point asymmetric bend

moment M: $Q = P(L_{2r} - L_{1r})$, $M = Qx$. The relationship between normal and shear stresses is as follows: $\sigma = 6\tau x/W$. For the 3PB scheme $-x = L/2 = 60/2 = 30$ mm. For 4PAB scheme, it is accepted: $L = 50$ mm, $L_1 = 10$ mm, $L_2 = 40$ mm, $L_{1r} = 0.2$, $L_{2r} = 0.8$.

The contribution of shear stresses is estimated through the lever coefficient $\gamma_{1b} = \sigma / (\tau k_f)$, where the coefficient k_f of the cross-sectional shape of is constant for the structural element.

The frequency of cyclic stresses was 3–15 Hz, and the stress ratio varied from $R_\sigma = 0.1$ to $R_\sigma \to 0.3$. A higher frequency was intended for the stage of crack formation. After the crack appeared, the frequency decreased. The appearance of a crack was recorded when its length was already more than 1 mm. Therefore, it was possible to obtain the upper sections of the crack growth curves with the definition of SIF ΔK_i* and ΔK_{ifc} (Fig. 1.1).

To be able to analytically integrate the Paris equation, it is necessary that the geometric function in determining the SIF was not dependent on the crack size. Therefore, the known solutions for SIF in the specimens [13–15] were approximated to reduce the number of variables. After that, they acquired the form that depends on the relative size of the crack $\varepsilon = a/W$:

$$K_{\rm I} = 0.37\sigma\sqrt{W}\exp(3.3\varepsilon), \quad K_{\rm II} = \tau(3.3\varepsilon - 0.09)\sqrt{W}. \tag{1.7}$$

As a result, taking into account Eqs. (1.6) and (1.7), we obtain the number of cycles of crack growth from size a_0 (ε_0) to size a_c (ε_c) (survival) for I and II modes, respectively:

$$N_{g\rm I} = 10^7\left(\frac{\Delta K_{\rm I}^*}{0.37\cdot\Delta\sigma\sqrt{W}}\right)^4\frac{W}{13.36}\left[\exp(-13.36\varepsilon_0) - \exp(-13.36\varepsilon_c)\right], \tag{1.8}$$

$$N_{g\rm II} = 5\cdot 10^6\left(\frac{\Delta K_{\rm II}^*\cdot\sqrt{W}}{\Delta\tau}\right)^3\left(\frac{1}{(3.3a_0 - 0.09W)^2} - \frac{1}{(3.3a_c - 0.09W)^2}\right). \tag{1.9}$$

Here it is accepted, as proposed at [2], $n_{\rm I} = 4$, $n_{\rm II} = 3$.

1.4 Experimental Results and Their Analysis

For the selected steels tested the 4PAB scheme, it was not possible to obtain fracture in the pure II mode. The crack immediately at an angle of approximately 45° grew from the notch toward the zone of tensile stresses caused by bending. Further breaking of the specimen occurs in the direction of support. This crack path is characteristic of metal tests in this scheme and indicates a mixed I+II mode. A certain specimens, which were tested according to the 4PAB scheme, collapsed into three parts due to an additional break of the I mode fracture by a straight (direct) crack (Fig. 1.3). Thus,

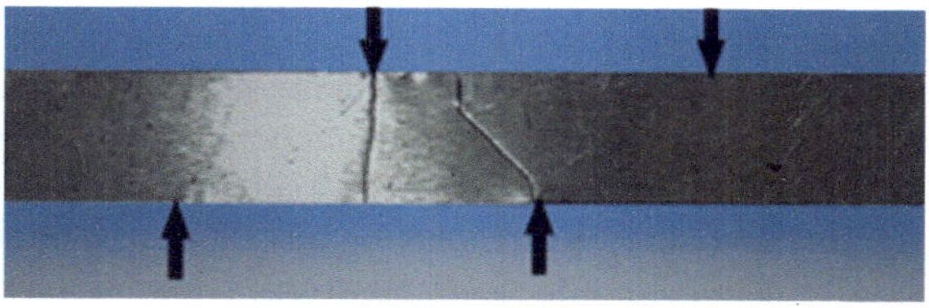

Fig. 1.3 Destruction of specimens during tests according to the 4PAB scheme

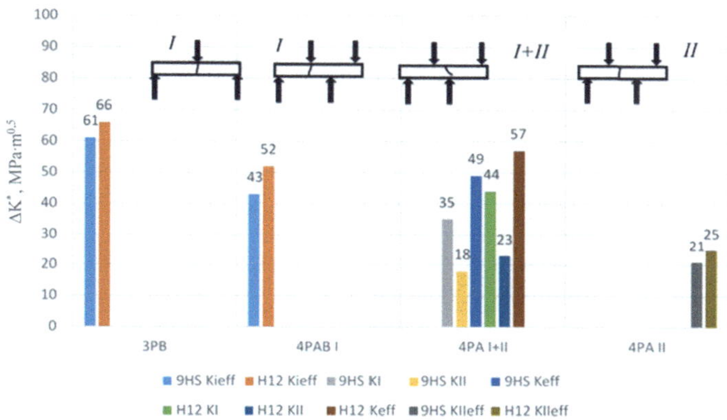

Fig. 1.4 Comparative diagram of the cyclic fracture toughness ΔK^* of steels 9HS (9HS), H12F1 (H12) at three-point bending (3PB) and at four-point bending for straight (4PAB I and 4PAB II) and oblique cracks (4PAB I+II)

destruction in the I mode occurred at the lever coefficients $\gamma_{1b} = 2$ (3PB) and $\gamma_{1b} = 0.8$ (4PB).

On the basis of the experimental FCG diagrams as the main indicator of resistance to cyclic fracture, the value of $\Delta K_{I(\mathrm{II})}{}^*$ was obtained (Fig. 1.4). For the mixed fracture of I+II mode, the value of the effective SIF $\Delta K_{\mathrm{I+II}}{}^*$ is given, which was obtained by the criterion of Richard Eq. (1.1).

A *computational* method for determining the parameters of the II mode of cyclic fracture based on the results of oblique crack growth kinetics was developed. It is also based on a reliable estimate of the parameters $\Delta K_{\mathrm{I}}{}^*$ and n_{I}, which were obtained in the 3PB scheme. Isolation of FCG parameters for mode II in conditions of mixed fracture I+II is possible due to the concept of amalgamation RSI Eqs. (1.2)–(1.5).

The tendency of deterioration of resistance to destruction at the 4PAB scheme of tests in the conditions of direct cracks is confirmed. For them, the value of ΔK^* is 25–45% less than the pure $\Delta K_{\mathrm{I}}{}^*$ obtained by the 3PB scheme. This leads to an increase in the rate of CG in 2.5–4.5 times, which reduces the durability and survivability. The deterioration of the resistance to the cyclic process in the conditions of the mode I for the 4PAB scheme in comparison with the 3PB scheme is explained by the increase in the contribution of the tangential shear stresses.

A decrease in the lever coefficient γ_{Ib} indicates an increase in the contribution of shear stresses and fractures of the mode II or an increase in the current ratio K_{II}/K_I. For the 3PB test scheme, the lever coefficient is higher, and the K_{II}/K_I ratio is 2.5 times smaller than for the 4PB scheme (direct cracks, I mode).

The factor γ_{Ib} has a contradictory effect on the resistance of materials to fatigue. At the stage of crack formation, its reduction leads to an increase in the cyclic strength and durability N. At the stage of crack growth, a decrease in the coefficient γ_{Ib} leads to an increase in the rate of FCG and, as a consequence, to a decrease in the value of ΔK_I^*.

Together with the experimental curves according to the developed algorithm, the model FCG curves were obtained (Figs. 1.5 and 1.6). It is the initial FCG curves ε –

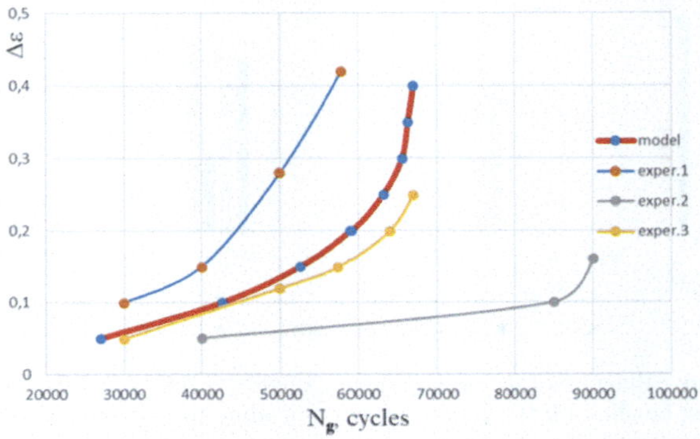

Fig 1.5 Growth kinetics of an oblique crack in a 9HS steel specimens for the shear stress $\Delta\tau = 130$ MPa, which was obtained by model and experimentally (exper. 1, 2, 3)

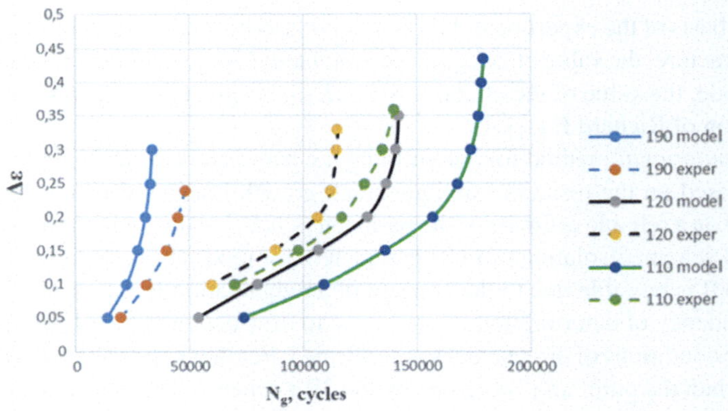

Fig. 1.6 Kinetics of oblique crack growth in H12F1 steel specimens for the shear stress $\Delta\tau = 110$, 120, 190 MPa, which was obtained by model (model, solid line) and experimentally (exper.110, 120, 190, dotted line)

Table 1.1 Parameters of the equation of survivability curves for $\Delta\tau$ in MPa

№	Test scheme	Path of crack	Mode	Steel	m_τ	C_τ
1	3PB	Direct	I	9HS	4.2	13.23
2	3PB	Direct	I	H12F1	4.2	13.34
3	4PAB	Direct	I	9HS	4.1	13.45
4	4PAB	Direct	I	H12F1	4.1	13.99
5	4PAB	Oblique	I+II	9HS	3.0	11.11
6	4PAB	Oblique	I+II	H12F1	3.1	11.64
7	4PAB	Direct	II	9HS	3.0	11.04
8	4PAB	Direct	II	H12F1	3.1	11.62

N_g (in contrast to the expected FCG in form diagrams v $- \Delta K$), or more precisely the survival period of N_g, served as a criterion for the convergence of model and experimental results. The run-up between them does not exceed 50% (Fig. 1.5), which is an acceptable accuracy for the fatigue process.

For the survivability curves used the equation:

$$\Delta\tau^{m_\tau} \cdot N_g = 10^{C_\tau}. \tag{1.10}$$

The transition from pure modes of fracture to mixed slope of the survivability curves increases. This can be seen from the fact that the parameter $m\tau$ decreases from 4.2 to 3 (Table 1.1).

It is theoretically proved that for a constant size of the finite crack ε_c, for the survivability curves at pure modes, there is an exact correspondence of the slope parameter of the survivability curve to the parameter of the Paris equation: $m_\sigma = n_{\mathrm{I}}$, $m_\tau = n_{\mathrm{II}}$. In mixed modes, the slope of the survivability curves changes due to the fact that the survivability of $N_{g(\mathrm{I+II})}$ for Eq. (1.5) is less than the survivability in pure modes: $N_{g(\mathrm{I+II})} < N_{g\mathrm{I}}$, $N_{g(\mathrm{I+II})} < N_{g\mathrm{II}}$.

The same features of the survivability curves were found in the study of mixed fracture of sheet rolls [2]. Therefore, we can assume that in mixed fracture, in which there are two modes, the parameter of the survivability curve $m_{\tau\mathrm{mix}}$ will be closed in the range: $\left(n_{\mathrm{II(III)}} - 1\right) \leq m_{\tau mix} \leq n_{\mathrm{II(III)}}$.

Given that $n_{\mathrm{II}} = n_{\mathrm{III}} = 3$, it is recommended to set $m_{\tau\mathrm{mix}} = 2\ldots 3$. That is, with the decrease in sensitivity to the influence of actual stresses, the influence of the non-stationary factor of the regime is leveled. The revealed tendency is a precondition of the conclusion, which was made at research of survivability of sheet rolls. It consists in the fact that the equivalent rate of FCG in mixed fracture, practically, does not depend on the level of tangential stresses. The obtained values of $\Delta K_{\mathrm{II}m}{}^*$ at mixed fracture from oblique crack can serve as initial data for determination of the pure value $\Delta K_{\mathrm{II}}{}^*$. Due to the identified dominance of the destruction of the II mode in the development of oblique cracks, you can use this as follows: $\Delta\kappa_{\mathrm{II}}^* = 1,1 \cdot \Delta\kappa_{\mathrm{II}m}^*$.

1.5 Conclusions

The validity of using the model of schematization of the kinetic diagram of fatigue fracture has been experimentally confirmed. For steels of rolling rolls, it is recommended to a priori assign the main indicators when predicting survivability as follows: $\Delta K_I^* = 60–70\ \text{MPa}\sqrt{\text{m}}$, $n_\text{I} = 4$, $n_\text{II} = 3$, $\Delta K_I^*/\Delta K_\text{II}^* = 2.5–3$.

The effectiveness of the developed algorithm for predicting survivability in mixed fracture, which does not require the determination of an effective SIF, has been confirmed. In the algorithm, the action of individual damaging processes (modes of fracture) is combined on the basis of the amalgamation rule of resource safety indices [1, 16].

References

1. Belodedenko SV, Hanush VI, Hrechanyi OM (2019) Rules for amalgamating the reliability of elements of powertrain systems at maintenance. J Mech Eng Autom 9:24–32
2. Belodedenko S, Grechany A, Yatsuba A (2018) Prediction of operability of the plate rolling rolls based on the mixed fracture mechanism. Eastern-Eur J Enterp Technol 1(91):4–11
3. Liebowitz H, Eftis J, Jones D (1977) Some recent theoretical and experimental developments in fracture mechanics. In: Fracture 1977: Proceedings of the fourth international conference on fractures. Waterloo, pp 695–723
4. Datsyshyn O, Panasyuk V (2020) Structural integrity assessment of engineering components under cyclic contact. Springer International Publishing
5. Troshchenko VT, Tsybanov GV, Khotsyanovsky AO (1994) Two-parameter model of fretting fatigue crack growth. Fatigue Fract Eng Mater Struct 17(1):15–23
6. Murakami Y, Hamada S (1997) A new method for measurement of Mode II fatigue threshold stress intensity factor range $K_{\tau th}$. FFEMS 20:863–870
7. Pokrovsky V et al (2006) Investigations of crack growth behavior under mixed mode loading. In: Proceeding of the 13 international colloquim mechanical fatigue of metals. TSTU, Ternopil, pp 259–265
8. Richard HA, Eberlein A (2016) 3D-mixed-mode-loading: material characteristic values and criteria's validity. Frattura ed Integrità Strutturale 37:80–86
9. Demir O, Ayhan AO (2016) Investigation of mixed mode-I/II fracture problems -Part 2: evaluation and development of mixed mode-I/II fracture criteria. Frattura ed Integrità Strutturale 35:340–349
10. Campagnolo A et al (2017) Mode II brittle fracture: recent developments. Frattura ed Integrità Strutturale 42:181–188
11. Miloud S et al (2012) Finite element model of crack growth under mixed mode loading. Int J Mater Eng 2(5):67–74
12. Broek D (1974) Elementary engineering fracture mechanics. Noordhoff International Publishing, Leyden
13. Murakami Y (1987) Stress intensity factors handbook. Pergamon Press
14. He MY, Hutchinson JW (2000) Asymmetric four-point crack specimen. J Appl Mech 67:207–209
15. Fett T (1998) Stress intensity factors and weight functions for special crack problems. FZKA 6025: Forschungszentrum Karlsruhe
16. Belodedenko SV et al (2019) Application of risk-analysis methods in the maintenance of industrial equipment. Procedia Struct Integrity 22:51–58

Chapter 2
Accelerated Fatigue Test Procedure for Leaf Springs Applied to Semi-Trailers

Rui Gustavo Lippert Schwanke, Joel Boaretto, Felipe Vannucchi de Camargo, Cristiano Fragassa, and Carlos Pérez Bergmann

Abstract The fatigue endurance assessment of leaf springs for trailers and semi-trailers is a non-standardized procedure often protected by industrial secrecy that can significantly vary depending on the manufacturer, thus, imposing a barrier to seek for more efficient and safe structures. Hence, the main objective of the present work is to create, starting from experimental evidences, a solid technical basis for the creation of a valid accelerated fatigue test procedure able to reproduce the damage to failure inferred to a semi-trailer leaf spring allowing an accurate prediction of its durability. For that, data was collected on public paved highways and off-road tracks by strain gauges mounted on a spring, which were then extrapolated for fatigue failure through a rainflow cycle counting, yielding a damage parameter that was reproduced in special test tracks. The accelerated fatigue test created had a good correlation to real operational conditions of the spring, represented by severity factors close to 1 along the longitudinal axis of the spring.

Keywords Design of experiment · Accelerated life testing · Rainflow cycle counting

2.1 Introduction

High reliability is of utmost importance in the automotive sector, where the relevance of the constant seek for improving energy efficiency of vehicles [1] is only overcome by the concern with ensuring safe designs. By their turn, these designs depend

R. G. L. Schwanke · J. Boaretto
Center of Technology Randon, Farroupilha, RS 95180-000, Brazil

J. Boaretto · F. V. de Camargo (✉) · C. P. Bergmann
PPGE3M, Federal University of Rio Grande Do Sul, Porto Alegre, RS 90035-190, Brazil
e-mail: felipe.vannucchi@ufrgs.br; felipe.camargo@senairs.org.br

C. Fragassa
Department of Industrial Engineering, University of Bologna, 40136 Bologna, Italy

F. V. de Camargo
SENAI Institute of Innovation in Polymer Engineering, São leopoldo, RS 93030-090, Brazil

G. Lesiuk et al. (eds.), *Structural Integrity and Fatigue Failure Analysis*, Structural Integrity 25, https://doi.org/10.1007/978-3-030-91847-7_2

on the thorough failure analysis of each individual structural auto-part, specifically those under dynamic loads, aiming to guarantee long-term durability [2]. Within this context, springs deserve special attention for their crucial role in damping systems, supporting the weight of the vehicle, providing safety and adherence for an improved propulsion, and smoothing the ride by isolating the passengers from vibrations and shocks [3, 4]. This is determinant to reduce the driver's fatigue in long drives and therefore the risk of accidents [5].

The present work focus on the trailer and semi-trailer segment, responsible for a major portion of transportation of goods worldwide, mainly in large emerging countries that do not possess developed railways. Knowing that among the transportation modals, highways stand as the one with higher number of accidents, cargo trailers comprise a pertinent category of vehicles that should be constantly under going safety improvements.

Although a component of elevated structural importance, trailer and semi-trailer leaf springs are not comprised within a specific technical standard that evaluates their durability taking into account real-world boundary conditions. Nowadays, the mechanical assessment procedures adopted generally depend on the manufacturer and are often covered by industrial secrecy non-disclosure agreements, hence imposing a barrier for researchers and engineers to push this technology forward based on a solid technical ground and aiming for more efficient structures.

The seek for continuous enhancement of dimensioning techniques for trailer and semi-trailer auto-parts has been approached by several recent works. Abdelkareem et al. [5] demonstrated an efficient study on how to improve the ride performance of a truck semi-trailer by a proper dimensioning of leaf springs with adjustable stiffness based on the load of the cargo carried, particularly effective when the semi-trailer is empty. The author underlined the key role played by the suspension system on improving ride comfort, truck performance, tractor acceleration, and the integrity of the goods carried through a comprehensive vibration analysis.

Similarly, Diwakar et al. [6] conducted an experimental study to determine the suspension settings that provide an optimal vibrational behavior of a trailer by adjusting the spring stiffness. Just like the present work, it considered different terrains in order to account for a model that represents the actual operational conditions of the trailer.

Tello et al. [7] have developed an analytic formulation applied in a semi-trailer axle bracing support that allows the prediction of fatigue failure of welded steel components based on the properties of the Thermally Affected Zone of the base material. This study has demonstrated to be quite useful specially for metals that do not have their fatigue parameters fully characterized.

Aiming at the constitution of an accurate accelerated fatigue test, the present work makes use of a solid experimental dataset collected on public paved highways and off-road tracks to calculate the expected durability of semi-trailer leaf springs through a cumulative linear damage model [8] for variable amplitude cyclic loads. The experimental data, obtained by strain gauges mounted on the spring and filtered by a rainflow cycle counting, yielded a certain damage that was successfully reproduced in an accelerated manner in the special test tracks of potholes, Belgian stones, body-twist, and off-road, thus providing a more efficient and yet highly representative fatigue test routine for leaf springs.

2.2 Materials and Methods

In-field data was collected by means of 6 transversal and 6 longitudinal strain gauges assembled on the foremost top leaf (Fig. 2.1), which is the spring under more severe loads, for 210 km of paved highways and 84 km of off-road tracks, with a load of 8.425 t of Total Gross Weight (TGW) above the axis of the spring studied, distributed 50% on each side of the axis. Strain gauge specifications are described in Table 2.1. Once the data was imported into the nCode software, a rainflow cycle counting [9] was used to set aside how many cycles of each specific stress amplitude (N_i) were suffered by the spring in each region of its length.

Then, once each stress amplitude points to a particular fatigue life on the SAE 5160 steel *S–N* diagram ($N_{f,i}$) [10], the damage (D) was calculated through the Palmgren–Miner relation [8] Eq. (2.1), where the closer D is to 1, the higher the probability of failure. The mechanical properties of this steel grade are reported in Table 2.2 [11].

$$\sum_{i=1}^{n} \frac{N_i}{N_{f,i}} = D \tag{2.1}$$

With the damage data of external tracks sorted out, the next step was to figure out how to reproduce the same damage on the company's internal test ground facilities in an accelerated manner. Adopting the conservative damage target that an actual

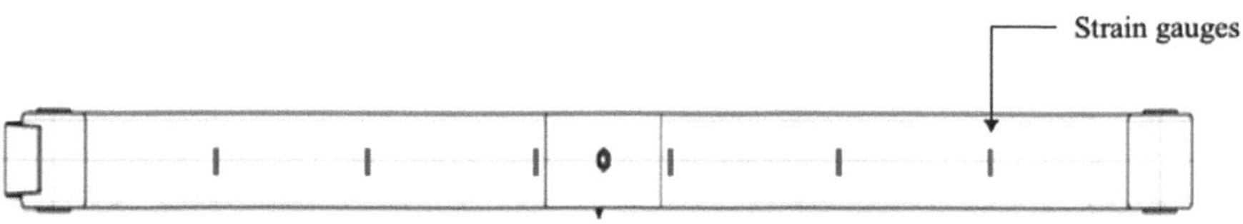

Fig. 2.1 Leaf spring top view: strain gauge setting

Table 2.1 Strain gauge specifications

Strain gauge	HBM 1LY41-6/120
Strain gauge factor	2.09
Equipment	MGC Plus HBM
Connection pattern	¼ Wheatstone
Data acquisition rate (Hz)	400
Filter	Low-pass 40 Hz

Table 2.2 Mechanical properties of the leaf spring studied [11]

Material	SAE 5160 steel
Ultimate tensile stress (MPa)	1360
Yielding tensile stress (MPa)	1240
Elastic modulus (GPa)	185
Poisson ratio	0.3

semi-trailer would circulate 50% on-road and 50% off-road during the operational life of a leaf spring, this respective accumulated damage has been extrapolated from the data collected on field, and the challenge was to set up a test procedure that reaches this same damage utilizing internal tracks.

Given that the internal paved tracks are in optimal conditions and infer way less damage than an actual highway with its erosion-related imperfections, the test procedure was elaborated aiming to submit the semi-trailer to operate for specific distances in each of the following special tracks: potholes, Belgian stones, body twist, and off-road (Fig. 2.2), at the respective velocities of 20, 15, 5, and 30–40 km/h. The latter did not have a specific constant velocity due to the curves of this track, where the truck had to reduce its speed. Even though circulating only on the off-road track would accelerate the achievement of the targeted damage, different courses were chosen to account for the variability of terrains that a semi-trailer could circulate on, attributing a higher level of accuracy to this study.

Starting from a test block with a particular number of laps in each of the aforementioned special tracks and assessing its respective damage over the spring by the same methodology considered for external tracks, the damage data was extrapolated reaching the ideal number of laps in each track that, altogether, the semi-trailer must

Fig. 2.2 Special test tracks: potholes (**a**), Belgian stones (**b**), body twist (**c**), and off-road (**d**)

go through to infer a damage as similar as possible to the target assumed before (D_{target}). Then, the severity factor (S), which is the coefficient between the damage of the internal tracks (D_{int}) by the target damage (Eq. 2.2) retrieved from the data collected on external (i.e. public) tracks, is calculated for each region of the spring to assess how faithful the internal test routine is to real operational conditions (i.e., how close the ratio is to 1).

$$S = \frac{D_{int}}{D_{target}} \tag{2.2}$$

2.3 Results

2.3.1 Public Road Test

The random signals with variable frequencies collected on external tracks were filtered by a Fast-Fourier Transform (FFT) procedure on the nCode software, generating a frequency spectrum power relation to allow the comprehension of which frequency bands are responsible for inferring most damage. Then, a rainflow cycle count methodology was used to determine how many of the cycles the spring was subjected to are actually significant for induction of damage. Both graphs are displayed in Fig. 2.3.

The frequency spectrum power evidenced how much more damage off-road tracks infer than paved highways, with the similarity of the fact that frequency bands in the order of approximately 1 and 4 Hz are particularly prejudicial in both cases.

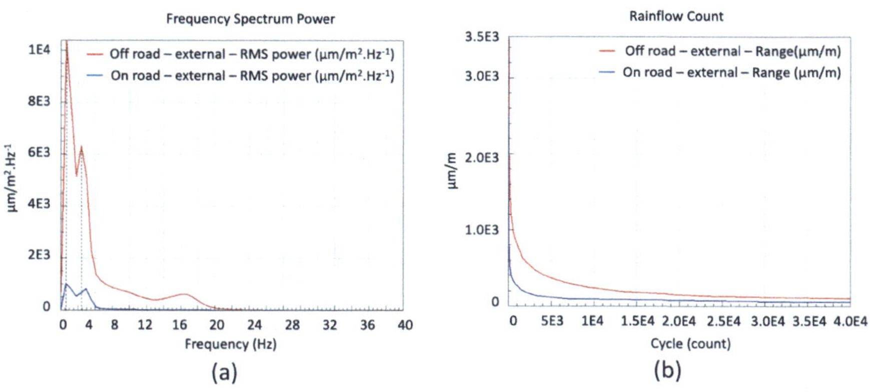

Fig. 2.3 Frequency spectrum power (**a**) and Rainflow cycle count (**b**) of the data collected on external tracks

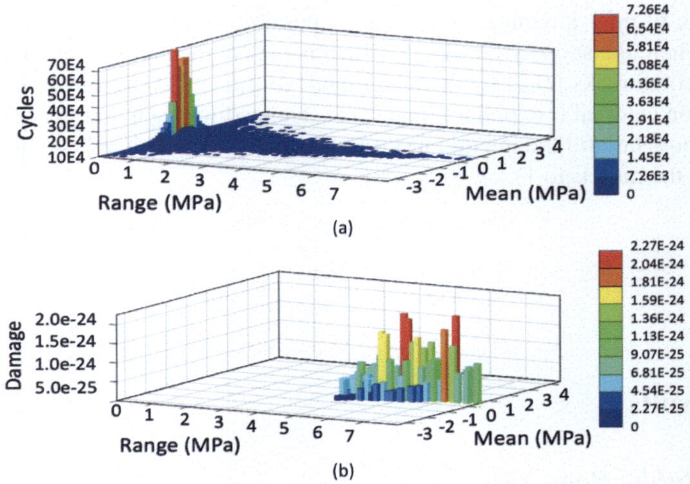

Fig. 2.4 Number of cycles at each stress amplitude (**a**) and the respective damage they infer (**b**) on the region of the 3rd foremost strain gauge in off-road tracks

The damage was numerically defined from the Palmgren-Miner [Eq. (2.1)], which is the sum of the ratios among the number of cycles at each stress amplitude by their respective number of cycles to failure considering those mean stresses. Figure 2.4 shows an example of the damage data obtained in the case of the 3rd foremost strain gauge of the spring on external off-road tracks.

Knowing that the expected durability of a semi-trailer leaf spring is 110.000 km, being 67% on paved roads and 33% on off-road tracks in the extreme case of a semi-trailer that is used in the countryside by the sugarcane industry [12]; the present work adopted the conservative design premise of estimating that leaf springs should be dimensioned to last the same 110.000 km, but on 50% of each kind of terrain. The damage calculated for the previous external tests ($D_{\text{on-road, ext}}$ for 210 km and $D_{\text{off-road, ext}}$ for 84 km) was extrapolated for the expected durability ($D_{\text{on-road, target}}$ and $D_{\text{off-road, target}}$, each for 55.000 km) to attain the targeted total damage value that each region of the spring must achieve to reach failure (D_{target}), as shown in Table 2.3.

2.3.2 Validation on Internal Test Ground

The same approach for damage calculation was used to evaluate 1 lap in each of the special tracks of off-road, potholes, body twist, and Belgian stones, which cover the distances of 3.2 km, 0.12 km, 0.18 km, and 0.15 km, respectively. Then, it was possible to calculate that 1 circuit formed by 1 lap in the body twist track plus 20 laps in each of the others yielded approximately 0.5% of the damage desired in each portion of the spring. Hence, given that 1 circuit covers 69.58 km, a complete test

Table 2.3 Damage of public road tests

Damage from external tracks				
$D_{\text{on-road, ext}}$	$D_{\text{off-road, ext}}$	$D_{\text{on-road, target}}$	$D_{\text{off-road, target}}$	D_{target}
0.00E+00	1.32E−25	0.00E+00	4.47E−23	4.47E−23
3.65E−25	4.41E−23	9.55E−23	1.50E−20	1.51E−20
7.16E−25	8.07E−23	1.87E−22	2.74E−20	2.76E−20
1.66E−25	7.84E−24	4.35E−23	2.66E−21	2.70E−21
1.69E−25	7.34E−24	4.43E−23	2.49E−21	2.53E−21
1.73E−25	8.63E−24	4.53E−23	2.93E−21	2.98E−21

Table 2.4 Damage and severity data

Damage from internal tracks						Severity
$D_{\text{off-road}}$ (20 laps)	D_{potholes} (20 laps)	$D_{\text{body twist}}$ (1 lap)	$D_{\text{belgian stones}}$ (20 laps)	D_{total} (1 circuit)	D_{total} (200 circuits)	
0.00E+00	0.00E+00	0.00E+00	0.00E +00	0.00E + 00	0.00E+00	0.00
3.28E−23	1.32E−23	2.32E−24	1.94E−23	6.77E−23	1.35E−20	0.90
5.50E−23	2.34E−23	6.73E−24	3.60E−23	1.21E−22	2.42E−20	0.88
1.23E−23	2.12E−24	0.00E+00	0.00E+00	1.44E−23	2.88E−21	1.07
1.11E−23	2.20E−24	0.00E+00	0.00E +00	1.33E−23	2.67E−21	1.05
1.10E−23	2.18E−24	0.00E +00	0.00E +00	1.32E−23	2.64E−21	0.89

routine of 200 circuits (13.916 km) is able to infer approximately the same damage to the leaf spring than 110.000 km on public roads.

Following Eq. (2.2), the severity factor can be calculated to estimate the level of accuracy that the accelerated test routine realized in the internal test ground has to reproduce the damage to failure of a leaf spring on real operational conditions (Table 2.4).

2.4 Conclusions

An efficient experimental accelerated fatigue test routine was created, allowing the reproduction of damage to failure of a semi-trailer frontal leaf spring by mounting it on a vehicle that covers 12.65% of the actual distance it needs to cover in public roads to fail. The calculation of the severity factor demonstrated the high precision of this novel test routine, where the closer it is to 1, the more precise the test is for each one of the 6 assessed regions of the spring. The resulting severity values ranged from 0.88 to 1.07. However, it is important to highlight that the damage and the severity of the leaf spring region corresponding to the foremost strain gauge were null. This can be explained by the fact that this region is too close from the chassis attachment point, which restrains the deformation in that region significantly.

Future studies include the adaptation of this fatigue endurance test to a laboratory procedure; the microstructural fracture analysis of a spring fractured from the elaborated test versus one from an actually operative semi-trailer; and a study of the vibrational resonant frequencies of the spring calculated numerically in comparison with the vibrational data collected on field and filtered by means of a FFT.

References

1. Mohamad B, Karoly J, Zelentsov A (2020) CFD modelling of formula student car intake system. Facta Univ Ser. Mech Eng 18(1):153–163
2. Beermann HJ (1986) The analysis of commercial vehicle structures. Mechanical Engineering Publications Ltd., London
3. de Camargo FV, Giacometti M, Pavlovic A (2017) Increasing the energy efficiency in solar vehicles by using composite materials in the front suspension. In: Campana G, Howlett RJ, Setchi R, Cimatti B (eds) International conference on sustainable design and manufacturing. Springer, Cham, pp 801–811
4. Minak G, Brugo T, Fragassa C, Pavlovic A, Zavatta N, de Camargo FV (2019) Structural design and manufacturing of a cruiser class solar vehicle. J Vis Exp 143:e58525
5. Abdelkareem MA, Makrahy MM, Abd-El-Tawwab AM, EL-Razaz ASA, Kamal Ahmed Ali M, Moheyeldein MM (2018) An analytical study of the performance indices of articulated truck semi-trailer during three different cases to improve the driver comfort. Proc Inst Mech En, Part K J Multi-body Dyn 232(1):84–102 (2018)
6. Diwakar N, Balaguru S (2020) Experimental study on vibration control of transportation trailers used for spacecraft. In: Yang LJ, Haq A, Nagarajan L (eds) ICDMC 2019, LNME. Springer, Singapore
7. Tello L, Castejon L, Malon H, Valladares D, Luque P, Mantaras DA, Ranz D, Cuartero J (2020) Development of a fatigue life prediction methodology for welded steel semi-trailer components based on a new criterion. Eng Fail Anal 108:104268 (2020)
8. Stephens RI, Fatemi A, Stephens RR, Fuchs HO (2001) Metal fatigue in engineering, 2nd edn. Wiley-Interscience, USA
9. Amzallag C, Gerey JP, Robert JL, Bahuaud J (1994) Standardization of the rainflow counting method for fatigue analysis. Int J Fatigue 16(4):287–293
10. Jaramillo SHE, de Sánchez NA, Avila DJA (2019) Effect of the shot peening process on the fatigue strength of SAE 5160 steel. Proc Inst Mech Eng Part C J Mech Eng Sci 233(12):4328–4335 (2019)
11. Gonzales MAC, Barrios DB, De Lima NB, Goncalves E (2010) Importance of considering a material microfailure criterion in the numerical modeling of the shot peening process applied to parabolic leaf springs. Latin Am J Solids Struct 7(1):21–40
12. Possamai ERF (2015) Study of fatigue behavior of SAE 5160 steel applied to automotive leaf springs. Masters dissertation, Federal University of Rio Grande do Sul

Chapter 3
Early Evidences on the Rotating Bending Fatigue Properties of Ductile and Vermicular Cast Irons

Nenad Radovic, Grzegorz Lesiuk, and Rodoljub Vujanac

Abstract Cast iron probably represents one of the most widely utilized constructive materials in our modern industrial world. This is due not only to the excellent mechanical and tribological properties, but also to a certain flexibility of the same. In fact, it can be produced in various types of metal alloy, which differ significantly in properties: from the poorly performing, but economical white cast iron, to the excellent resistant nodular cast iron. In this study, several preliminary results on the experimental measurement of fatigue resistance are proposed for two distinct families of cast iron, the ductile and the vermicular. While the former is well-known, with a great variety of applications, vermicular cast iron is almost unfamiliar at the moment in terms of utilization, despite different potentials. This experiment compares the results of the Rotating Bending Fatigue test for cast iron specimens made under foundry process conditions as identical as possible in the way to reduce bias.

Keywords Metal alloys · Foundry · Spheroidal cast iron · Compact graphite cast iron · Mechanical properties · Fatigue behavior · Fatigue limit

3.1 Introduction

3.1.1 Cast Iron

Cast iron is a ferrous alloy consisting mainly of iron and carbon with a relatively high carbon content (2.06% < C < 6.67% which is the saturation limit) obtained by reduction or heat treatment of iron [1, 2].

N. Radovic
University of Bologna, Bologna, Italy

G. Lesiuk
Wroclaw University of Science and Technology, Wroclaw, Poland

R. Vujanac (✉)
University of Kragujevac, Kragujevac, Serbia
e-mail: vujanac@kg.ac.rs

G. Lesiuk et al. (eds.), *Structural Integrity and Fatigue Failure Analysis*, Structural Integrity 25, https://doi.org/10.1007/978-3-030-91847-7_3

Not much appreciated until the end of the nineteenth century for its lower quality and worse workability compared to mild steel, it has taken on its importance over the years thanks to the many benefits offered by this material. As a result, despite still many years have passed, cast iron continues to be a widely used material [3].

Compared to mild steel ($C < 1.5\%$), cast iron has greater hardness, therefore greater resistance to abrasion, and less resilience, therefore greater brittleness. It also has a lower thermal expansion coefficient (−20%) which makes this alloy suitable for couplings where there are variations in temperature. Finally, the higher carbon content protects the cast iron products from rust [4].

3.1.2 Cast Production

Several advantages in terms of production can be also included as process economy; excellent fusibility; good workability by machine tooling; possibility of creating very complex shapes by simple casting; and so on [5–7].

But, maybe, one of the most interesting aspects regarding the cast iron (respect to the contents of the present article) is related to the huge flexibility offered by this family of metallic alloys: with relatively few process changes (i.e., additives, temperatures…), it is possible to obtain a wide range of materials, very different from each other in terms of characteristics and usability [8–10].

All those who deal with foundry alloys certainly know the various forms of cast iron such as:

- Gray cast iron;
- White cast iron;
- Malleable cast iron;
- Gray lamellar cast iron;
- Ductile nodular cast iron (SGI).

Together with their main characteristics and differences.
Less known is instead:

- Vermicular cast iron, or compact .graphite (CGI).

Although it was probably first produced at the beginning of the last century, even before SGI, it was often confused as a production error [11, 12] and identified as an independent form of cast iron only around the 1960s [11–13].

CGI shows interesting mechanical and productive properties such as, e.g., a lower hardness than SGI which favors its use in applications where it becomes useful to reduce vibrations (e.g., cylinder blocks). At the same time, it is not so resistant, serving as a middle ground between SGI and less performing cast irons [14–17].

With the aim of extending CGI use in the design of components, it becomes essential to know its main characteristics [17–20]. Several investigations, available in literature, already offer measures from the experiments regarding the CGI essential properties. At the same time, they show a remarkable variability. This study, part of a broader research dealing with the CGI experimental characterization [21–27], intends to propose an initial overview of CGI fatigue behavior (also compared with SGI one) in the case of rotating bending testing configuration. These results complement the ones already published on CGI fatigue in push–pull testing configuration [27].

3.2 Materials and Methods

3.2.1 Samples

The fatigue behavior of the two cast irons was investigated using a four-point rotating bending test in accordance with ISO 1143 [27] and ISO 12107 [28]. Standardized specimens were obtained by tool machining in accordance with ISO starting from metal castings (Fig. 3.1). The sandcasting process was performed in SCM Foundry in Rimini, Italy.

Special actions were taken to reduce the metallurgical variability of samples and, therefore, the results comparability.

Two castings were produced within the same casting process and using the same metal alloy as a base: its chemical composition was properly modified by adding additives in the way to obtain SGI (firstly) and CGI (secondly). Particular attention was paid to maintaining all the other process conditions unchanged as much as possible, especially regarding the casting temperature (1400 °C). Table 3.1 reported the chemical composition (in % of weight) of ductile (SGI) and vermicular (CGI) irons.

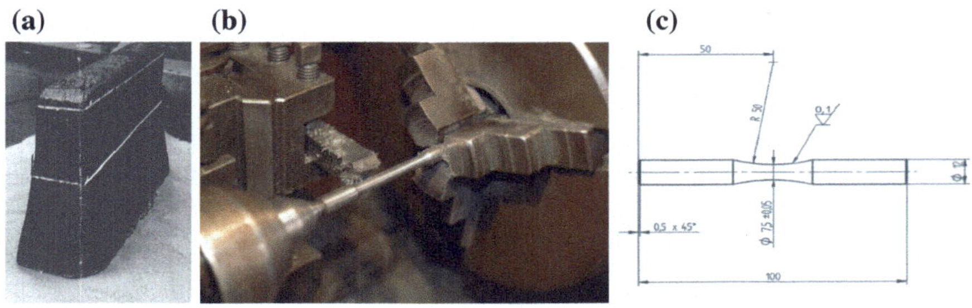

Fig. 3.1 Production of standardized specimens for rotating bending fatigue tests (ISO 1143): **a** metallic block from sand casting; **b** tool machining; **c** "dog-bone" shape

Table 3.1 Chemical composition (in % of weight) of ductile (SGI) and vermicular (CGI) irons

Cast iron	FE	C	SI	MN	P	S	NI	CR
SGI	93.00	3.70	2.65	0.276	0.036	0.002	0.060	0.083
CGI	93.13	3.65	2.57	0.272	0.034	0.005	0.060	0.082
Δ%	0.14%	−1.35%	−3.02%	−1.45%	−5.56%	150.0%	–	– 1.20%
	CU	MO	MG	SN	TI	PB	AL	ZN
SGI	0.077	0.002	0.049	0.011	0.033	0.007	0.011	0
CGI	0.075	0.002	0.012	0.011	0.074	0.006	0.011	0
Δ%	−2.60%	–	−75.51%	0.00%	124.24%	−14.29%	–	–

3.2.2 Modified Staircase Method

The fatigue tests were carried out with a sequence of loads defined using the so-called *modified staircase method*. This method, as detailed in the ISO 12107 standard, permits to estimate the fatigue limit of the materials by carrying out a reduced number of tests. For an effective application, a draft estimation of the fatigue strength, including its variability, for each material under investigation should be needed. Having such an information, testing procedure could start by assuming this entity (S_M) as the value of the applied stress during the first test and its standard deviation ($\Delta\sigma$) as the step of stress increase/decrease during the following ones. The procedure is repeated until all samples are tested. Not knowing these values, the fatigue strength can be imagined as part of the ultimate tensile stress (UTS) and variation range of about 5–10%.

Specifically, S_M, the load value to be used during the first fatigue test, was estimated as 40% and 30% of UTS in the case of, respectively, SGI and CGI. This difference between percentages is mainly related to the better confidence of mechanical properties coming from literature for SGI respect to CGI. Suggested values for UTS are ~600 and ~475 MPA for, respectively, SGI and CGI [1–9, 13, 22, 24]. Thus, the related values S_M, to be adopted during tests, are ~240 and ~140 MPa. Finally, as standard deviation (Δ_σ), as stress increase/decrease step, it was used 8.8 MPa in both cases. These parameters are reported in Table 3.2. Test limit was equal to 10^7 cycles (*run-out*): after that, the test was *terminated,* and the *running time* considered as *censored data.*

Table 3.2 Parameters definition for modified staircase method

Parameter	SGI	CGI
UTS (MPa)	601.7	476
Fraction (%)	40	30
S_M (MPa)	240.7	142.8
Δ_σ (MPa)	8.8	8.8

3.2.3 Bias Reduction

As a way to reduce bias, samples were chosen randomly. Moreover, every precaution was introduced to reduce the risk of metallurgical differences between the samples such as, for example, the extraction of specimens from adjacent areas.

3.3 Results and Discussion

3.3.1 Experimental Data

Table 3.3 reports the experimental data from the rotating bending fatigue tests. For each material, SGI and CGI, 6 samples were tested.

In the table it is possible to follow the test progression. In the case of SGI, for instance, the first test took place at ~241 MPa of load (S_M), equal to about 40% of the UTS for this material. The fatigue test was performed up to $3.9{\cdot}10^5$ cycles when the specimen showed a break (marked as "*F*" of failure in the table). The data was acquired as a *failure*. Precisely for this failure, the subsequent test was carried out reducing the testing load for a value equivalent to the standard deviation (Δ_σ), therefore, equal to ~231 MPa. At this new load level, the test continued up to the maximum expected number of cycles, $10{\cdot}10^6$, when it was terminated (market as "*R*" of run-out in the table). The data was acquired as *censored.*

Figure 3.2 shows in a graph the experimental values for the two materials. Unlike the table, here the samples are not reported in the exact order in which they were tested but taking into account an increasing trend of stress. The filled circles indicate breaks while the others indicate run-outs. It is possible to observe, for example, that as it was logical to expect, the run-outs are concentrated at lower load values.

Table 3.3 Results from rotating bending fatigue tests

Sample	SGI			CGI		
	Stress	Cycle		Stress	Cycle	
1	240.7	390,000	F	142.8	4,400,096	F
2	231.3	10,000,000	R	133.6	10,000,000	R
3	240.7	470,000	F	142.8	10,000,000	R
4	231.3	509,000	F	151.4	10,000,000	R
5	222.5	10,000,000	R	160.0	1,650,000	F
6	231.3	10,000,000	R	151.4	1,153,000	F
Mean	233.0			147.0		

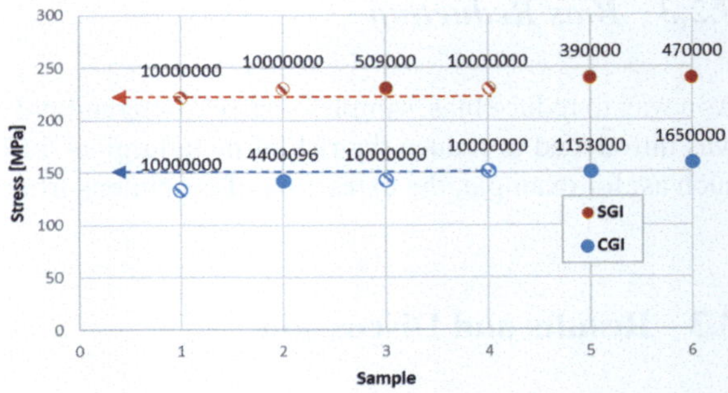

Fig. 3.2 Graphical representation of results from rotating bending fatigue tests

3.3.2 Fatigue Limits

Fatigue properties were estimated considering the probabilities of failure and confidence levels in accordance with ISO 12107.

Specifically, the so-called *Mean Fatigue Strenght* (MFS) was assumed as the mean values between the experimental cycles as directly available in Table 3.3. It also means MFS is 233 and 147 MPa for, respectively, SGI and CGI.

The Lower Limit Fatigue Strength (LFS) was estimated as

$$\text{LFS} = \text{MSF}--k \cdot \Delta_{\sigma} \tag{3.1}$$

with k a parameter able to consider the probabilities of failure and confidence level respect to the number of samples. Considering 10% as probability of failure (*PF*), 95% as confidence level (*CL*), and 6 as samples (v),

$$k = 2.755 \tag{3.2}$$

and, in the case, LFS is equal to 208 and 123 MPa for, respectively, SGI and CGI. These values are reported in Fig. 3.2 by horizontal lines and represent an early evaluation of the maximum level of stress the materials can sustain in infinite.

3.4 Conclusions

This experimental study provides several preliminary results regarding the measure of fatigue resistance for two different cast irons, the ductile (SGI) and the vermicular (CGI). In particular, it detects and compares results of the 4-points rotating bending fatigue test of specimens made under an identical sand cast foundry process. Several standards were used as a reference in the tests and for statistical processing of the results. As expected, SGI exhibits a better fatigue behavior, but both alloys remain special observations in the case of situations in which load conditions with high number of cycles are considered.

References

1. Minkoff I (1983) The physical metallurgy of cast iron. Wiley, New York
2. ASM International (1996) Specialty handbook: cast irons, pp 33–267
3. Angus HT (2013) Cast iron: physical and engineering properties. Elsevier, London
4. Elliott R (1988) Cast iron technology. Butterworth-Heinemann, London
5. Emerson PJ, Simmons W (1976) Final report on the evaluation of the graphite form in ferritic ductile iron by ultrasonic and sonic testing, and the effect of graphite form on mechanical properties. AFS Trans 76(26):109–128
6. Fuller AG, Emerson PJ, Sergeant GF (1980) A report on the effect upon mechanical properties of variation in graphite form in irons having varying amounts of ferrite and pearlite in the matrix structure and the use of nondestructive tests in the assessments of mechanical properties of such Irons. AFS Trans 80(9):21–50
7. Radiša R, Dučić N, Manasijević S, Marković N, Ćojbašić Ž (2017) Casting improvement based on metaheuristic optimization and numerical simulation. Facta Univ Ser Mech Eng 15(3):397–411
8. Blaszczyk C, Melon M, The crucial role of material selection in machine design. Misumi
9. Fragassa C (2017) Material selection in machine design: the change of cast iron for improving the high-quality in woodworking. Proc Inst Mech Eng C J Mech Eng Sci 231(1):18–30
10. Lucisano G, Stefanovic M, Fragassa C (2016) Advanced design solutions for high-precision woodworking machines. Int J Qual Res 10(1):143–158
11. Itofuji H, Kawano Y, Inoyama N (1983) The formation mechanism of compacted/vermicular graphite in cast irons. Trans Am Foundrymen's Soc 91:831–840
12. Pan EN, Ogi K, Loper CR Jr (1982) Analysis of the solidification process of compacted/vermicular graphite cast iron. Trans Am Foundrymen's Soc 90:509–527
13. Murthy VSR, Seshan S (1985) Vermicular graphite cast iron—current state of the art. Sadhana 8(4):361–372
14. Lampic-Oplander M (2001) Cast iron with vermicular graphite GJV. Giesserei-Praxis (Ger) 1:17–22
15. Lim CH, Goo BC (2011) Development of compacted vermicular graphite cast iron for railway brake discs. Met Mater Int 17(2):199–205
16. Wang YQ, Hu F, Shi HW, Fan XM (2010) Application and development of vermicular iron to cylinder block castings. Modern Cast Iron 6:23–26
17. Guzik E, Dzik S (2009) Structure and mechanical properties of vermicular cast iron in cylinder head casting. Arch Foundry Eng 9(1):175–180
18. Ma ZJ, Tao D, Yang Z et al (2016) The effect of vermicularity on the thermal conductivity of vermicular graphite cast iron. Mater Des 93:418–422
19. Guzik E (2010) Structure and mechanical properties as well as application of high quality vermicular cast iron. Arch Foundry Eng 10(3):95–100
20. Laneri K, Bruna P, Crespo D (2008) Microstructural characterisation and kinetics modelling of vermicular cast irons. Mater Sci Technol 24(10):1214–1221
21. Radovic N, Morri A, Fragassa C (2012) A study on the tensile behavior of spheroidal and compacted graphite cast irons based on microstructural analysis. In: 29th Danubia Adria symposium on experimental solid mechanics. Belgrade, Serbia, pp 164–170
22. Fragassa C, Radovic N, Pavlovic A, Minak G (2016) Comparison of mechanical properties in compacted and spheroidal graphite irons. Tribol Ind 38(1):49–59
23. Fragassa C, Minak G, Pavlovic A (2016) Tribological aspects of cast iron investigated via fracture toughness. Tribol Ind 38(1):1–10

24. Fragassa C, Babic M, Bergmann CP, Minak G (2019) Predicting the tensile behaviour of cast alloys by a pattern recognition analysis on experimental data. Metals 9(557):1–21. https://doi.org/10.3390/met9050557
25. Fragassa C, Babic M, Pavlovic A, do Santos ED (2020) Machine learning approaches to predict the hardness of cast iron. Tribol Ind 42(1):1–9
26. Babic M, Fragassa C, Lesiuk G, Marinkovic D (2020) New method for complexity determination by using fractals and its applications in material surface characteristic. Int J Qual Res 14(3):705–716
27. Fragassa C, Zigulic R, Pavlovic A (2016) Push-pull fatigue test on ductile and vermicular cast irons. Eng Rev 36(3):269–280
28. ISO 1143 (2010) Metallic materials—rotating bar bending fatigue testing

Chapter 4
Rotating Bending Fatigue of Spheroidal Cast Irons

Nenad Radovic, Dragan Marinkovic, and Nenad Miloradovic

Abstract Cast iron probably represents one of the most widely utilized constructive materials in our modern industrial world. This is due not only to the excellent mechanical and tribological properties, but also to a certain flexibility of the same. In fact, it can be produced in various types of metal alloy, which differ significantly in properties: from the poorly performing, but economical white cast iron, to the excellent resistant nodular cast iron. In this study, several preliminary results on the experimental measurement of fatigue resistance are proposed for two distinct families of cast iron, the ductile and the vermicular. While the former is well known, with a great variety of applications, vermicular cast iron is almost unfamiliar at the moment in terms of utilization, despite different potentials. This experiment compares the results of the Rotating Bending Fatigue test for cast iron specimens made under foundry process conditions as identical as possible in the way to reduce bias.

Keywords Metal alloys · Foundry · Spheroidal cast iron · Mechanical properties · Fatigue behavior · Fatigue limit

4.1 Introduction

4.1.1 Spheroidal Cast Iron

The spheroidal cast iron (SGI), also known as ductile cast iron, is a cast alloy in which graphite, instead of in the form of lamellae, occurs in nodules in the shape of spheroids. The nodules are located in a metal matrix whose structure is a function of

N. Radovic
Alma Mater, Studiorum University of Bologna, Bologna, Italy

D. Marinkovic
Technical University of Berlin, Berlin, Germany

N. Miloradovic (✉)
University of Kragujevac, Kragujevac, Serbia
e-mail: mnenad@kg.ac.rs

G. Lesiuk et al. (eds.), *Structural Integrity and Fatigue Failure Analysis*,
Structural Integrity 25, https://doi.org/10.1007/978-3-030-91847-7_4

the chemical composition of the specific type of cast iron, of the cooling rate at the time of solidification and of any subsequent heat treatments [1, 2].

The spheroidal form of graphite produces a lower concentration of tension than the lamellar one; moreover, the spherical shape is the one that, with the same volume, has the smallest surface and the matrix is therefore less damaged, thus being able to better exploit its characteristics [3, 4].

Furthermore, in spheroidal cast iron, graphite nodules exert a stopping action for cracks, unlike lamellar graphite which offers a preferential way for their propagation. Spheroidal cast iron shows a notable improvement of all mechanical characteristics including the fatigue behavior [5–7].

4.1.2 Spheroidal Cast Production

Spheroidal cast iron is produced starting from a base cast iron and adding appropriate magnesium alloys that can favor the spheroidization of the cast iron (Fig. 4.1). Concentrations of chemical elements in molten alloys and process temperatures must be kept under strict control to ensure the correct formation of the microstructure [1, 3].

This accuracy in process and products is important in general for every cast iron, but even more relevant in the case of SGI. For example, a variation of only 2% of Si leads to an increase in the temperature of the stable eutectic and the lowering of the metastable one from the initial 6 to 40 °C with effects on spheroidization. Each modification in the structure affects the main characteristics of the alloy, but even

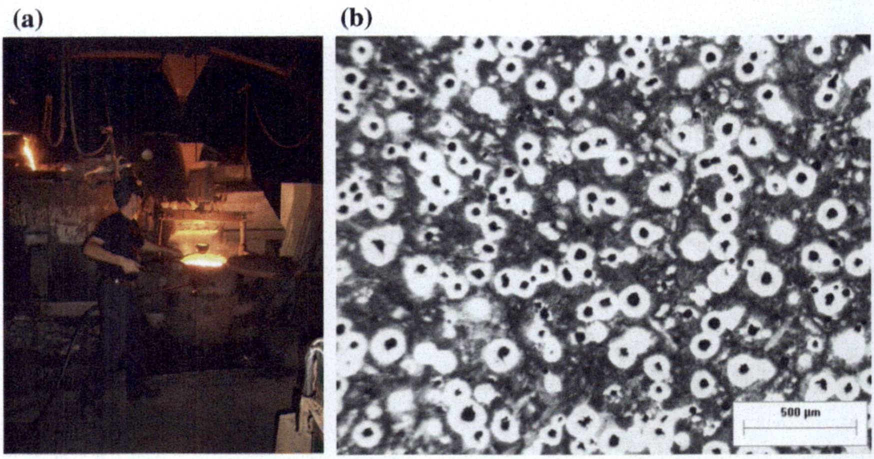

Fig. 4.1 Production of spheroidal cast iron: **a** sampling for process control; **b** microstructure with evidence of the spheroids

more so for those properties that are intrinsically linked to the benefits offered by the spheroids, such as fatigue strength [2–4, 6, 8].

Thus, the study of those properties such as fatigue strength, linked to the propagation of cracks across the microstructures, deformed by nodules, becomes even more fundamental in terms of appearance such as, for example, tensile strength.

4.1.3 Aim and Scope

In line with what just said, this article describes an experimental study aimed at measuring the fatigue strength of a ductile iron produced in a cupola furnace at the SCM Foundry, Italy. This work is part of a broader investigation that led the research team to study some of the main mechanical [9, 10] and tribological properties [11] of nodular cast iron under carefully controlled process conditions, including the development of advanced techniques for data interpretation based on machine learning [12–14]. In particular, regarding the fatigue resistance, a characterization study of SGI is already available, but based on a push–pull load configuration [15]. The present work complements these results by adopting a rotating load configuration.

4.2 Materials and Methods

4.2.1 Samples Production

The fatigue behavior of SGI was investigated by a four-point rotating bending test in accordance with ISO 1143 [16] and ISO 12107 [17]. Standardized specimens were extracted from a metallic block and shaped by tool machining (Fig. 4.2).

Table 4.1 reported the chemical composition (in % of weight) of ductile (SGI) and vermicular (CGI) irons.

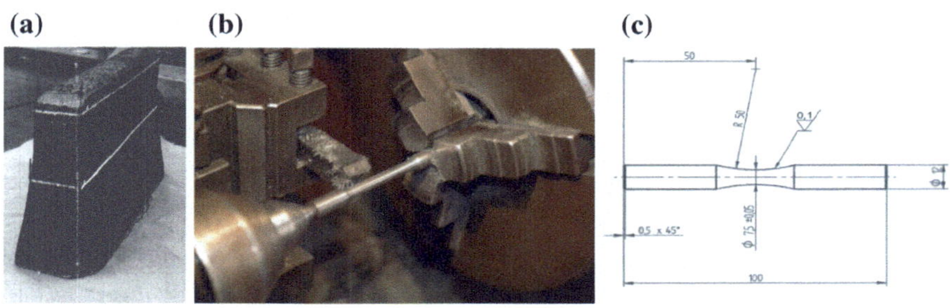

Fig. 4.2 Production of standardized specimens for rotating bending fatigue tests (ISO 1143): **a** metallic block from sand casting; **b** tool machining; **c** 'dog-bone' shape

Table 4.1 Chemical composition (in % of weight) of ductile cast iron (SGI)

Fe	C	Si	Mn	P	S	Ni	Cr
93.1	3.66	2.65	0.218	0.032	0.004	0.069	0.062
Cu	Mo	Mg	Sn	Ti	Pb	Al	Zn
0.062	0.002	0.055	0.013	0.034	0.007	0.011	0

4.2.2 *Fatigue Life*

ASTM considers the 'fatigue life' as the number of stress cycles applied before a failure in the specimen occurs. For some materials, such as cast iron, there is a stress value below which the material will not fail regardless of the number of cycles.

The standards also define, as known, different methods for determining the fatigue behavior of a material, including the identification of the fatigue strength.

In this investigation, fatigue tests were carried out by applying a sinusoidal stress. The tests were performed at different stress levels (S) by recording at how many cycles the failure occurs (N). Each completed test leads to a known failure time (F). In the absence of failure, the test can be terminated without this information. Anyway, for these samples that have finished early (*run-out*), the data (R) can be also considered for evaluation and analyzed using typical statistical techniques.

However, the results may depend on the number of specimens used and on the sequel of loads applied to these specimens.

For this reason, special attention was paid to applying the regulations in the present study. Specifically, the fatigue tests were implemented with a sequence of loads defined using the *modified staircase method* in accordance with the ISO 12107 standard. This method aims to reduce the number of tests and samples.

In brief, it is based on the procedure to decrease the test stress (σ_0) when a previous test has led to failure and increase it if not. The step of change of load (d) is the same, both in the case of increase or decrease, possibly equal to the uncertainty of this kind of measure. The procedure is repeated until all samples are tested.

In the present case, from the knowledge of the general properties of SGI and assumptions of similar investigation, it was decided to use an initial test stress of $\sigma_0 = 215$ MPa, equivalent to 1/3 (33%) of a characteristic value of the Ultimate Tensile Stress (UTS) of SGI. As modification value, $d = 10$ MPa was chosen. Test limit was set to 10^7 cycles (*run-out*). These parameters are also summarized in Table 4.2.

Table 4.2 Parameters definition for modified staircase method

Parameter	SGI
UTS	650 MPa
Fraction	33%
σ_0	215 MPa
D	10 MPa
Test time	10^7 cycles

Table 4.3 Results from rotating bending fatigue tests

Sample	Stress	Cycle		Sample	Stress	Cycle	
1	215	3,000,000	F	9	235	10,000,000	R
2	205	10,000,000	R	10	245	434,000	F
3	215	10,000,000	R	11	235	239,000	F
4	225	10,000,000	R	12	225	1,175,000	F
5	235	574,000	F	13	215	10,000,000	R
6	225	10,000,000	R	14	225	945,800	F
7	235	10,000,000	R	15	215	10,000,000	R
8	245	495,000	F				

F–failure; ***R***–run-out

4.3 Results and Discussion

4.3.1 Experimental Data

N. 15 samples were tested by rotating bending fatigue tests considering a stress level between 205 and 245 MPa. *N*, 7 of them failed (‘*F*’) while *N*. 8 samples run out to 10^7 cycles (‘*R*’). Table 4.3 reports the experimental data.

The same table also exhibits details regarding the test progression. For instance, the first test took place at 215 MPa of load (σ_0) leading to a fail after $3.0 \cdot 10^6$ cycles. Then a 10 MPa reduction in the stress level was implemented and new test showed no fail up to $10 \cdot 10^6$. Thus, the stress level was increased again for additional 10 MPa, but, also in this case, no failure occurred. Then, stress level was increased once again, for additional 10 MPa. And so on.

Figure 4.3 graphically shows the same experimental values. In this case, samples are not reported in the testing order but respect to the level of stress. The filled circles, as symbols, indicate failures while other symbols indicate run-outs. Data show, for instance, that the run-outs are mainly concentrated at lower load values.

4.3.2 Fatigue Limit

Fatigue properties were estimated considering the probabilities of failure of 10% and confidence levels of 90% in accordance with ISO 12107, as shown in Table 4.4. With the scope it was necessary to evaluate the following parameters:

$$A = \sum_{i=1}^{l} i f_i \tag{4.1}$$

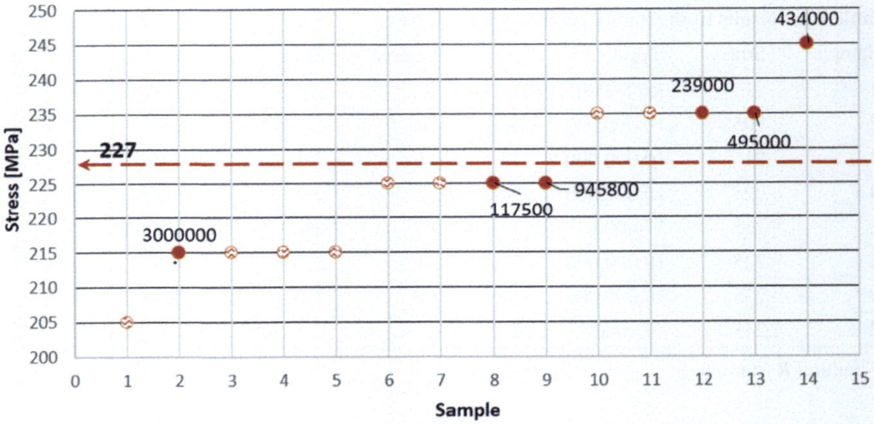

Fig. 4.3 Graphical representation of results from rotating bending fatigue tests

Table 4.4 Scheme for fatigue limit determination at 50% probability of failure

Stress (Mpa)	Sample															Event	
	1	2	3	4	5	6	7	8	9	10	11	12	13	14	15	F	R
245								F		F						2	0
235					F		R		R		F					2	2
225				R		R						F		F		2	2
215	F		R										R		R	1	3
205		R														0	1
																7	8
																SUM	
d =	10 Mpa																
σ_0	205 Mpa																

i	ni	i*ni	i²*ni
1	0	1	0
4	2	8	16
3	2	6	12
2	2	4	8
1	1	1	1
0	0	0	0
	7	19	37
	N	A	B

F–failure; *R*–run-out

$$B = \sum_{i=1}^{l} i^2 f_i \tag{4.2}$$

$$C = \sum_{i=1}^{l} f_i \tag{4.3}$$

$$D = \frac{BC - A^2}{C^2} \tag{4.4}$$

As a consequence, the fatigue limit at 50% probability of failure can be determined as:

$$\sigma_{D(50\%)} = \sigma_o + d\left(\frac{A}{N} \pm 0.5\right) = 227\,\text{MPa} \tag{4.5}$$

4.4 Conclusions

This experimental study provides results regarding the measure of fatigue resistance for ductile cast iron (SGI). Several standards were followed as a reference in the testing and data analysis. Specifically, using 4-points rotating bending fatigue test, up to 10^7 cycles, and 15 specimens, it detects 227 MPa as fatigue limit. This value, much lower than other material resistances (e.g., ultimate tensile stress) makes evident the attention of designers in using cast iron in the presence of high number of load cycles.

References

1. Minkoff I (1983) The physical metallurgy of cast iron. Wiley, New York
2. ASM International (1996) Specialty handbook: cast irons, pp 33–267
3. Angus HT (2013) Cast iron: physical and engineering properties. Elsevier, London
4. Elliott R (1988) Cast iron technology. Butterworth-Heinemann, London
5. Emerson PJ, Simmons W (1976) Final report on the evaluation of the graphite form in ferritic ductile iron by ultrasonic and sonic testing, and the effect of graphite form on mechanical properties. AFS Trans 76(26):109–128
6. Fuller AG, Emerson PJ, Sergeant GF (1980) A report on the effect upon mechanical properties of variation in graphite form in irons having varying amounts of ferrite and pearlite in the matrix structure and the use of nondestructive tests in the assessments of mechanical properties of such irons. AFS Trans 80(9):21–50
7. Radiša R, Dučić N, Manasijević S, Marković N, Ćojbašić Ž (2017) Casting improvement based on metaheuristic optimization and numerical simulation. Facta Univ Ser Mech Eng 15(3):397–411
8. Fragassa C (2017) Material selection in machine design: the change of cast iron for improving the high-quality in woodworking. Proc Inst Mech Eng C J Mech Eng Sci 231(1):18–30
9. Radovic N, Morri A, Fragassa C (2012) A study on the tensile behavior of spheroidal and compacted graphite cast irons based on microstructural analysis. In: 29th Danubia Adria symposium on experimental solid mechanics, Belgrade, Serbia, pp. 164–170
10. Fragassa C, Radovic N, Pavlovic A, Minak G (2016) Comparison of mechanical properties in compacted and spheroidal graphite irons. Tribol Ind 38(1):49–59
11. Fragassa C, Minak G, Pavlovic A (2016) Tribological aspects of cast iron investigated via fracture toughness. Tribol Ind 38(1):1–10
12. Fragassa C, Babic M, Bergmann CP, Minak G (2019) Predicting the tensile behaviour of cast alloys by a pattern recognition analysis on experimental data. Metals 9(557):1–21. https://doi.org/10.3390/met9050557
13. Fragassa C, Babic M, Pavlovic A, do Santos ED (2020) Machine learning approaches to predict the hardness of cast iron. Tribol Ind 42(1):1–9
14. Babic M, Fragassa C, Lesiuk G, Marinkovic D (2020) New method for complexity determination by using fractals and its applications in material surface characteristic. Int J Qual Res 14(3):705–716
15. Fragassa C, Zigulic R, Pavlovic A (2016) Push-pull fatigue test on ductile and vermicular cast irons. Eng Rev 36(3):269–280
16. ISO 1143 (2010) Metallic materials—rotating bar bending fatigue testing
17. ISO 12107 (2003) Metallic materials—Fatigue testing—Statistical planning and analysis data

1.4 Conclusions

References

Chapter 5
Microstructural and Mechanical Characterization of Riveted 19th-century Steels in Terms of Fatigue Lifetime Prediction

B. Szymczyk, G. Lesiuk, A. Zięty, and J. Warycha

Abstract In the paper, the problem of long-term operated materials from nineteenth-century steel was discussed. The detailed analysis was focused on materials aspect and strength degradation based on the experimental results from rivets and metal plate from materials extracted from Sand Bridge (1861) located in Wrocław, Poland. As it is reported, the most sensitive parameters for degradation are cyclic crack growth properties. However, there is no direct link from degradations symptoms to the engineering fatigue lifetime calculation—the presented numerical analysis aimed to bridge this gap using FM-TOOL software. Obtained results suggest significant fatigue lifetime decrease caused by degradation processes. Also, crucial role plays fracture resistance and thus, higher fracture toughness properties expressed in cyclic stress intensity factor K_{fc} obtained for steel in post-operated state.

Keywords Puddle iron · Rivets · Microstructure · Fatigue crack growth

5.1 Introduction

There are a significant number of exploited bridges built in the nineteenth century in the world. There are over 120 bridges in Wroclaw alone, many of which remember the nineteenth century, such as the Sand Bridge, Tumski Bridge or Młyńskie Bridges. Without these crossings, efficient communication between the islands and the city centre would not be possible. Therefore, proper diagnosis of the condition of the bridge is crucial. The visual examinations are the first step to determine the wear condition of the bridge. However, these are only preliminary studies that provide information about the general condition of the examined object. Only taking a sample and carrying out full material test—microstructure, impact strength, hardness and tensile tests—can assess the bridge's wear condition. However, these studies are

B. Szymczyk · G. Lesiuk (✉) · A. Zięty · J. Warycha
Faculty of Mechanical Engineering, Wroclaw University of Science and Technology, Smoluchowskiego st. 25, PL50370 Wrocław, Poland
e-mail: Grzegorz.lesiuk@pwr.edu.pl

G. Lesiuk et al. (eds.), *Structural Integrity and Fatigue Failure Analysis*, Structural Integrity 25, https://doi.org/10.1007/978-3-030-91847-7_5

mostly insufficient. It should be remembered that the crossings created in the nineteenth century were built of puddled or early mild steel, and these could be subject to significant ageing processes. Therefore, if possible, fatigue tests are performed, and on their basis, the limit number of cycles to failure is determined.

5.2 Plate Steel and Rivet Steel from Nineteenth-Century Bridges

The nineteenth century is famous for the spread of steel as a construction material. This is particularly evident in bridge construction, which was booming in the nineteenth century. All over the world, you can still find crossings made at that time. To a large extent, these are beautiful riveted constructions, which are symbols of the past.

5.2.1 Puddled and Early Mild Steel

At the end of the eighteenth century, Henry Cort invented the puddling kiln, which revolutionized steels production and made it possible to produce puddle steels. The puddling furnace was the flaming coal-fired furnace, consisted of three zones: the hearth, working zone and chimney. In such a furnace, the coal was burned on a horizontal grate, and the exhaust gases passed into the working zone, heating the charge located there and then escaped through the chimney. During the process, iron oxide was added, which contributed to the oxidation of silicon, manganese and phosphorus. This process raised the temperature in the furnace and caused the "charge to boil". The charge had to be stirred all the time, hence the name of steel. [1, 2]

The melting point of iron was not reached in the pudding furnaces; therefore, the product obtained had a spongy consistency. This contributes to numerous nonmetallic inclusions formed into sprits after the rolling process. Therefore, this steel is characterized by a low carbon content—$0.018\% \div 0.3\%$, and anisotropy, which is the property resulting from the streaking structure and a variable Young's modulus. In addition, numerous inclusions of sulphur and phosphorus contribute to the increase of tensile strength at the expense of elongation and plasticity, improve corrosion and abrasion resistance slightly, and simultaneously tend to segregation and cracking of steel. This steel is not suitable for hardening [1, 3, 4].

Another breakthrough invention was the Bessemer furnace, invented in 1856. Its design allowed for achieving full penetration of the charged material, which contributed to better properties of the steel. In this way, liquid steel was free of slag impurities, which contributed to the homogeneous structure and properties in all directions. The steel obtained in Bessemer furnaces was suitable for hardening as well as heat and cold treatment. It was the material with good tensile strength and good ductility and sufficient isotropy. The last, but not at least advantage of this steel is the possibility of obtaining 3–4 times more material from one process than from a pudding furnace [1, 3, 5].

5.2.2 *Riveting*

Riveting is one of the oldest joining methods, known already in the Middle Age. This technology flourished most in the eighteenth and nineteenth centuries when industrial revolutions took place. Riveted joints are intermediate permanent joints where the joint is a rivet—a cylindrical mandrel with ahead. They are used to connect flat elements e.g. steel, wooden or polymer sheets. To combine two elements, holes of the appropriate diameter should be made in both of them. Most often, holes are drilled in the assembly to ensure their concentricity. Then the rivet is placed in the prepared hole, the rivet head is supported with a special bracket and finally crimped to form a rivet joint [1, 6, 7].

The rivet diameter and size of the production define the tools and the rivets joint process temperature. Manual or pneumatic hammers are used for small diameter rivets. However, if the rivet diameter is large, and large-scale production, hydraulic or mechanical riveting machines are used. The situation is similar to the process temperature. If the rivet diameter is smaller than 9 mm, it rivets cold, if it is larger, it rivets hot [6, 7].

Due to the various applications, space availability and tools used, there are many types of rivets. The most popular rivets are ball-headed solid rivets. There are also solid rivets with trapezoidal or lenticular heads. Additionally, it can be found hollow or tubular rivets with a turned head. The latter can be found in some ships [1, 6, 7].

5.3 Microstructural Analysis

Steels produced in the nineteenth century are characterized by a low carbon content—in the range of $0.018 \div 0.3\%$. Therefore, their structure is dominated by ferrite with slight precipitation of perlite. Due to such a structure, the degradation processes occurring in the puddled and/or early mild steels consist in the decomposition of ferrite into perlite and carbides, precipitation of carbides and nitrides inside and at the grain boundaries (tertiary cementite is formed at the boundaries) and perlite degradation called degenerate perlite. [1, 8] Any change in the microstructure of the material and any degradation reduces its mechanical properties.

First, the chemical composition of the collected rivet from the Sand Bridge was tested. The results are presented in Table 5.1.

The research shows that the Sand Bridge was built of puddled steel. The content of ordinary steel additives (silicon, manganese, sulphur and phosphorus) does not exceed the values typical for steel produced in the nineteenth century.

Table 5.1 The chemical composition of the Sand Bridge rivet

C	Si	Mn	P	S	Cr	Ni	Cu	Al	Mo	V
0.07	0.01	0.36	0.013	0.021	0.08	0.48	0.23	0.01	0.04	< 0.005

Source Own research

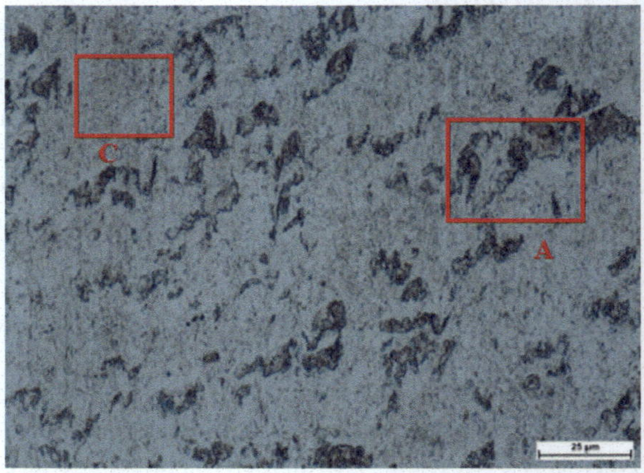

Fig. 5.1 Pin of the rivet from the Sand Bridge–etched condition, A–degenerate perlite, C–brittle precipitations inside the grains. (*Source* Own research)

Significant amounts of nickel and copper are puzzling as they exceed the accepted ranges for puddled steels. Nickel dissolved in ferrite may cause its strengthening, while cooper increases corrosion resistance, and its content increases with the increased consumption of scrap in the processing process. Hence, it can be concluded that in the nineteenth century, previously processed material was used for the production of puddle steel [1].

The microscopic tests were carried out for the rivet head, mandrel and the bridge part. For the observation, the rivet was cut in a longitudinal section. The specimens were subjected to microscopic analysis in the non-etched state and with 5% HNO_3 etching, using the Nikon Ma200 light microscope (Figs. 5.1, 5.2 and 5.3).

Hardness tests were also carried out using the Vickers Hardness Tester FV–800. The samples were loaded with a value of 0.5 kgf. Five measurements were made on different parts of the samples, each lasting 10 s. The entire test procedure was run at room temperature. The hardness test results for the rivet head, pin and for the bridge fragment are presented in Table 5.2.

5.4 Mechanical Characterization and Degradation Processes–Fatigue Lifetime Reduction

In papers [9, 10] were presented the studies of fatigue crack growth in welded steel extracted from Sand Bridge in Wroclaw. In both papers, the tests were conducted

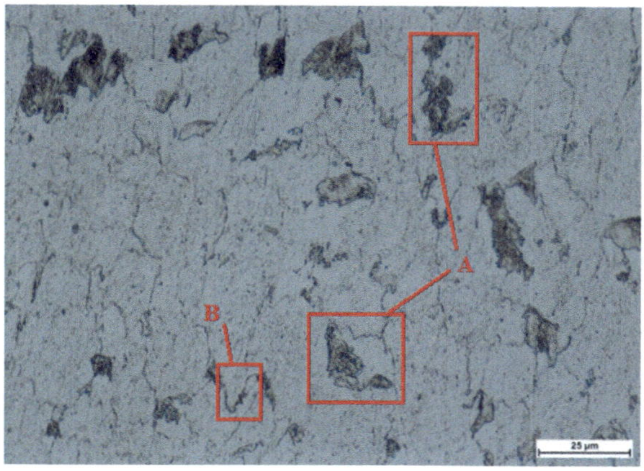

Fig. 5.2 The head of the rivet from the Sand Bridge-etched condition; A–degenerate perlite, B–cementite at the grain boundaries, C–precipitions inside the grains. (*Source* Own research)

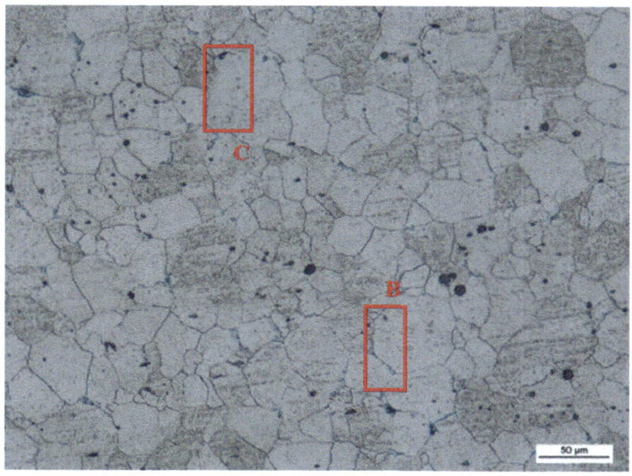

Fig. 5.3 Base metal of the Sand Bridge-etched condition, B–cementite at the grain boundaries, C–precipition inside the grains. (*Source* Own research)

Table 5.2 Hardness measurement results for the rivet head and pin and the bridge fragment

Measurement	1	2	3	4	5	Average
Head rivet [HV1]	108.0	110.1	108.2	114.1	105.4	109.2
Pin rivet [HV1]	124.3	121.1	110.0	116.3	125.4	119.4
Structural element from bridge [HV1]	97.5	100.5	97.4	101.2	100.2	99.4

Source Own research

following ASTM E647 on $M(T)$ specimens (Middle Cracked Tension Specimen). The dimensions of specimens were following: $W = 40$ mm, $L = 160$ mm, $t = 5$ mm. The initial crack length was $a_0 = 4$ mm. The specimen was loaded with a sinusoidal waveform keeping constant stress ratio $R = 0.1$, force load ranging from 1.65–16.5 kN at a frequency of 10 Hz.

In the paper [10], the specimens tested were in the operational condition, while in the paper [9] they were in both the operating and normalized condition. Comparison of both states allows determining the degree of microstructure degradation. Both papers present similar results of the Paris model constants, which are shown in Table 5.3. In addition, the approximate number of cycles to material failure was also determined.

To better illustrate such changes, a simulation of fatigue lifetime was performed using FM-TOOL software delivered by Nobo Solutions company (Fig. 5.4).

Stress intensity factor was analytically calculated based on solutions available in handbook [11]:

$$K_I = F_m \sigma_m \sqrt{\pi a}. \tag{5.1}$$

$$F_m = 0.265(1-\lambda)^4 + \frac{0.857 + 0.265\lambda}{(1-\lambda)^{1.5}}. \tag{5.2}$$

$$\lambda = \frac{a}{t} \tag{5.3}$$

where:

K_I Stress intensity factor (SIF)
F_m Geometrical factor (nondimensional)
a Crack length

Table 5.3 Kinetic fatigue fracture properties of tested steel

	Post-operated state [9]	Normalized state [9]
K_{th} [MPa $\sqrt{m}$]	14	17
K_{fc} [MPa $\sqrt{m}$]	24	28
Paris law exponent m	5.34	5.11
Paris law constant C	$10^{-11.52}$	$10^{-11.74}$

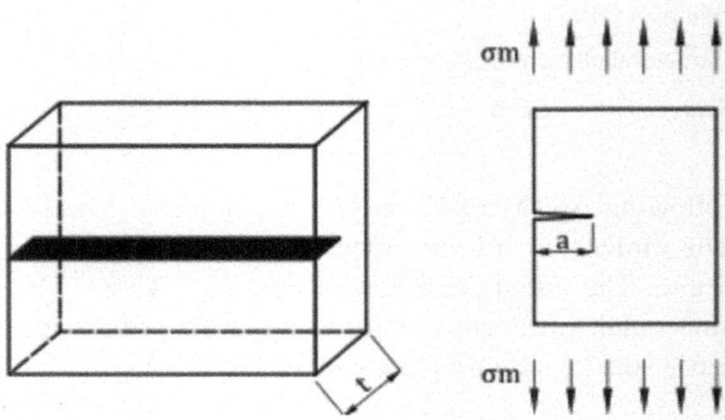

Fig. 5.4 Crack model for simulation ($t = 0.015$ m, $a = 0.003$ m, membrane stress $\sigma_m = 100$ MPa)

σ_m Membrane stress (tension).

Graphical representation of F_m (2) function is shown in Fig. 5.5.

For simulation, Paris law was selected as the main crack propagation rule. It is worth to note that the stooping criterion was based on critical stress intensity factor K_{fc} experimentally determined (Table 5.3). Numerical results are plotted as fatigue curves in Figs. 5.6 and 5.7. As it was expected for material in post-operated state fatigue lifetime does not exceed 222,591 cycles on the contrary to the fatigue life in the material after heat treatment (normalization) $Nf = 691{,}745$. It allows for the calculation of the fatigue lifetime improvement expressed in relative fatigue lifetime ratio:

$$\eta = \frac{N_{\text{normalized}} - N_{\text{postoperated}}}{N_{\text{postoperated}}} \cdot 100\% \approx 211\% \tag{5.4}$$

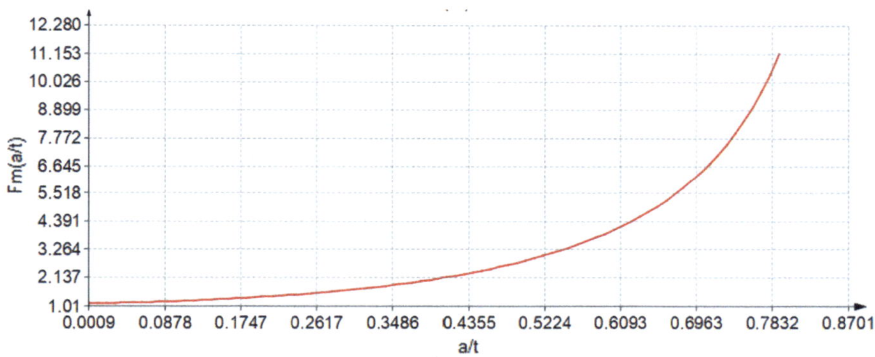

Fig. 5.5 Nondimensional geometric factor F_m as a function of normalized crack length (*a/t*) for assumed load case. (*Source* Own research)

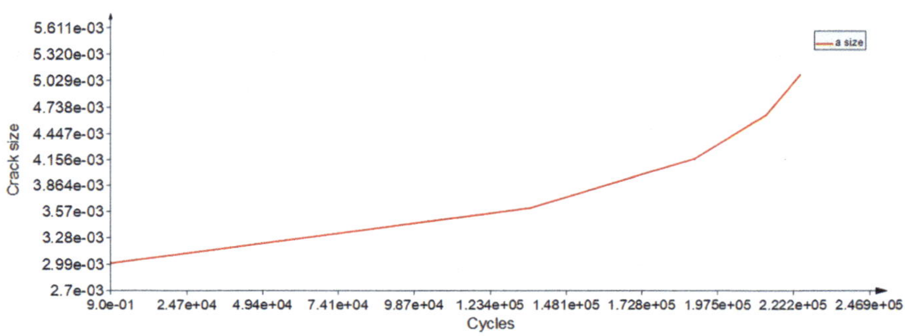

Fig. 5.6 Fatigue crack growth curve for material from Sand Bridge (post-operated state, crack size in m) with a critical number of cycles to failure $N_f = 222{,}591$, (*Source* Own research)

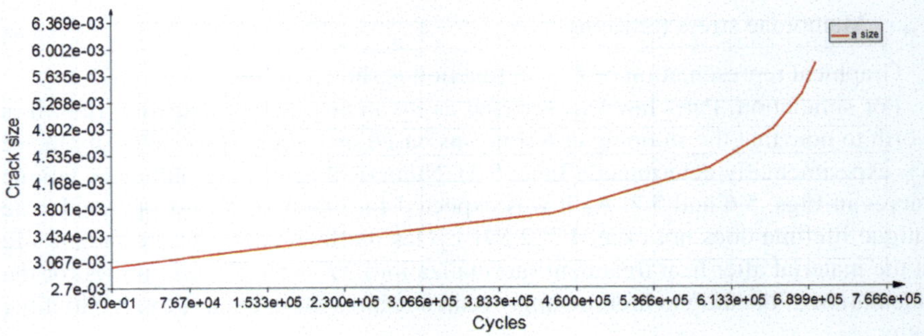

Fig. 5.7 Fatigue crack growth curve for material from Sand Bridge (normalized—after heat treatment state, crack size in m) with a critical number of cycles to failure $N_f = 691\ 745$, (*Source* Own research)

5.5 Conclusions

Based on the performed materials investigations and literature data analysis combined with numerical simulations, the following conclusions can be drawn:

- Microscopic analysis of the tested materials from nineteenth-century bridge show undoubtedly the brittle nature of such materials,
- Hot riveting is one of the effective joining methods referred to in the old nineteenth-century structures. It is more likely that cracks preexisted in the metal base will propagate much faster than in modern (replaced) new materials.
- Higher fatigue crack growth rate can be associated with microstructural degradation processes
- Heat treatment (normalization) as a recovery material treatment, demonstrates much slower fatigue crack growth rate in such material—at least fatigue lifetime was extended (for assumed load case) more than 200% compared to the post-operated state.

Acknowledgements The publication has been prepared as a part of the Support Programme of the Partnership between Higher Education and Science and Business Activity Sector co-financed by City of Wroclaw.

References

1. Szymczyk B (2020) Ocena nośności XIX—wiecznych połączeń nitowych w świetle degradacji mikrostrukturalnej zachodzących w długotrwale eksploatowanych stalach, Bachelor Thesis, Wroclaw University of Science and Technology
2. https://fantasticoss.wordpress.com/2012/09/02/hutnictwo-w-wieku-xviii/. Accesed 10 Feb 2021

3. Czapliński K (2009) Dawne wyroby ze stopów żelaza. Dolnośląskie Wydawnictwo Edukacyjne
4. Gierczak J, Ignatowicz LR, Sawicki M (2016) Właściwości materiałowe stali stosowanej w konstrukcjach balkonów i logii pochodzących z XIX i XX w. Builder s, pp 68–69
5. Gierczak J (2016) Przebudowy konstrukcji stalowych część 2; analiza stali. Builder, s. pp 88–91
6. Osiński Z (2002) Podstawy konstrukcji maszyn. Wydawnictwo Naukowe PWN, Warszawa
7. Majdański Z (2020) Wykonywanie połączeń nierozłącznych—połączenia nitowe. Instrukcja laboratoryjna ZSM
8. Lesiuk G, Szata M (2010) Degradacja mikrostruktur elementów konstrukcyjnych pochodzących z mostów wzniesionych na przełomie XIX i XX wieku. Zesz Naukowe WSOWL, nr 1(155)
9. Lesiuk G, Szata M (2011) Aspects of structural degradation in steels of old bridges by means of fatigue crack propagation. Mater Sci 47(1):82–88
10. Lesiuk G, Szata M (2014) The mathematical model of the fatigue crack growth rate in 19th puddled steel. Górnictwo Odkrywkowe 55
11. Tada H, Paris PC, Irwin GR (1973) The stress analysis of cracks, handbook

Part II
Fatigue Life Prediction

Chapter 6
Acoustic Emission Characteristics for Determining Fatigue Damage Behaviour

N. Md Nor, S. N. Mat Saliah, S. Abdullah, S. S. K. Singh, and N. A. Yahya

Abstract In reality, damage in the reinforced concrete structure under fatigue causes poor performance under service loading. Hence, this study aims to investigate the acoustic emission characteristics of the reinforced concrete beam subjected to fatigue loading at different sensors. Laboratory experiments were performed on the 150 mm × 150 mm × 750 mm reinforced concrete beam under fatigue loading in conjunction with acoustic emission monitoring. Then, the acoustic emission parameters on four sensors such as absolute energy, amplitude and average frequency were analysed. It is found that a good correlation between acoustic emission characteristics and the normalised cycles for all sensors. Hence, the response on the acoustic emission characteristic can be used to predict fatigue damage prognosis of the reinforced concrete beam.

Keywords Acoustic emission · Fatigue · Fatigue damage

6.1 Introduction

Fatigue in a concrete structure is a process of damage accumulation owing to the repetitive application of loads. The repetition of the loads may not produce any effect at an earlier stage. Over several cycles, as the loads continuously applied, the damage generates intrusions and extrusions that resemble a crack, which led to a disaster when the damage becomes larger. It generally occurs on concrete structures such as buildings, bridges and dams. Such a problem led to a recommendation of a proper

N. Md Nor (✉) · S. N. Mat Saliah
Civil Engineering Studies, Universiti Teknologi MARA, Cawangan Permatang Pauh, Kampus Permatang Pauh, Permatang Pauh, Malaysia
e-mail: ida_nsn@uitm.edu.my

S. Abdullah · S. S. K. Singh
Department of Mechanical and Manufacturing Engineering, Faculty of Engineering & Built Environment, Universiti Kebangsaan Malaysia, Bangi, Malaysia

N. A. Yahya
School of Civil Engineering, College of Engineering, Universiti Teknologi MARA, Shah Alam, Malaysia

G. Lesiuk et al. (eds.), *Structural Integrity and Fatigue Failure Analysis*, Structural Integrity 25, https://doi.org/10.1007/978-3-030-91847-7_6

inspection technique so that rational decisions regarding maintenance, repairing, rehabilitation and replacement can be made. Yuyama et al. [1] recommend using acoustic emission technique to inspect fatigue damage of reinforced concrete slab. Wang et al. [2] apply acoustic emission for the investigation of the fatigue damage process of rubberised concrete and plain concrete. Acoustic emission has been used by Md Nor et al. [3] for inspection of fatigue damage behaviour of reinforced concrete beam. Mohammad et al. [4] has applied acoustic emission to predict the fatigue life of SAE 1045 carbon steel. It is also has been used by Md Nor et al. [5] on fatigue damage classification of the precast reinforced concrete beam.

Acoustic emission is defined as the class of phenomena whereby elastic waves are generated by the rapid release of energy from the localised source or sources within a material, or the transient elastic waves generated [6]. The acoustic emission has been generally used for detection of fatigue damage in a concrete structure employing sensors that fixed on the selected positions. From the signal captured by the sensors, it gives information on the condition of the structure, where the increase in acoustic emission characteristic is generally associated with the rise of new structural damage. Investigation of different types of sensors in terms of their frequency, signal and sensitivity has been investigated by Meserkhani et al. [7] and Bhuiyan et al. [8]. Meserkhani et al. [7] found that the general-purpose sensor with the frequency range of 35–100 kHz produced better absorbing acoustic emission energies than the other sensors. Bhuiyan et al. [8] found that all sensors represent a similar pattern of acoustic emission hit. Although types of sensors are essential in acoustic emission monitoring, acoustic emission characteristic at certain sensor positions also needs to be taken into consideration. However, the effect of the sensor's position was not investigated in-depth primarily related to the acoustic emission energy on concrete either absolute energy or signal strength. In the present study, the absolute energy, amplitude and average frequency are highlighted. The absolute energy is true energy which is calculated by squaring the digitised acoustic emission signal and integrating the results during a hit [9]. To develop a comprehensive understanding of the acoustic emission activities, acoustic emission signal collected at different locations of sensors becomes the key that needs to be taken into account. In doing so, the present study investigates the acoustic emission parameters of a reinforced concrete beam subjected to fatigue loading at different sensors.

6.2 Experimental Programme

6.2.1 Preparation of Reinforced Concrete Beam

A total of ten reinforced concrete beams were prepared with the concrete grade of C40. The concrete proportion of 1: 0.43: 2.16: 2.60 was designed for cement: water: fine aggregate: coarse aggregate. To boost the workability of the fresh concrete mix, a retarder with 1% of cement weight was added. The beams are prismatic standard size with an effective length, Le of 630 mm, having an actual length, $L = 750$ mm, width, $b = 150$ mm and thickness, $h = 150$ mm. The beams were designed as singly reinforced concrete and procured by two high yield steel reinforcements of 16 mm diameter on the tension part. The nominal cover of 20 mm was used for this beam. In order to fix the stirrups, two mild steel reinforcements of 8 mm diameter as hanger bars were set on the compression part. Then, 12 mm diameter of mild steel as stirrups with a spacing of 100 mm centre to the centre were utilised.

In the beam preparation, the reinforcement was submerged in the mould, and the concrete mix was cast. After ±24 h, the beams were demoulded and cured in water for 28 days. The compressive strength of the concrete at the age of 28 days was 44.65 N/mm^2, which found to be greater than the design concrete.

6.2.2 *Laboratory Test*

A constant-amplitude sine wave load cycle (frequency of 1 Hz) was applied to the beam size 150 mm × 150 mm × 750 mm. Four sensors (CH1 to CH4) were used, and the positions of the sensors on the beam are illustrated in Fig. 6.1.

For fatigue test, maximum load, P_{max} and minimum load, P_{min} were based on the ultimate load, P_{ult} value. The P_{ult} was taken by averaging the ultimate load of five beams under monotonic loading. The average P_{ult} was 158.85 kN. Hence, the P_{max} and the P_{min} were 0.8 P_{ult} and 0.2 P_{ult}, respectively. Wang et al. [2] stated that acoustic emission signals for data analysis obtained from high maximum loading are relatively small. The fatigue test in conjunction with acoustic emission monitoring was performed with this load range until failure, as shown in Fig. 6.2. Four sensors (CH1 to CH4) were used during acoustic emission monitoring. Two sensors were fixed on top of the beam with a distance of 175 mm from the edge of the beam. Meanwhile, the other two sensors were set at the centre of the cross section of the

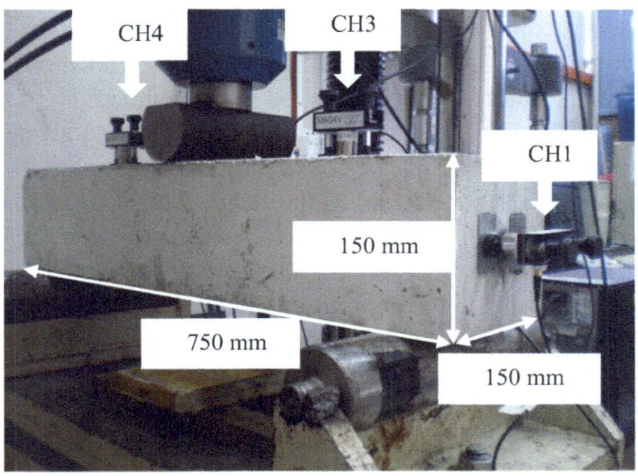

Fig. 6.1 Size of the beam and the location of sensors

Fig. 6.2 Set up of the beam

beam, as shown in Fig. 6.1. Then, the acoustic emission absolute energy, amplitude and average frequency were analysed and compared for all sensors (CH1, CH2, CH3, and CH4).

6.3 Results and Discussion

6.3.1 Absolute Energy

Figure 6.3 shows the absolute energy, cumulative absolute energy with respect to the normalised cycle (N/Nf) that collected from sensors 1 to 4 and designated as CH1 to CH4, respectively. Generally, the fatigue process in a structure divides into three stages; initiation stage, steady stage (propagation) and final stage (unstable fracture). This process can be seen clearly in CH1, as presented in Fig. 6.3a. CH1, which located at the edge of the beam, received the highest absolute energy at the initiation stage of fatigue damage compared to other sensors. It is because the amplitude of the $P_{\max}$ and $P_{\min}$ applied to the beam-induced formation of the first fatigue crack, which propagated to the neutral axis of the beam, close to CH1. The process of fatigue damage can be clearly observed through the graph of cumulative absolute energy in Fig. 6.3a. From this graph, it is a fact that high absolute energy is presented in the beginning and the end period of fatigue loading. Xu [10] stated that early microcracks generate a higher signal as the flexural cracks form. The absolute energy was then reduced instantly and almost constant throughout the cycle, produced lower energy in the steady stage. A similar finding has been found by Wang et al. [2] on the fatigue damage process of rubberised concrete and plain concrete.

In the fatigue damage process, high absolute energy at the initiation stage is closely related to the formation of microcrack in the concrete rather than the reinforcement in the beam. Yuyama et al. [1] stated that this is because crack density increases rapidly at the beginning of fatigue life of concrete structure due to formation of

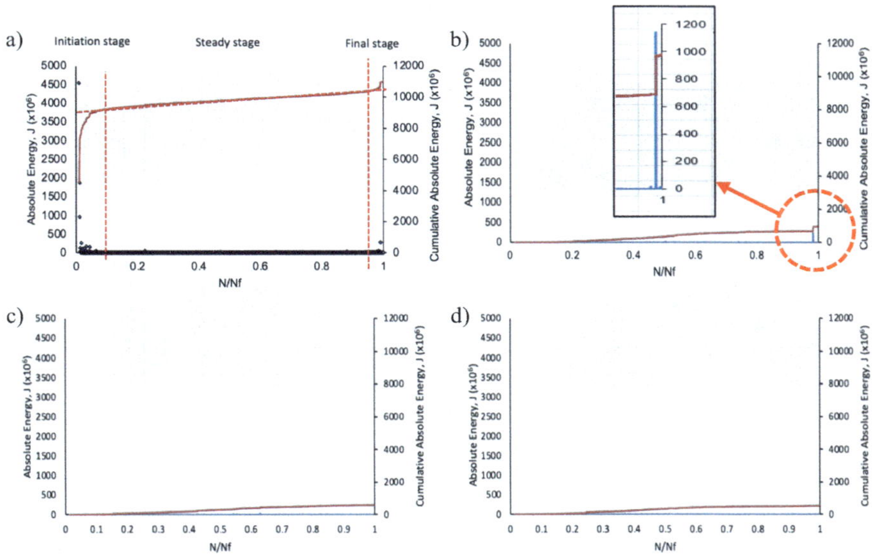

Fig. 6.3 The relationship between absolute energy and cumulative absolute energy and N/Nf for **a** CH1, **b** CH2, **c** CH3 and **d** CH4

microcracking. However, the crack extension starts to slow down after this. As the fatigue loading increases, no significant crack growth can be observed in the steady stage. However, from the experimental work, the other sensors CH2 to CH4 collected lower acoustic emission data and lower absolute energy compared to CH1, as shown in Figs. 6.3b–d. A similar pattern of acoustic emission characteristics can be found for CH2 to CH4 for absolute energy and their accumulation.

Related to high absolute energy, Muhammad Sauki et al. [11] imply that accumulation of damage in the reinforced concrete structure generated high acoustic emission activity. As stated by Abdul Hakeem et al. [12], the cumulative energy is closely related to the formation of crack as the beam subjected to load. However, the appearance of a crack in a structural system depends on the load rate applied during testing [3]. Carpinteri et al. [13] interpreted that the acoustic emission energy correlates to the surplus of elastic energy concerning the dissipated one, which correspondence of unstable behaviours. The more unstable behaviours or instability in the structure, the higher the emitted energy and thus produces higher acoustic emission activities [13].

Although the CH2 represents low absolute energy at the initiation stage and steady stage of fatigue process, high absolute energy can be observed at the final stage as the beam failed, as shown in Fig. 6.3b. Wang et al. [2] imply that the energy of acoustic emission signals also related to the loading and unloading process. Fewer acoustic emission events happen at the unloading stage, and the energy of these acoustic emission signals is much lower when compared with those at reloading stage. The

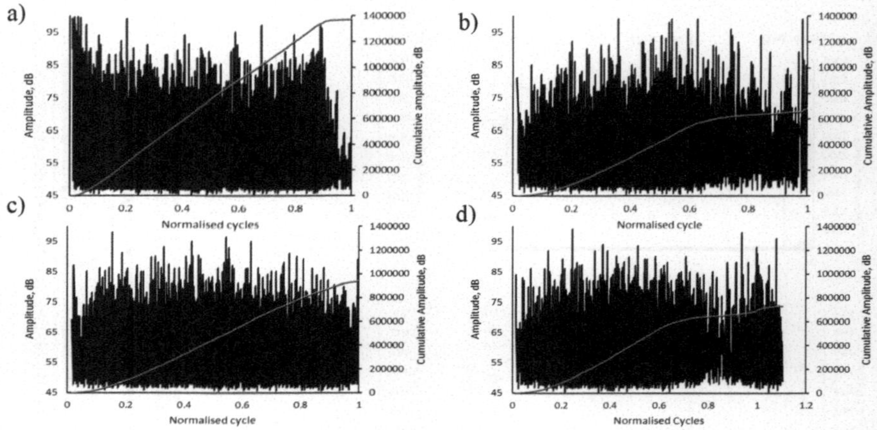

Fig. 6.4 The amplitude and cumulative amplitude for **a** CH1, **b** CH2, **c** CH3 and **d** CH4

peak load generally produces high acoustic emission energy [2]; therefore, acoustic emission in CH2 is.

6.3.2 Amplitude, Cumulative Average Frequency and Cumulative Amplitude

Figure 6.4 shows one of the acoustic emission parameters of amplitude that generally used for evaluation of a structure. CH1 represents a higher amplitude at the beginning of the fatigue process compared to other sensors. This increased amplitude indicates the formation of crack on the concrete as the fatigue load and unloads applied to the beam. The position of the sensors is also played an essential role in collecting the acoustic emission signals. A sensor closer to the acoustic emission source will capture high amplitude than the farther sensor. Since the CH1 is the more relative sensor to the formation of the first crack, it is relatively received the high amplitude at the beginning of the fatigue process. Yuyama et al. [1] added that the microcrack initiation in the beam induces the crack density increases. However, the crack extension starts to slow down after that. However, an almost similar pattern can be observed for CH2 to CH4 when the fatigue process reaches a steady stage. At this stage, the sensors collected high amplitude with the value of more than 95 dB. It pronounces to the progression of crack that occurred in the beam. The amplitude relies on the severity of the fatigue damage during loading and reloading process.

The analysis of the linear correlation between two variables of cumulative average frequency and cumulative amplitude and the normalised cycle is analysed to quantify the damage degree inside the reinforced concrete beam. It also has been performed by Wang et al. [2] to quantify the degree of damage of concrete and rubberised

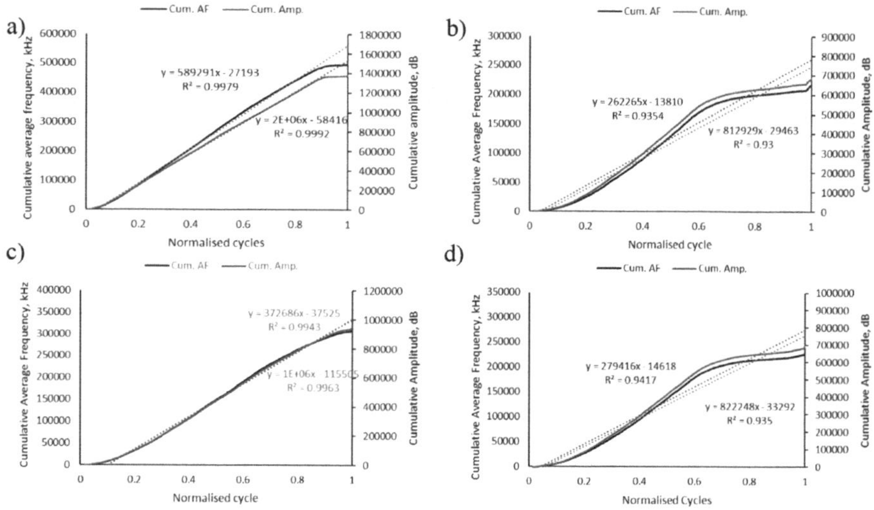

Fig. 6.5 Linear correlation of **a** CH1, **b** CH2, **c** CH3 and **d** CH4

concrete. Figure 6.5 shows the relation between the cumulative average frequency and cumulative amplitude and normalised cycles obtained during the fatigue test. A good linear correlation exhibited between the cumulative average frequency and the cumulative amplitude and the normalised cycle, by presenting the R-squared for all sensors is more than 0.9. Cui et al. [14] stated that a good correlation is indicated if the R-squared value approaches 1.0. Figure 6.5 also presents the equation corresponding to fatigue trendline in terms of the linear-law. From these correlations, the response on the average frequency and amplitude—time series can be used to predict fatigue damage prognosis of the reinforced concrete beam. The fatigue damage prognosis primary aims to predict load-induced fatigue damage to the beam in the future according to the current fatigue damage state of the beam. It can be inferred that the accumulated average frequency and amplitude can be used as a damage variable to quantify damage process of the reinforced concrete beam.

6.4 Conclusions

In a nutshell, acoustic emission characteristics analysed on four sensors of CH1 to CH4 to investigate the fatigue damage of the reinforced concrete beam, which includes the absolute energy, amplitude and average frequency. From the four sensors, the CH1 represents the higher absolute energy compared to other sensors as the sensors located nearer to the acoustic emission source. From the analysis, it is found that high absolute energy was observed at the initiation stage as the microcrack formed and the final stage as the reinforced concrete beam failed. However, the energy is much lower at the steady stage as the crack propagation slows down. Acoustic emission amplitude is higher at the initiation stage of the fatigue process for CH1 compared to other sensors. However, as the fatigue loading increases, high amplitude is noticed at CH2, CH3 and CH4 with the value of more than 95 dB.

From the correlation analysis, a good linear correlation presented for all sensors between the cumulative average frequency and the cumulative amplitude and the normalised cycle with the value of R-squared is more than 0.9. Hence, the response on the average frequency and amplitude related to time series can be used to predict fatigue damage prognosis of the reinforced concrete beam.

Acknowledgements The authors would like to express their thanks to Universiti Teknologi MARA, Cawangan Pulau Pinang, Universiti Teknologi MARA, Malaysia and Universiti Kebangsaan Malaysia for providing financial support towards the publication of this manuscript.

References

1. Yuyama S, Li ZW, Yoshizawa M, Tomokiyo T, Uomoto T (2001) Evaluation of fatigue damage in reinforced concrete slab by acoustic emission. NDT&E Int 34:381–387
2. Wang C, Zhang Y, Ma A (2011) Investigation into the fatigue damage process of rubberised concrete and plain concrete by the AE analysis. J Mater Civ Eng 23(7):953–960
3. Md Nor N, Ibrahim A, Muhamad Bunnori N, Mohd Saman H, Mat Saliah SN, Shahidan S (2014) Diagnostic of fatigue damage severity on reinforced concrete beam using acoustic emission technique. Eng Fail Anal 41:1–9
4. Mohammad M, Abdullah S, Jamaludin N, Innayatullah O (2014) Predicting the fatigue life of the SAE 1045 steel using an empirical Weibull-based model associated to acoustic emission parameters. Mater Des 54:1039–1048
5. Md Nor N, Ibrahim A, Muhamad Bunnori N, Mohd Saman H (2013) Acoustic emission signal for fatigue crack classification on reinforced concrete beam. Constr Build Mater 49:583–590
6. ASTM E976–10 (2010) Standard guide for determining the reproducibility of acoustic emission sensor response. ASTM International 5
7. Meserkhani A, Jafari SM, Rahi A (2020) Experimental comparison of acoustic emission sensors in the detection of outer race defect of angular contact ball bearings by artificial neural network. Measurement
8. Bhuiyan Y, Lin B, Giurgiutiu V (2019) Characterisation of piezoelectric wafer active sensor for acoustic emission T sensing. Ultrasonics 92:35–49
9. Vallen Systeme (2012) AMSY-6, Operational manual, Vallenm Systemer GmbH
10. Xu J (2008) Nondestructive evaluation of prestressed concrete structures by means of acoustic emissions monitoring. Ph.D. thesis, Auburn, Alabama. ProQuest LLC
11. Muhammad Sauki MIH, Md Nor N, Juhari N, Endut MZ (2019) Acoustic emission characteristic for different thicknesses of RC slabs. J Phys Conf Ser 1349
12. Abdul Hakeem Z, Noorsuhada MN, Azmi I, Noor Syafeekha MS, Soffian Noor MS (2017) Analysis of acoustic emission cumulative signal strength of steel fibre reinforced concrete, 2017) (SFRC) beams strengthened with carbon fibre reinforced polymer (CFRP) In: AIP Conference Proceedings 1901:1–7
13. Carpinteri A, Lacidogna G, Corrado M, Di Battista E (2016) Cracking and crackling in concrete-like materials: a dynamic energy balance. Eng Fract Mech 155:130–144
14. Cui C, Xu YL, Zhang QH, Wang FY (2020) Vehicle induced fatigue damage prognosis of orthotropic steel decks of cable-stayed bridges. Eng Struct 212

Chapter 7
Predicting the Fatigue Life of Hydraulic Unit Under Variable Operating Conditions

Evgeniia Georgievskaia

Abstract The project lifetime of most hydraulic units is 30–40 years. However, the actual operational conditions at HPP can significantly prolong or significantly reduce this value. If you want to correctly assess the real unit's lifetime, it is very important to take into account the whole operating range. The announced approach to calculating the total accumulated damage proposes dividing the whole operating range into specific loading blocks with a characteristic frequency and a characteristic amplitude. The fatigue damages are calculated separately for each block take into account the peculiar fatigue features of material behavior at characteristic frequencies and the total number of loading cycles. The total exhausted lifetime is the sum of the damages on each load block. The residual lifetime is the difference between allowed damage and exhausted damage. The proposed approach allows us to make reliable individual lifetime predictions for various scenarios under real operation at HPP correctly and quickly. It helps to capture the most dangerous modes and realize smart lifetime management.

Keywords Hydraulic unit · Lifetime · Fatigue damage · Variable operating conditions

7.1 Introduction

Now hydropower is about 20% of total world energy production and about 75% of total renewable energy. The project lifetime of most hydraulic units is 30–40 years. As practice shows, some of them have premature damages after 10 years or less.

Figure 7.1 shows the typical picture: comparison of the exhausted lifetime as of 2011, the forecast lifetime to 2025, and critical life for the main hydro equipment in the example of a large Russian energy company [1].

In world practice, the following gradation is accepted: the critical level of equipment wear is about 30–35%, the unreliable level is more than 30–35%, and the wear

E. Georgievskaia (✉)
Center of Design and Technological Innovation» LLC, Saint Petersburg, Russia
e-mail: sciencedir@cdti.ru

G. Lesiuk et al. (eds.), *Structural Integrity and Fatigue Failure Analysis*, Structural Integrity 25, https://doi.org/10.1007/978-3-030-91847-7_7

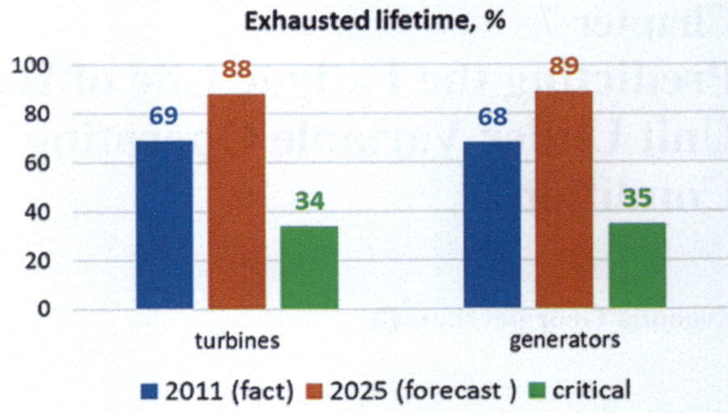

Fig. 7.1 Comparison of the exhausted, forecast, and a critical lifetime of hydraulic units [1]

of more than 60% is dangerous. As you see, the design lifetime of the most hydraulic turbines is already exhausted now. Their wear is classifieds as dangerous but some of them successfully work 60, 70, 80, and even 100 years.

Many examples of hazardous damages for the main unit's components you can find in open access, for example [2–5]. To return the unit to good condition, expansive and longtime overhaul or their replacement are required.

The risk of failure increases with every year of operation and especially dangerous for large high-head units. The reliable lifetime prediction helps to increase the reliability and safety of HPP and makes unit operation more effective. So, it is a very actual problem for modern hydropower.

7.2 The Impact of Operating Modes

One of the main reasons for significantly contrasting the values of a real lifetime is modern trends of essential increasing the operational range of hydraulic units [6].

In the past two decades, new renewable energy sources such as wind, solar, tidal, and wave are wide and rapid spread. They are characterized by very nonstable levels of power output. Therefore, the hydraulic units are used for ensuring the stability of a power grid. The choice in favor of hydraulic units is due to their high maneuverability in comparison with other traditional power generating ones.

Figure 7.2 shows a comparison of the actual (green columns) and design (red lines) modes [7]. Such operation causes long-term work at off-design modes (transient and non-stationary) and a large number of starts/stops.

The actual operating conditions play a decisive role in the real lifetime of hydraulic units. The off-design modes make a significant contribution to the accumulation of the unit's fatigue damage [8–11].

Figure 7.3 shows the characteristic distribution of relative values for the dynamic stress amplitudes under the whole operating range as the ratio of the average measured amplitude of dynamic stresses to the value of the static stress component at the same operating mode [12] for the most stressed point of the hydraulic runner (the junction

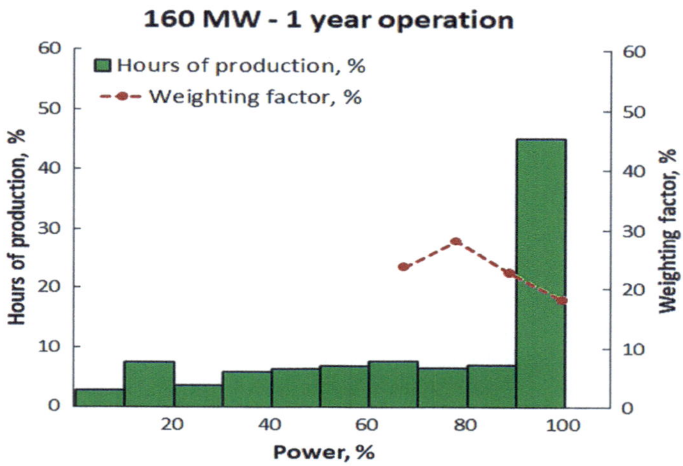

Fig. 7.2 Comparison of the actual (*green* columns) and design (*red* lines) modes [7]

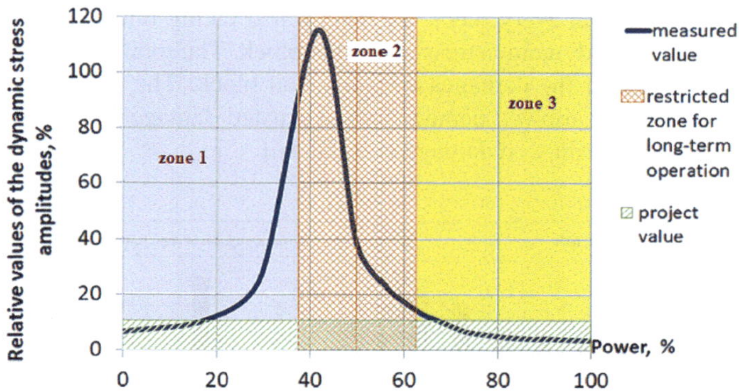

Fig. 7.3 The characteristic distribution of relative values for the dynamic stress amplitudes under different operating range (the ratio of the average measured amplitude of dynamic stresses to the value of the static stress component at the same operating mode [12])

zone of the blade output edge and the runner crown). The scale of power is normalized to the nominal value.

The division of the range into characteristic zones is made in accordance with the established rules in hydropower. Zone 1—a low-efficiency zone. Zone 2—a zone that is not intended for long-term operation of the equipment. Zone 3—a zone of optimal efficiency.

As you see, dynamic stresses at low and middle load are very significant in comparison to the project values near the best efficiency point. So, despite their short duration, the off-design modes make a significant contribution to the accumulation of fatigue damage due to the high-stress level in the unit's components and significantly accelerates the growth of defects. So, the correct estimation of the residual life requires taking into account the actual working hours at each operating mode.

7.3 Proposed Approach

The proposed approach requires to divide the whole operating range presented at hill chart into characteristic loading blocks for calculating the total accumulated damage. Each such block has a characteristic frequency (ultra-low, low-, high- and very high-) of the external load and a characteristic amplitude as you see in Fig. 7.3. The example of such a division and correspondent stresses at every block you can see in Fig. 7.4.

The fatigue damages are calculated separately for each block take into account the special fatigue features of material behavior at the characteristic frequencies. The fatigue damage is calculated as the ratio of allowable safety factor to the actual one for each load block. The concreate value of an allowed safety factor depends on the type of turbines, operating mode, and the accepted industry standard. The total amount of load cycles at every block are determined taking into account the characteristic frequency of dynamic stresses for this block. The total exhausted lifetime is the algebraic sum of the damages at every load block. The residual lifetime is the difference between allowed damage and exhausted damage. With this problem statement, the value of allowed damage is equal to 1.

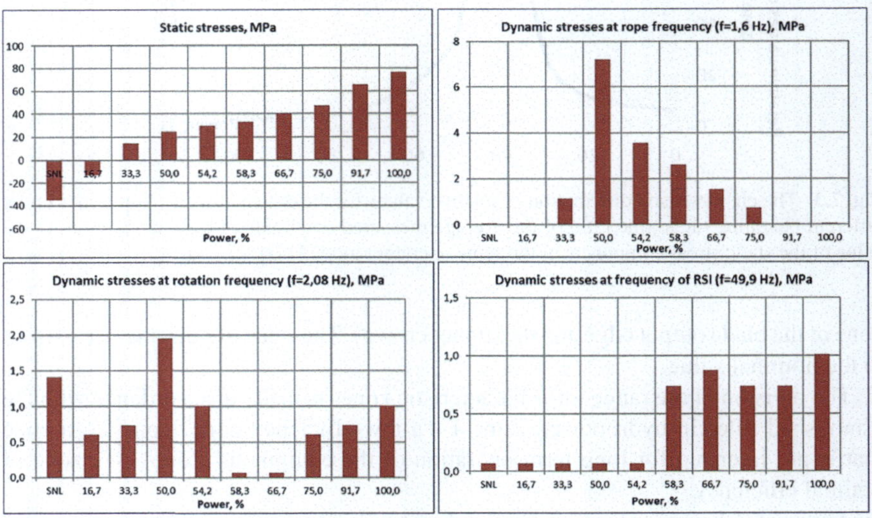

Fig. 7.4 Example of static and dynamic stresses corresponding to the division of the whole operating range to specific load blocks

It should be noted that the fatigue limit under operating in water is permanently decreased, unlike the air environment.

7.4 Results

To demonstrate the capabilities of the proposed approach some of the results are presented here.

Figure 7.5 shows the distribution of lifetime by operating zones in comparison to operating time at the different characteristic zones. Operation at speed-no-load is extracted as a separate zone.

As you see from the presented results, the distribution of the lifetime does not correspond to the distribution of operational time. The operating time at zone 2 is only 0.06% of the total time. However, the exhausted lifetime is about 13%. The same can be said about the start/stop modes. The total operating time on them is negligible, and the exhausted lifetime exceeds 12.6%.

Another example is the comparison of the lifetime distribution in different operating modes for two scenarios: base and peak modes under equal operating time (see Fig. 7.6).

Base mode supposes the long-term operating near the best efficiency point and about 10–12 start/stops per year.

Peak mode means frequent changes in power ranges. The peak modes are characterized by long-term operation at off-design modes and a large number of starts/stops. It causes a significant decrease in residual life, as you see in Table 7.1. The residual lifetime in this example will differ more than 3 times for base and peak modes.

Figure 7.6 helps to identify nonoptimal operating zones from the standpoint of life-time management. The impact of the off-design operational modes is clearly visible, especially in the power range of 33–67% of the nominal.

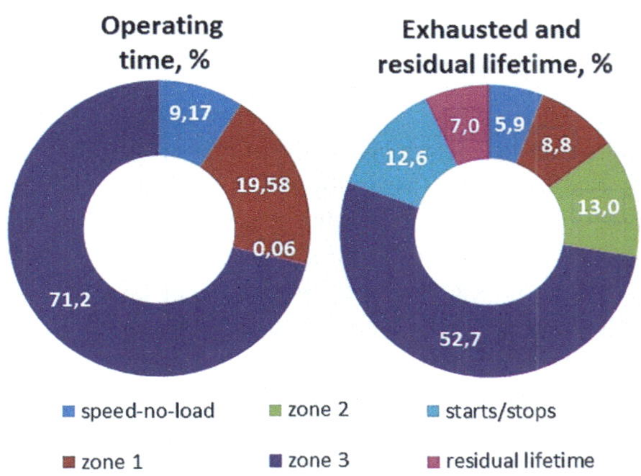

Fig. 7.5 Comparison of operating time and corresponding lifetime

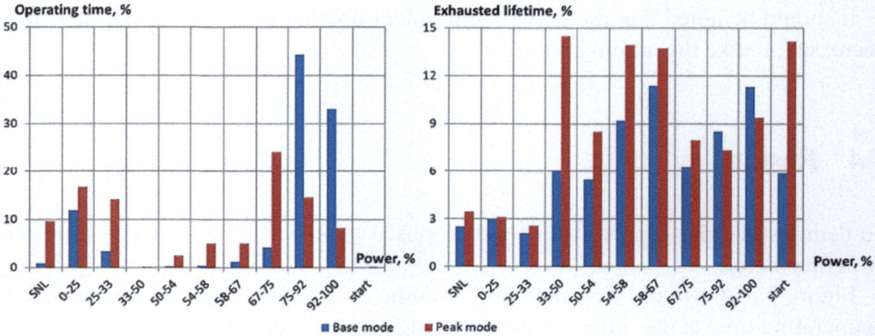

Fig. 7.6 Distribution of operating time and lifetime under base and peak scenarios

Table 7.1 Predicting lifetime

Parameters	Base mode	Peak mode
Total operating time for the prognostic period (until the next overhaul), hours	37,230	37,230
Average number of start-up and shutdown per year for the prognostic period, pc	29	214
Forecast of exhausted service life until the next overhaul, %	94.9	97.9
Residual lifetime at the end of the prognostic period, years	32.9	10.3
Residual lifetime at the end of the prognostic period, thousand hours	175.0	54.8

The proposed approach makes it possible to solve correctly the problem of reliable prediction of the lifetime for various scenarios under real operation at HPP. It allows us to capture the most dangerous modes from the point of view of the rapid exhaustion of the service life and realize smart lifetime management.

7.5 Conclusion

The individual features and the actual operating conditions play a decisive role in the real lifetime of hydraulic units. The correct prediction of residual lifetime after long operation under variable operating modes is very important for providing the right level of reliability, safety, and effectiveness of HPP.

The proposed approach allows us to make reliable individual lifetime predictions for various scenarios under real operation at HPP correctly and quickly. It helps to capture the most dangerous modes and realize smart lifetime management.

It allows us:

- to capture the most dangerous modes from the point of view of the rapid exhaustion of the service life,

- to catch the crack appearance at the earliest stage when the standard diagnostic system does not detect any changes yet,
- to optimize the unit's service life under real operation at HPP,
- to extend significantly the life of the units,
- to get an economy on the unit's maintenance only by the realization the optimal scenarios of equipment usage,
- to make more profit from its operation and realize smart lifetime management.

Funding This research was funded by the Foundation for Assistance to Small Innovative Enterprises in Science and Technology (FASIE), Russia, grant number 3288ГС1/55639, 27 November 2019, and the Skolkovo Innovation center, grant number 40113/07002/0329–2020, 22 April 2020.

References

1. Bogush B (2019) About modernization of equipment and introduction of modern information technologies to ensure energy security and energy efficiency of HPP and HAPP operation. In: Conference: Rusenergyweek-2019. Energy efficiency and energy security of hydropower facilities in the context of modernization and digital transformation. Moscow, Russia (2019) (RUS)
2. Momcilovic D, Motrovic R, Antanasovska I, Vuherer T (2012) Methodology of determination the influence of corrosion pit on the decrease of hydro turbine shaft fatigue failure. Mach Des 4(4):231–236
3. Pejovic S, Gajic A, Zhang Q (2014) Smart design requires updated design and analysis guidelines. IOP Conf Ser Earth Environ Sci 22(4):42008–42017.https://doi.org/10.1088/1755-1315/22/4/042008
4. Egusquiza E, Valero C, Xingxing H, Jou E, Guardo A, Rodriguez C (2012) Failure investigation of a large pump-turbine runner. Eng Fail Anal 23:27–34
5. Liu X, Luo Y, Wang Z (2016) A review on fatigue damage mechanism in hydroturbines. Renew Sustain Energy Rev 54:1–14. https://doi.org/10.1016/j.rser.2015.09.025
6. Kougias I, Aggidis G, Avellan F, Deniz S, Lundin U, Moro A, Muntean S, Novara D, Ignacio Pérez-Díazh J, Quaranta E, Schild P, Theodossiou N (2019) Analysis of emerging technologies in the hydropower sector. Renew Sustain Energy Rev 113:109257. https://doi.org/10.1016/j.rser.2019.109257
7. Parkinson E (2017) Advanced technologies of works on repair and restoration of objects in hydropower. In: 10th scientific and technical conference "hydropower. Hydraulic engineering. New developments and technologies". Reports and speeches, 95–104. Publishing house "VNIIG them. B.E. Vedeneeva", St. Petersburg, Russia (2017) (RUS)
8. Zouhar J, Obrovsky J, Feilhauer M, Skotak A (2016) Case study and numerical analysis of vibration and runner cracks for the Lipno I hydroelectric project. IOP Conf Ser Earth Environ Sci 49:072011. https://doi.org/10.1088/1755-1315/49/7/072011
9. Duparchy F, Brammer J, Thibaud M, Favrel A, Lowys PY, Avellan F (2017) Mechanical impact of dynamic phenomena in Francis turbines at off design conditions. IOP Conf Ser J Phys Conf Ser 813:012035. https://doi.org/10.1088/1742-6596/813/1/012035
10. Trivedi C, Gandhi B, Cervantes M (2013) Effect of transients on Francis turbine runner life: a review. J Hydraul Res 51(2):112–132. https://doi.org/10.1080/00221686.2012.732971

11. Seidel U, Mende C, Hübner B, Weber W, Otto A (2014) Dynamic loads in Francis runners and their impact on fatigue life. IOP Conf Ser Earth Environ Sci 22(3):32054–32062. https://doi.org/10.1088/1755-1315/22/3/032054
12. Georgievskaia E (2019) Effect of cracks on dynamic parameters and lifetime of hydraulic units. Procedia Struct Integrity 22:229–234. https://doi.org/10.1016/j.engfailanal.2019.08.003

Chapter 8
Predicted Distribution in Measured Fatigue Life from Expected Distribution in Cyclic Stress–strain Properties Using a Strain-Energy Based Damage Model

P. J. Huffman, J. A. F. O. Correia, A. Mourão, T. Bittencourt, and R. Calçada

Abstract In practice, fatigue properties such as stress life and strain life are often fit from a relatively small amount of data, making accurate statistical analysis a challenge. Using a strain energy based damage model to predict strain-life and stress-life curves from individual specimens (Huffman in Int J Fatigue 88:197–204, 2016 [1]), a set of curves can be created to more accurately reflect the range of strain-life values that would come from more extensive testing. This work examines how the expected variation in stress–strain properties could be used to predict the set of curves if the strain hardening behavior of the individual specimens is unknown.

Keywords Probabilistic analysis · Strain energy · Fatigue damage

8.1 Probabilistic Fatigue

The distribution in fatigue life has been studied in a variety of ways [2–5]. There have been statistical analyses of fatigue based on stress-life and strain-life curves, as well as fatigue crack growth rates. Many of these examine the results of testing and apply descriptive statistics such as arithmetic means and standard deviations based on normal distributions in order to describe the probability of a specimen failing at a particular number of cycles at a particular load amplitude. Unfortunately, these techniques are not particularly predictive beyond being able to describe the data that has already been collected [2–5]. While curves based on a probability of failure can be created from these data and used to predict component fatigue life, they do little to help explain the physics or enable interested parties to make predictions from limited data sets.

P. J. Huffman
Successful Failure Consulting Corp, Ankeny, IA 50023, USA

J. A. F. O. Correia (✉) · A. Mourão · R. Calçada
Faculty of Engineering, University of Porto, 4200-465 Porto, Portugal
e-mail: jacorreia@inegi.up.pt

T. Bittencourt
University of São Paulo, Tiergartenstr. 17, São Paulo 69121, Brazil

G. Lesiuk et al. (eds.), *Structural Integrity and Fatigue Failure Analysis*,
Structural Integrity 25, https://doi.org/10.1007/978-3-030-91847-7_8

By using more advance statistical techniques such as Weibel distributions, better predictions can be made [2–6]. These statistics can be applied to damage parameters instead of the more basic stress or strain measurements for added benefit.

In addition, there have been statistical examinations of parameters which are related to stress–strain properties. The elastic modulus, cyclic strength coefficient, and cyclic strain exponent have been studied for their influence on the variability of the stress–strain behavior of specimens [3]. That work uses measured variability in the strain-life parameters along with the variability in strain hardening parameters to explain measured variability in fatigue life. The work presented here uses those variabilities in the cyclic stress–strain properties, *K*', the cyclic strain hardening coefficient, and *n*', the cyclic strain hardening exponent, as well as *E*, the elastic modulus, in order to predict the variability in the strain life.

8.2 Strain-Energy Density Damage

The strain-energy density method uses the Ramberg–Osgood relationship to calculate the ratio of elastic and complimentary plastic strain-energy density each to the strain-energy density from dislocations expressed as U_{d}, the strain energy in terms of energy per unit length, and ρ_{c}, the dislocation density.

Using the strain energy from loading and the strain energy from dislocations, the damage is formulated as

$$\left(\frac{2}{U_{\mathrm{d}}^2\rho_{\mathrm{c}}^2 E}\right)\left(\frac{n'}{1+n'}\right)\left(\frac{1}{K'}\right)^{(1/n')}\left(\sigma_{\max}^2\right)(\sigma_a)^{\left(\frac{1+n'}{n'}\right)} = D = \frac{1}{2N_f}. \tag{8.1}$$

where D is the damage and $2N_f$ is the number of reversals until failure. Using the Ramberg–Osgood relationship to convert stress to elastic strain, the fatigue strength coefficient, σ_f' ., is derived to be

$$\sigma_f^{'} = E \times \left(\frac{2}{U_{\mathrm{d}}^2\rho_{\mathrm{c}}^2}\left(\frac{1}{K'}\right)1,^{1/n'}\left(\frac{n'}{1+n'}\right)E^{\frac{1+2n'}{n'}}\right)^{\frac{-n'}{1+3n'}} \tag{8.2}$$

the fatigue ductility exponent, $\varepsilon_f^{'}$, is

$$\varepsilon_f^{'} = \left(\frac{2(K')^3}{U_{\mathrm{d}}^2\rho_{\mathrm{c}}^2 E}\left(\frac{n'}{1+n'}\right)\right)^{\frac{-1}{1+3n'}} \tag{8.3}$$

the fatigue strength exponent is

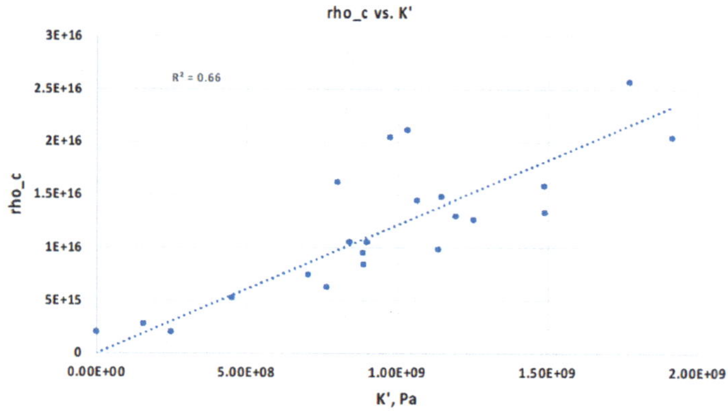

Fig. 8.1 The critical dislocation density versus the cyclic hardening coefficient. The dislocation densities and cyclic hardening coefficients are from Ref. [1]

$$b = \frac{-n'}{1 + 3n'} \tag{8.4}$$

and the fatigue ductility exponent is

$$c = \frac{-1}{1 + 3n'}. \tag{8.5}$$

The variability of the cyclic stress–strain parameters used in

$$E \pm 5\% \text{ , } K' \pm 5\% \text{ , } n' \pm 5\% \text{ , } \rho_c \pm 20\% \tag{8.6}$$

The variability of the cyclic stress–strain parameters are taken as the lower end values from [3] and the variability of ρ_c is 20% because the relationship between ρ_c and K' indicates that 20% of the variation in ρ_c is explained by the variation in K'. Since the two covary, this work used low values of ρ_c with low values of K', and high values of ρ_c with high values of K' (Fig. 8.1).

A total of 125 curves are constructed for each measured data set taking 5 different evenly distributed values each for E, K' and n'. Five different values were also taken for ρ_c, however they are assumed to covary with K' and don't add to the number of curve parameter combinations. The range of the values is described by Eq. (8.6).

8.3 Results

The data used is adapted from the Society of Automotive Engineers Fatigue Design and Evaluation database, a repository from several other data sources including measurements made at the University of Waterloo, Ontario, for that database [7–9]. Figure 8.2 shows the set of 125 strain-life curves based on the variations of the input parameters indicated by Eq. (8.6). Figure 8.3 shows the curves predicted for a 4130 alloy. Figure 8.4 is for an AA 1100 aluminum alloy, and Fig. 8.5 is for a Ti6Al4V titanium alloy.

In Fig. 8.2 it can be seen that some of the variability is accounted for, however there are deviations of the data from the group of curves. In Fig. 8.3, the measured data does not deviate significantly from the predicted curves, however the data set is somewhat limited. Figure 8.4 shows good prediction of the scatter except at long

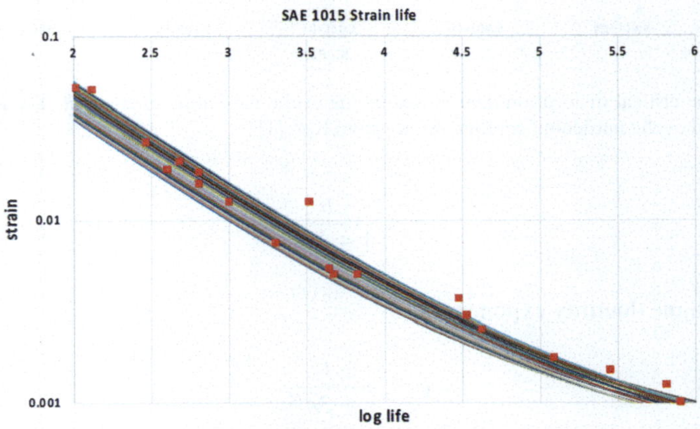

Fig. 8.2 Predicted strain-life curves of 1015 steel. Measurement data from Ref. [7]

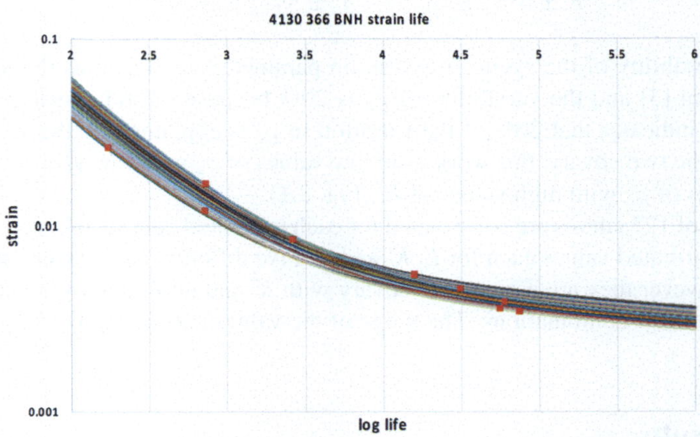

Fig. 8.3 Predicted strain-life curves of 4130 steel heat treated to BHN 366. Measurement data from Ref. [8]

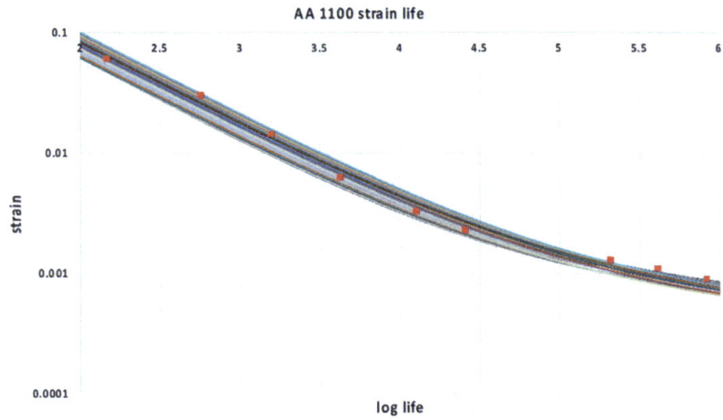

Fig. 8.4 Predicted strain-life curves of AA 1100 aluminum. Measurement data from Ref. [8]

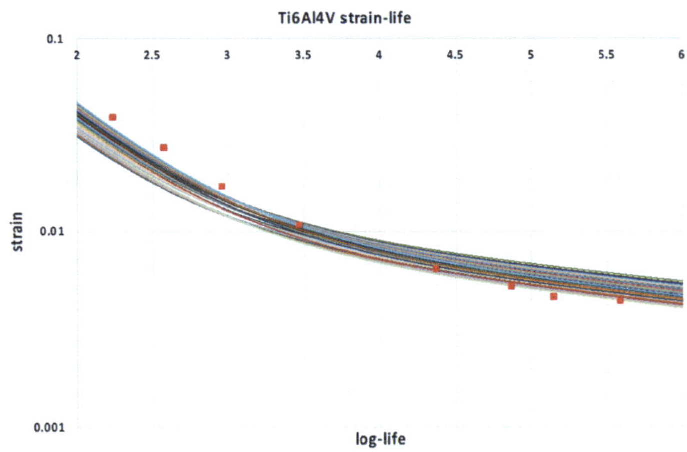

Fig. 8.5 Predicted strain-life curves of Ti6Al4V titanium alloy. Measurement data from Ref. [9]

lives. The predicted scatter in Fig. 8.5 would cover the scatter in measured data, but the curve itself is a lower quality match to the measurement.

8.4 Conclusion

8.4.1 Discussion

The results show a variation in predicted fatigue behavior based on expected variation in cyclic properties, calculated from a strain-energy density damage model. Although some of the variability in measurement appears to be accounted for, more work is needed to gather statistical information about the material properties to be used as inputs. The variation used for the cyclic elastic and plastic properties were on the lower end of what was presented by Jakubczak et al. Increasing those values to the mean or upper bounds would increase the predicted variation and may capture more of the variation in measurement. However, care should be taken as overprediction of the variation should also be avoided. In addition, little is known about the meaning of the critical dislocation density, ρ_c, and the variation of that variable may need to be calculated instead of measured. At present, there is no demonstrated way to measure this property.

A weakness of the damage model used can be seen in Figs. 8.2 and 8.4, where at low strains and long lives the measured life diverges from the predicted life, the prediction being conservative [1]. While it might be advantageous to use a better fitting model, one is not known which can predict the strain life from the cyclic properties. The use of the strain life and cyclic stress strain to predict variation in fatigue crack growth rates is discussed in Ref. [3].

8.4.2 Future Work

More accurate results are desired, which could come from better statistical inputs. If the statistics of the cyclic stress–strain parameters were calculated based on experimental results from a particular set of test data, the predictive power could be more accurately assessed. Given the promising results of the preliminary work, statistical inputs based on the statistical properties of a material could yield a usefully accurate prediction of the variation in fatigue life that could be expected from the experiment.

Extrinsic data about the specimens could be incorporated as well, including variation in specimen geometry, and surface finish. Attributes such as grain size and chemistry, however, are assumed to be accounted for by the relationship between those and the cyclic stress–strain properties [1].

References

1. Huffman PJ (2016) A strain energy density based damage model for fatigue crack initiation and growth. Int J Fatigue 88:197–204

2. Castillo E, Fernández-Canteli A (2009) A unified statistical methodology for modeling fatigue damage. Springer
3. Jakubczak H, Sobczykiewicz W, Glinka G (2006) Fatigue reliability of structural components. Int J Mater Prod Technol 25:64–83
4. Caiza PDT, Ummenhofer T, Correia JAFO (2020) Applying the Weibull and Stüssi methods that derive reliable Wöhler curves to historical German bridges. Pract Periodical Struct Des Constr 25(4):04020029
5. Pedrosa B, Correia JAFO, Rebelo C, Veljkovic M (2020) Reliability of fatigue strength curves for riveted connections using normal and weibull distribution functions. J Risk Uncertainty Eng Syst Part A Civil Eng 6(3)
6. Caiza PDT, Ummenhofer T (2020) Probabilistic relationships between strain range, stress range and loading cycles. Application on ASTM A969 steel. Int J Fatigue 137:105626
7. SAE Fatigue Design and Evaluation Database (2009) https://fde.uwaterloo.ca/Fde/Materials/dindex.html
8. Hirschberg MH, Manson SS, Smith RW (1963) Fatigue behavior of materials under strain cycling in low and intermediate life range. NASA-TN-D-1574
9. Nachtigal AJ (1947) Strain-cycling fatigue behavior of ten structural metals tested in liquid helium 4k, in liquid nitrogen 78K, and in ambient air 300K. NMASA-TN-D7532

Chapter 9
Modification Optimization-Based Fatigue Life Analysis and Improvement of EMU Gear

Yonghua Li, Chi Zhang, Hao Yin, Yang Cao, and Xiaoning Bai

Abstract This paper proposes an improved fatigue life analysis method for optimal design of EMU gear, which aims at the defects of traditional Miner fatigue cumulative damage theory. A fatigue life analysis method by modifying S–N curve and considering material difference is presented to improve the fatigue life of EMU gear based on shape modification optimization. A corrected method for stress amplitude, average stress, and S–N curve is proposed, which considers low stress cycle, material difference, and other factors. The fatigue life prediction of EMU gear is carried out by corrected S–N curve and transient dynamic analysis. Moreover, the gear modification technology combined with intelligent optimization method is adopted to investigate the approach of fatigue life analysis and improvement. It is found that the fatigue life analysis and improvement method proposed in this paper has higher prediction accuracy than the traditional method in most cases. It is concluded that the fatigue life is significantly improved by applying the proposed approach to the EMU gear. The results show that it is more corresponded to engineering practice by using the improved fatigue life analysis method than the traditional method. The function of stress and modification amount established by response surface method meets the requirement of precision. The fatigue life of EMU gear based on the intelligent algorithm for seeking the optimal modification amount is significantly improved compared with that before the modification.

Keywords Fatigue life prediction · Transient structural analysis · Gear modification · Particle swarm optimization

---

Y. Li (✉) · C. Zhang · H. Yin · Y. Cao
School of Locomotive and Rolling Stock Engineering, Dalian Jiaotong University, Dalian, P.R. China
e-mail: yonghuali@163.com

X. Bai
School of Mechanical Engineering, Dalian Jiaotong University, Dalian, P.R. China

G. Lesiuk et al. (eds.), *Structural Integrity and Fatigue Failure Analysis*, Structural Integrity 25, https://doi.org/10.1007/978-3-030-91847-7_9

9.1 Introduction

The gear mechanism is widely used in the transmission system because of its high efficiency, stable transmission characteristics, and precise transmission ratio. The meshing of helical gear is a process in which two teeth mesh gradually and separate gradually compared with that of spur gear. Therefore, helical gear is widely used in high speed because of its good meshing performance. The transmission gear is easy to produce unbalanced loadbearing in the meshing process due to the complex load cases of EMU, which reduces the fatigue life [1, 2]. Premature failure of gear leads to great loss of persons and property. Therefore, it is necessary to analyze the fatigue life of the EMU gear based on the modification optimization, aiming to improve the meshing state of the gear and determine the failure time.

Excessive predicted life will lead to premature failure, which causes safety hazards. Too little predicted life will lead to premature replacement of parts and equipment and waste of resources, which leads to waste of resources [3]. Accordingly, accurate fatigue life prediction of the transmission gear is beneficial to the operation and safety improvement of the vehicle [4]. Some scholars have studied the fatigue life analysis method at present.

Liu et al. [5] proposed a double-crack gear pair model and obtained the relationship between the dividing circle and the propagation life of the tooth root crack. Dai et al. [6] established gear engagement equation according to the limitation of traditional gear optimization, which considers the contact width and other factors. Find that the contact width has obvious influence on fatigue life. Li et al. [7] determined the damage model and obtained the root crack initiation life by comparing the experimental results with the simulation results. Zhang et al. [8] proposed a tooth profile correction method for over-meshing and under-meshing problems, so as to improve the fatigue life. Li et al. [9] converted crack closure into traction force, which considers the influence of crack closure effect on crack propagation. Find that crack closure effect could prolong the life of gear. Li et al. [10] established a probabilistic life prediction model based on the least order statistics, which solves the problem of long fatigue test time of gear. Zhu et al. [11] established the life prediction model and the fuzzy fatigue life calculation formula considering the small load action and the fuzzy damage. Huang et al. [12] revised the Corten-Dolan theoretical model, taking into account the degradation caused by material and historical loading of loads. The accuracy of fatigue life prediction is improved. Zhi et al. [13] calculated fatigue life of welding robot, which is optimized by fuzzy theory and agent model. Matsubara [14] proposed a method to predict high cycle fatigue life under multi-axis loads. This fatigue limit estimation method is superior to other standards. Li [15] presented a concise methodology to predict the flexural fatigue life of tooth V-belt through a combination of material property test and finite element analysis method. The prediction accuracy of this method is high. However, few scholars combine gear modification with intelligent optimization method and discuss the improved method of fatigue life analysis based on the simulation results of transient dynamics.

This paper takes the EMU gear as the research object. The model is simplified twice in fatigue life analysis to improve computational efficiency. The fatigue life analysis method is improved by considering the deficiency of the traditional linear Miner fatigue cumulative damage theory. The fatigue life of modified gear is analyzed more accurately compared with the traditional method, which modification amount is obtained by particle swarm optimization.

9.2 The Establishment of EMU Gear Model

The gear material of EMU is 20CrNi3Mo. The input end is the pinion end, which is connected with traction motors of the passive vehicle through a coupling. The pinion transmits torque to engage with the driven gear, which is mounted on the axle. Original model of EMU gear in this paper is shown in Fig. 9.1.

The speed of the EMU is v. The diameter of wheel is D. If the wheel and the driven gear are concentric, the speed of the driven gear n_2 is

$$n_2 = \frac{60v}{3.6\pi D} \tag{9.1}$$

The gear meshing mode is external meshing, then the pinion speed is

$$n_1 = \frac{60v}{3.6\pi D} \cdot \frac{z_2}{z_1} \tag{9.2}$$

Basic parameters of EMU gear pair are shown in Table 9.1.

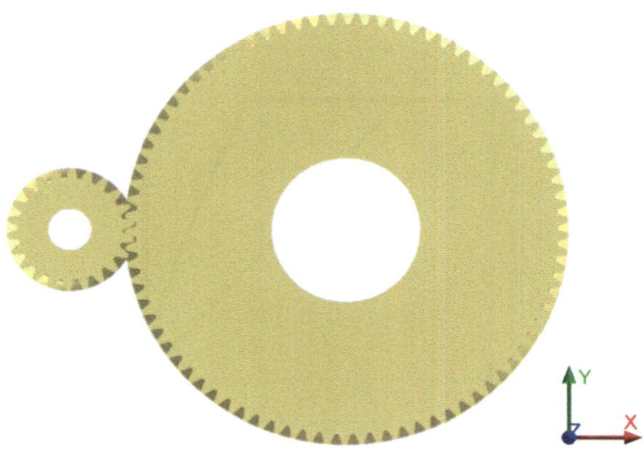

Fig. 9.1 Original model of EMU gear

Table 9.1 Basic parameters of EMU gear pair

Gear pair parameters	Driving gear	Driven gear
Number of teeth z_i	22	81
Tooth width b_i (mm)	55	50
Speed n (r/min)	4542.5	1233.8
Normal module m_n (mm)	7	
Normal pressure angle α_n (°)	20	
Normal tip height coefficient h_{an}^*	1	
Top clearance coefficient c_n^*	0.25	
Helix angle β (°)	18	
Normal center distance a (mm)	379.05	

9.3 Gear Modification Theory

The gear modification can improve the contact state of tooth surface and increase the fatigue life of the transmission gear. Gear meshing produces impact and excitation when the distortion of gear causes changes in angular velocity. The commonly used gear modification methods are divided into three categories, which is the tip and profile chamfering and rounding, the tooth profile modification and the tooth alignment modification. In this paper, the linear tip and root relief method is used to optimize the EMU gear.

The linear modification methods of tooth tip and tooth root are shown in Fig. 9.2. The top and root of the gear cross section are modified linearly to remove the interference part, so as to ease the meshing process and optimize the contact state of the tooth surface.

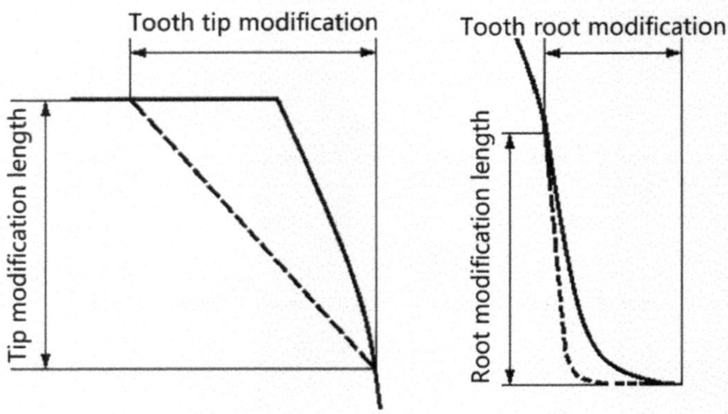

(a) Gear tip modification. (b) Tooth root modification.

Fig. 9.2 Linear modification of gear

The modification amount of tooth tip can be described with the following equation according to the elastic deformation equation.

$$\delta = \frac{K_A K_{mp} F_t \cos\beta}{b_i \cos\alpha_n c_r} \tag{9.3}$$

where K_A is the service coefficient, K_{mp} is the branch coefficient, F_t is the tangential force, c_r is the engagement stiffness.

The EMU gear is modified based on intelligent algorithm in this paper. The modification process based on intelligent algorithm is shown in Fig. 9.3. Firstly, it is necessary select the appropriate gear modification amount as variables and carry out experimental design. Secondly, the quadratic response surface is fitted to the experimental design, which obtains the functional relationship between the modification amount and the maximum contact stress. Finally, in order to minimize the contact stress of gears, the particle swarm optimization algorithm is used to find the optimal solution of the response surface function.

9.4 Transient Dynamic Analysis of EMU Gear

It is assumed that small loads generated by non-meshing teeth are ignored. Since the gear speed is constant, the whole process can be reflected by simulating the meshing process of two teeth. If the analysis time is too short, the calculated fatigue life will be inaccurate. Therefore, it is necessary to accurately calculate the meshing time.

If the rotation speed of the driving gear is $n_1(t)$, the time of the driving gear rotation t_1 can be described by integrating with the following equation.

$$\int_0^{t_1} n_1(t)\,\mathrm{d}t = \frac{1}{60} \tag{9.4}$$

The accumulative damage of driving gear is greater under the same time because of its higher speed. Therefore, take a complete meshing of the tooth surface as a cycle. Assuming that the driving gear rotates at a uniform speed, the calculation time is

$$t = \frac{t_1}{\varphi_1} = \frac{60/n_1}{360/z_1} = \frac{6z_1}{n_1} \tag{9.5}$$

Transient dynamic analysis is also known as time-history response analysis, which can determine the structural response at any time. The input data is load changing. Output data is displacement, stress, and strain [16]. The governing equation of transient dynamics is

$$[M]\{\ddot{u}\} + [C]\{\dot{u}\} + [K(u)]\{u\} = \{F(t)\} \tag{9.6}$$

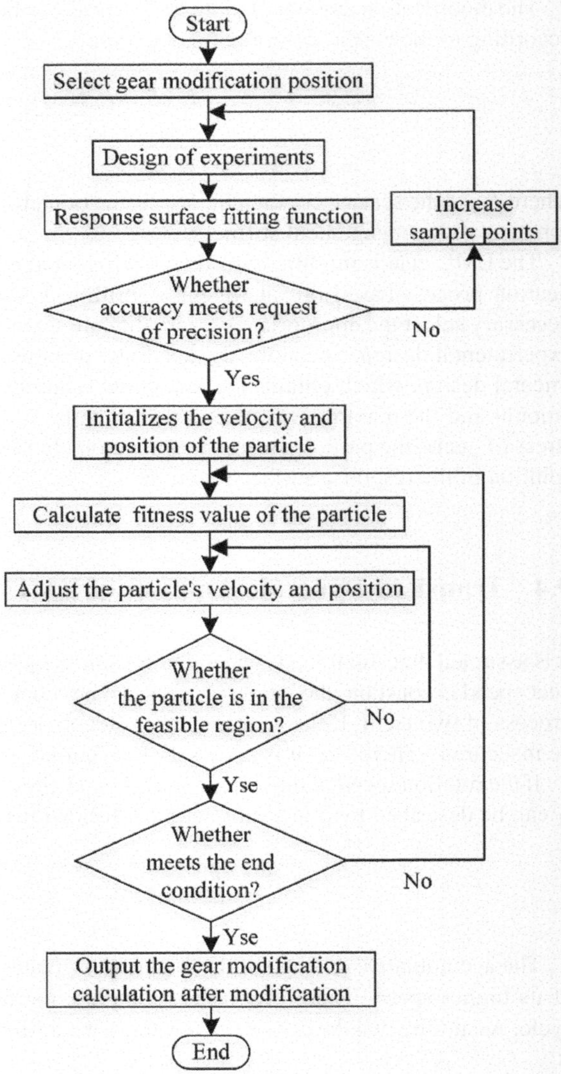

Fig. 9.3 Modification process based on intelligent algorithm

where $[M]$ is the overall structure mass matrix. $[C]$ is the overall structure damping matrix. $[K]$ is the overall structural stiffness matrix. $\{u\}$ is the displacement vector of the structure. $\{F\}$ is the vector matrix of the external load of the structure. $\{F(t)\}$ is a function of the load in terms of time.

The method of solving motion equation can be divided into direct integral method and modal superposition method. Although the modal superposition method has high computational efficiency, it is only applicable to the case of zero velocity and displacement and cannot be used to analyze nonlinear problems. Therefore, the direct integral method is adopted to solve equation. Equilibrium equation is established by linear combination of two continuous steps. The direct integral method can be obtained as follows:

$$\left\{F^a_{n+1-\alpha_f}\right\}=[M]\left\{\ddot{u}_{n+1-\alpha_m}\right\}+[C]\left\{\dot{u}_{n+1-\alpha_f}\right\}+[K(u)]\left\{u_{n+1-\alpha_f}\right\} \tag{9.7}$$

where

$$\begin{aligned} \left\{\ddot{u}_{n+1-\alpha_m}\right\} &= (1-\alpha_m)\{\ddot{u}_{n+1}\}+\alpha_m\{\ddot{u}_n\} \\ \left\{\dot{u}_{n+1-\alpha_f}\right\} &= \left(1-\alpha_f\right)\{\dot{u}_{n+1}\}+\alpha_f\{\dot{u}_n\} \\ \left\{u_{n+1-\alpha_f}\right\} &= \left(1-\alpha_f\right)\{u_{n+1}\}+\alpha_f\{u_n\} \\ \left\{F^a_{n+1-\alpha_f}\right\} &= \left(1-\alpha_f\right)\left\{F^a_{n+1}\right\}+\alpha_f\left\{F^a_n\right\} \end{aligned} \tag{9.8}$$

All variables can be defined by γ.

9.5 Fatigue Life Prediction of EMU Gear

The fatigue life prediction of gear is carried out in this paper based on the linear Miner cumulative damage theory. Linear Miner cumulative damage theory is widely used in fatigue life calculation because of its simple calculation and relatively accurate results. The linear Miner cumulative damage theory is shown as follows:

$$D=\sum_{i=1}^{n}\frac{n_i}{N_i} \tag{9.9}$$

where N_i is the fatigue life corresponding to the current level. n_i is the cycles number in the corresponding level. Fatigue damage occurs when $D=1$.

The following three points are optimized in this paper aiming at the defects of the traditional fatigue life analysis method.

9.5.1 Correction of S–N Curve

The stress-life curve can be described with the following equation, which assumes that the stress amplitude and life in the original S–N curve is a power function.

$$\sigma_a^m N = C \tag{9.10}$$

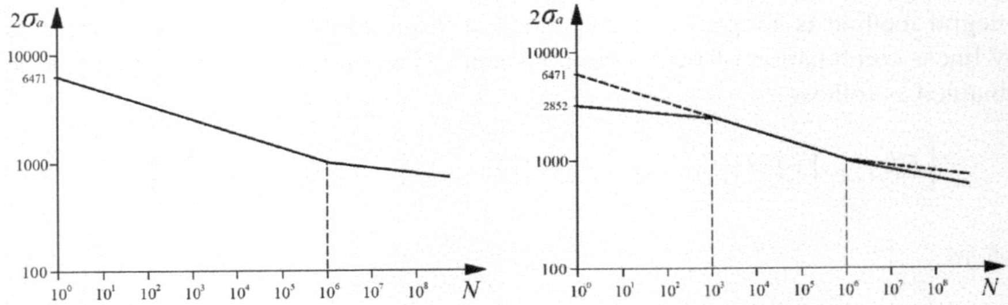

(a) Standard S-N curve. (b) Revised S-N curve.

Fig. 9.4 S–N curves

where S is the stress amplitude. N is the number of cycles. m and C are a constant.

The standard S–N curve of materials selected in this paper is shown in Fig. 9.4a. It can be seen from the figure that the curve is divided into two parts with different slopes bounded by 10^6 cycles. When the number of cycles is 1, the cut-off stress is much greater than the ultimate tensile strength, which is not consistent with the actual situation. Therefore, it is necessary to improve the slope of S–N curve under high stress state. Meanwhile, it is necessary to improve the S–N curve under low stress conservatively based on Haibach method, which considers the effect of low cycle.

The corrected S–N curve is shown in Fig. 9.4b. Assuming that the slopes of the two lines in Fig. 9.4a are k_1 and k_2, respectively, the slope of corrected S–N curve under the low stress state k_3 can be obtained as follows:

$$k_3 = \frac{k_1}{k_1 + k_2} \tag{9.11}$$

9.5.2 *Correction of Stress Amplitude*

The converted stress amplitude $\sigma_{ij}(t)$ can be expressed as the following equation by assuming that the material is ideal.

$$\sigma_{ij}(t) = \sum \frac{(P_k(t) \cdot \text{Scalefactor} + Off) \cdot \sigma_{ij,\text{static}}}{\text{Divider}_k} \tag{9.12}$$

where $P_k(t)$ is load spectrum. *Scalefactor* is the stress amplification factor. *Off* is the offset due to the residual stress. $\sigma_{ij,\text{ static}}$ is the result of stress calculation. $Divider_k$ is the stress reduction coefficient.

The proportionality factor and bias factor should be introduced into the material if the actual material is not ideal. The stress amplitude $\sigma(t)$ can be expressed as

$$\sigma(t) = Off_{\text{mat}} + SF_{\text{mat}} \cdot \sigma_{ij}(t) \tag{9.13}$$

where Off_{mat} is the material bias factor. SF_{mat} is the material proportion coefficient.

9.5.3 Correction of Mean Stress

Average stress also affects fatigue life in the actual situation. When the average stress is greater than 0, the fatigue life of the object is less than that of compression. The standard S–N curve is measured when the average stress is zero, so it is necessary to convert the actual average stress level to the stress amplitude under the symmetric cycle for calculation.

In this paper, the average stress is modified based on FKM method, which considers safety and economy. As shown in Fig. 9.5, three stress ratios are used to divide the mean stress and stress amplitude diagram into four sections. Four coefficients are used to define the sensitivity of the mean stress.

The stress ratio ranges corresponding to the four coefficients are

$$\begin{aligned} &-\infty \leq R < 0 \\ &0 \leq R < 0.5 \\ &0.5 \leq R < 1 \\ &R > 1 \end{aligned} \tag{9.14}$$

The slopes of the four lines are, respectively, M_1, M_2, M_3, and M_4.

$$\begin{aligned} M_1 &= 0 \\ M_2 &= -M_\sigma \\ M_3 &= -\frac{M_\sigma}{3} \\ M_4 &= 0 \end{aligned} \tag{9.15}$$

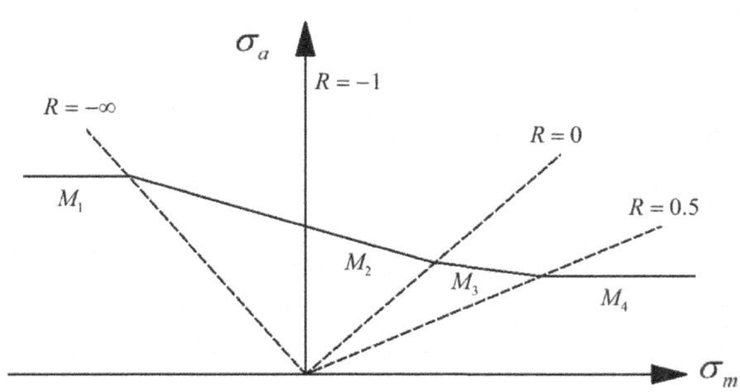

Fig. 9.5 Average stress correction method based on FKM

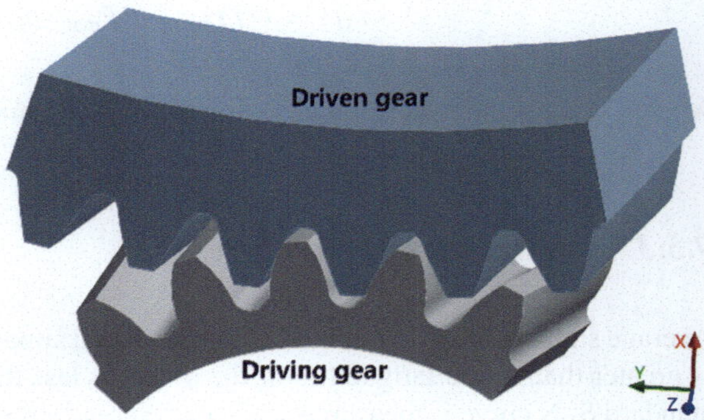

Fig. 9.6 The simplified EMU gear model

where

$$M_\sigma = a_M \cdot 10^{-3} \cdot R_M + b_M \tag{9.16}$$

a_M and b_M are material coefficients. R_M is the ultimate tensile strength of the material.

9.6 Case Analysis

9.6.1 Simplification of Gear Model

The model is simplified to improve the computational efficiency. Assume that the stress of non-meshing teeth is 0. The transient dynamic analysis process is simplified as a single tooth contact process. At the same time, the geometric model is simplified.

The simplification of the finite element model reduces the number of finite element meshes obviously, which makes each substep converge quickly. The simplification of the motion process reduces the number of calculated substeps. Two simplifications significantly improve the computational efficiency. The simplified EMU gear model is shown in Fig. 9.6.

9.6.2 Transient Dynamic Analysis and Modification Design of Gear

Set the friction coefficient of contact surface as 0.05. The driving gear rotates at 4542.5 r/min. The load torque of driven gear is 1000 N·m. The calculation time is 0.02906 s by substituting the data into Eq. 5. The time of each substep is 0.002906 s by setting 10 load steps for calculation.

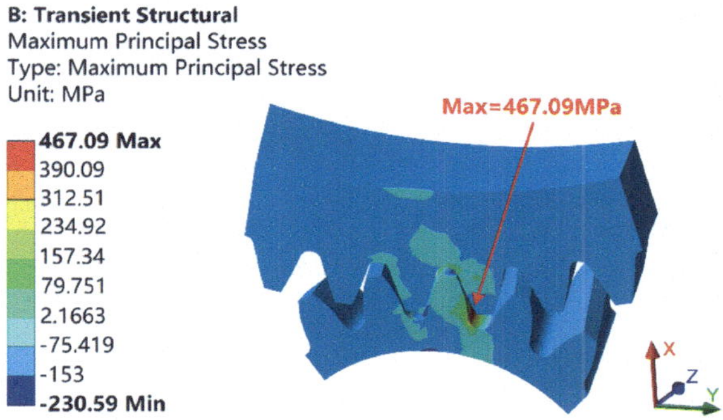

Fig. 9.7 Counter plot of the first principal stress before modification

Counter plot of the first principal stress before modification is shown in Fig. 9.7. It can be seen from the figure that the maximum stress of gear meshing is 467.09 MPa, appearing near the tooth root, which is relatively large. Therefore, it is necessary to improve the contact state and reduce the stress by gear modification. The stress of non-meshing teeth can be ignored, which also verifies the rationality of the finite element model and the movement process in this paper.

In order to solve the problem of high contact stress when meshing, it is necessary to seek the optimal value of the tooth tip and tooth root modification amount of gear with the aim of minimizing the contact stress.

First, it is necessary to determine the modification position which has great influence on contact stress. Because there are only four factors, the full-factor experimental design method is adopted to carry out the design of experiments. The simulation results of the four parameters are obtained.

Secondly, the experimental results are functionally fitted based on the secondary response surface. This paper uses particle swarm algorithm to find the optimal solution. It is found that the two modification amounts are the optimal solutions when the value is zero. Therefore, the tooth root of the driving gear and the tooth tip of the driven gear are identified as gear modification objects.

Finally, the appropriate amount of modification is selected to repeat the process described above. The tooth root amount of the driving gear and the tooth tip amount of the driven gear are taken as variables to design the experiments. The simulation results of gear modification are shown in Table 9.2.

Through the quadratic response surface fitting function, the mathematical model of optimal design can be expressed as

$$\begin{cases} \min\ f(X) = 507.41 - 25.77x_1 - 36.06x_2 + 2.77x_1^2 + 3.92x_2^2 \\ \text{Find } X = \{x_1, x_2\}^T \\ \text{s.t.} \quad \begin{cases} 0 \le x_1 \le 6 \\ 0 \le x_2 \le 6 \end{cases} \end{cases} \tag{9.17}$$

Table 9.2 Simulation results of gear modification

Plan	Tooth root modification of driving gear/μm	Tooth tip modification of driven gear/μm	Maximum stress/MPa
1	9.20	4.00	381.71
2	6.00	0.00	384.59
3	8.40	6.00	390.00
…	…	…	…
58	1.20	1.20	463.12
59	0.00	1.20	523.23
60	1.20	0.00	542.15

where f is the maximum stress. x_1 is the tooth root modification amount of driving gear. x_2 is the tooth tip modification amount of driven gear. The accuracy of the fitting function meets the requirements.

The optimal modification amount of the driving gear tooth root is 4.6516 μm. The optimal modification amount of the driven gear tooth tip is 4.5995 μm. Contact stress is 364.545 MPa.

9.6.3 *Fatigue Life Analysis of Modification EMU Gear*

The stress time diagram of the modified gear is shown in Fig. 9.8. The maximum stress occurs at 0.008718 s and the stress is 384.61 MPa, which is less than the yield limit of the material and meets the strength requirements. Two maximum stress points appear near the tooth root of the driving gear. The stress transition is smooth in the meshing process. Therefore, the contact condition of tooth surface is good.

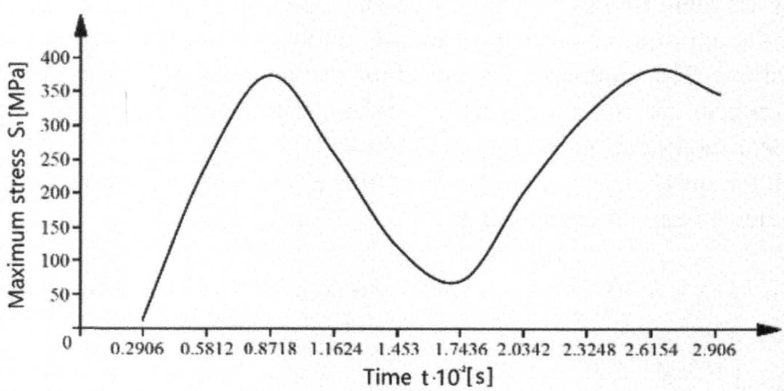

Fig. 9.8 The stress time diagram of the modified gear

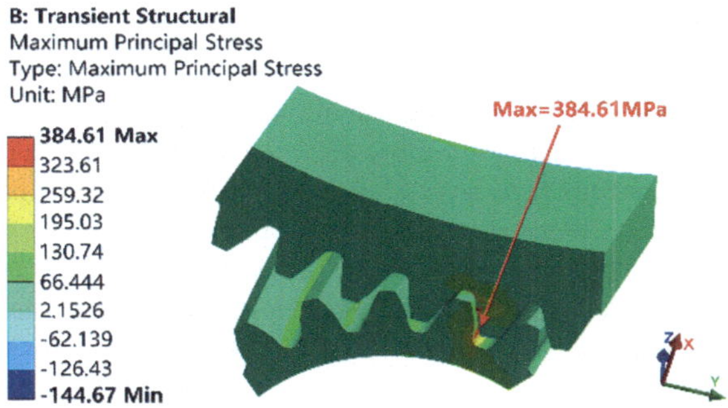

Fig. 9.9 Counter plot of the first principal stress after modification

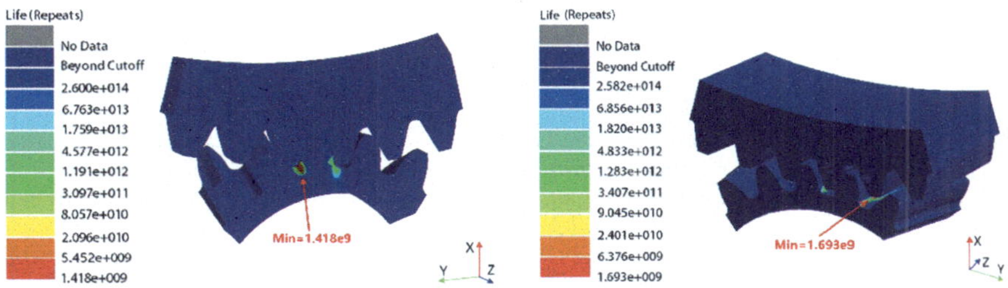

(a) Contour plot of gear fatigue life before modification.

(b) Contour plot of gear fatigue life after modification.

Fig. 9.10 Fatigue life comparison

Counter plot of the first principal stress after modification is shown in Fig. 9.9. The maximum stress point is at the root of the meshing tooth, which causes the fatigue fracture of the tooth root.

The results of transient dynamic analysis with the modified mean stress and stress amplitude are substituted into the modified S–N curve to calculate the fatigue life. The fatigue life comparison of EMU gear is shown in Fig. 9.10. It can be seen from the figure that the minimum life position appears at the maximum contact stress. The fatigue life of the gear before modification is 1.418E9, and the fatigue life of the gear after modification is 1.693E9. After conversion, the vehicle can travel a mileage of 56.47 million kilometers after gear modification, which meets the design life requirements of the vehicle. The fatigue life was improved by 19.4% compared with that before modification.

9.7 Conclusion

Compared with traditional methods, particle swarm algorithm can obtain optimal correction calculation more accurately. The correction of S–N curve, stress amplitude, and average stress level solves the problem that the traditional fatigue life prediction method does not conform to the reality. The correction of S–N curve, stress amplitude, and average stress solves the problem that traditional fatigue life prediction methods are inconsistent with reality. The fatigue life of EMU gear is predicted more systematically and completely by considering the difference between test materials and actual materials under the premise of ensuring safety and economy.

References

1. Vivek P, Vaibhav C, Yogesh P (2019) Geometrical complexity and crack trajectory-based fatigue life prediction for a spur gear having tooth root crack. Eng Fail Anal 105:444–465
2. Fan N, Tian H, Pei B et al (2017) Modification and optimization design of motor train unit transmission gear based on KISSsoft. J Mech Trans 41(3):83–87
3. Liu D, Li Q, Wang X et al (2016) Life prediction method for EMU axle box bearings based on actual measured loadings. J Mech Eng 52(2):45–54
4. Guillaume V, Jean-Philippe N, Fabrice V et al (2019) From Hertzian contact to spur gears: analyses of stresses and rolling contact fatigue. Mech Ind 20(6):1–16
5. Liu W, Dai H, Feng X (2019) Analysis on tooth crack growth life of a gear pair with double cracks. Lubr Eng 44(12):52–57
6. Dai L, Pu W, Tian X et al (2018) Effect of short teeth on contact fatigue in planetary gears under mixed lubrication. Tribology 38(2):121–128
7. Li X, Li G, Ren J et al (2018) Numerical Simulation of helical gear tooth root crack initiation life of high-speed EMUs. China Mech Eng 29(9):1017–1024
8. Zhang N, Zhang Y, Li Y (2019) Optimization method for double-arc profile of harmonic reducer. J Xi'an Jiaotong Univ 53(12):31–37
9. Li Y, Ding H, Zhang Q et al (2018) Influence of tooth-root crack closure effect on gear fatigue life. J Lanzhou Univ Technol 44(2):170–174
10. Li M, Xie L, Zhang Y et al (2017) Probabilistic life prediction and bending fatigue test for gear. J Mech Eng 53(18):169–175
11. Zhu S, Huang H, Xie L (2009) Prediction of fuzzy fatigue life under low amplitude loading strengthening. Chin J Aeronaut 30(6):1048–1052
12. Huang H, Zhu S, Wang Z et al (2011) Nonlinear fatigue damage cumulative rule based on strength degradation and its application to fatigue life reliability analysis. J Basic Sci Eng 19(2):323–334
13. Zhi P, Li Y, Chen B et al (2020) Fuzzy design optimization-based fatigue reliability analysis of welding robots. IEEE Access 8:64906–64917
14. Matsubara G, Hayashida et al. (2018) Predicting the multiaxial fatigue limit and the multiaxial high-cycle fatigue life based on the unified equivalent shear stress from axial strength characteristics with various waveforms. Int J Fatigue 112:52–62
15. Li W, Xin Z (2018) Flexural fatigue life prediction of a tooth V-belt made of fiber reinforced rubber. Int J Fatigue 111:269–277
16. Wang Z, Huo M (2019) Transient dynamics analysis of ANSYS workbench in belt conveyor reducer. Coal Mine Mach 40(2):75–77

Chapter 10
Fatigue Life Analysis of Strain Loadings Using Effective Strain Damage Approach with the Consideration of Cycle Sequence Effects

A. A. A. Rahim, C. H. Chin, S. Abdullah, S. S. K. Singh, and N. Md Nor

Abstract This paper presents fatigue life determination of automobile coil spring subjected to variable amplitude loading by considering the loading sequence effect. Fatigue failure of automobile coil spring is common occurrence that is subjected to random loading due to road conditions. Load sequence effect was found to have more significant effects in the automobile coil spring strain signals with many high amplitude events. Strain time histories of coil spring were acquired via road tests on various road conditions. To determine the effect of load sequence on fatigue life, effective strain damage model was also applied in fatigue life assessment and compared with strain-life models. Results illustrated that the lowest fatigue life of between 1.19 and 2.93 $\times$ 10^3 blocks was found in the industrial road signal. This owes to the high amplitude events contributing to high fatigue damage. The fatigue life of highway signals predicted using this significantly effective model is close to the predicted value using strain-life models. This is mainly because the load sequence effect was not dominant from the highway signal due to the absence of high amplitude events. Finally, the load sequence effect was found to have more significant effects in strain signals that had many high amplitude events.

Keywords Effective strain · Load sequence effect · Variable amplitude

10.1 Introduction

In engineering structures' durability analysis, fatigue failure is the most critical failure mode caused by cyclic loading. Other types of failure modes are buckling, ductile fracture, impact, creep, wear, corrosion and stress corrosion cracking to name a few. Various cyclic loads experienced by the structure during service are taken into account

A. A. A. Rahim · C. H. Chin · S. Abdullah (✉) · S. S. K. Singh
Department of Mechanical and Manufacturing Engineering, Faculty of Engineering and Built Environment, Universiti Kebangsaan Malaysia, UKM, 43600 Bangi, Selangor, Malaysia
e-mail: shahrum@ukm.edu.my

N. M. Nor
Faculty of Civil Engineering, Universiti Teknologi MARA, Cawangan Pulau Pinang, Malaysia

G. Lesiuk et al. (eds.), *Structural Integrity and Fatigue Failure Analysis*,
Structural Integrity 25, https://doi.org/10.1007/978-3-030-91847-7_10

in fatigue life assessment. For example, in automobile industry, strain signal of lower arm in suspension system caused by road surface condition during vehicle movement may fail due to fatigue [1]. In oil and gas industry, pipeline components experiencing accumulated fatigue may be damaged during service [2]. In addition, wall-thinning and stress corrosion cracking occurred in material degradations accelerate the fatigue damage phenomena. Therefore, the risk of fatigue failure happening increases as the structure had been in service for a long time.

Three fatigue life models usually used for local strain-based approaches are Coffin-Manson, Morrow and Smith–Watson–Topper (SWT). These models are widely used in fatigue life analysis and known as non-conservative method in assessing fatigue life [3]. Based on these non-conservative models, a plane strain state is assumed to predict strains in the crack tip region [4]. As a result, the fatigue life value is overestimated in long loading histories. The researchers improve fatigue life prediction method by using non-linear finite elements analysis [5]. However, this method needs advanced computational analysis, especially for complex geometry structures, loading conditions and residual stress state.

Fatigue failure commonly occurs in components that are subjected to vibration loadings, especially suspension coil spring in automobiles. Coil spring is an important component for the suspension system that provides good vehicle control and riding quality. The consequence of the coil spring failure can be catastrophic which could result in the loss of lives. Therefore, durability test of the coil spring is necessary to predict the useful fatigue life under different loadings and conditions. Time-domain methods including stress-life and strain-life approaches were conventionally used for durability prediction. These methods were found to be the conservative prediction of fatigue life as the load sequence effect of the loading histories was not considered. The main issue is in determining the fatigue life of coil spring subjected to variable amplitude loading by considering the loading sequence effects. Thus, this study aims to assess the durability of coil spring using effective strain approach by considering the load sequence effect in variable amplitude strain histories.

10.2 Methodology

10.2.1 Finite Element Analysis

Figure 10.1 shows the methodology flow of this study. Prior to on-road tests, static finite element analysis was conducted to estimate the stress distribution of coil spring under preloading. Geometrical model of the coil spring was built according to its real geometric dimension. Using static analysis, the coil spring was subjected to a total load of 3600 N corresponding to a quarter of the total weight of the car with three passengers [6]. From the stress distribution, the location of the highest stress was determined. It is important to determine this location as the strain gauge should

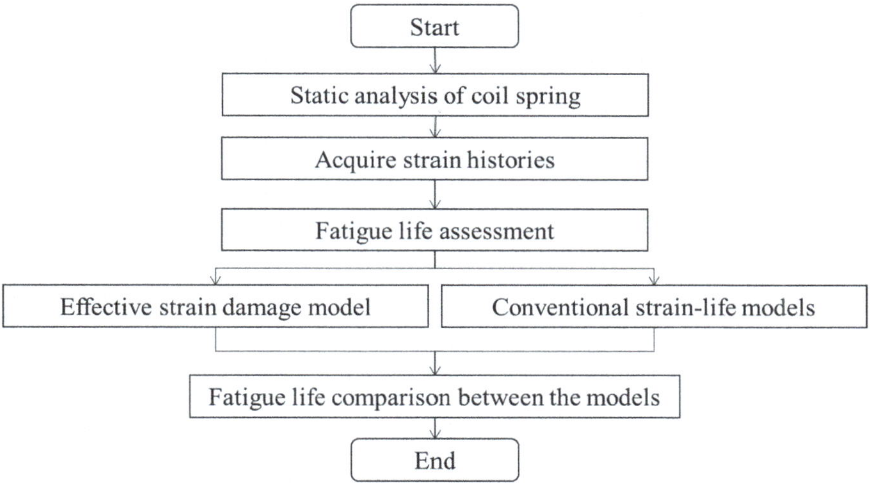

Fig. 10.1 Process flow of fatigue life assessment using effective strain damage model

be located on the suitable point to acquire the optimum strain response for fatigue analysis.

10.2.2 Road Tests and Strain Histories Acquisition

In this study, strain time histories of coil spring subjected to different road conditions were acquired. During road test, the test car was driven across different roads in industrial area, residential area and highway. A strain gauge was affixed on the coil spring where the hotspot was found through static analysis. The strain gauge was connected to data acquisition instrument that can record strain histories in real time. Strain histories were acquired at a sampling frequency of 500 Hz to ensure that the recorded signals can sufficiently represent the real situation of road surfaces [7]. Figure 10.2 shows the experimental setup of the road tests to acquire strain histories.

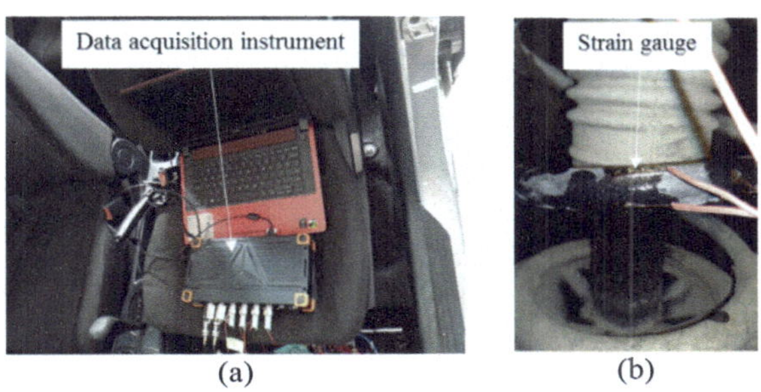

Fig. 10.2 Experimental setup of road tests: **a** data acquisition instrument, **b** strain gauge locations

10.2.3 Fatigue Life Assessment Using Effective Strain Damage (ESD) and Strain-Life Models

After the strain histories of coil spring were collected in road tests, rainflow cycle counting method was applied to determine fatigue cycles in the strain histories. Strain-life models including Coffin-Manson, Morrow and Smith–Watson–Topper (SWT) models were then used to compute the fatigue damage of each fatigue cycles. The conventional Coffin-Manson, Morrow and the SWT models were expressed in the following equations:

$$\varepsilon_a = \frac{\sigma_f^{'}}{E}(2N_f)^b + \varepsilon_f^{'}.(2N_f)^c \quad (10.1)$$

$$\varepsilon_a = \frac{\left(\sigma_f^{'} - \sigma_m\right)(2N_f)^b}{E} + \varepsilon_f^{'}.(2N_f)^c \quad (10.2)$$

$$\sigma_{max}\varepsilon_a = \frac{\left(\sigma_f^{'}\right)^2}{E}(2N_f)^{2b} + \sigma_f^{'}\varepsilon_f^{'}.(2N_f)^{b+c} \quad (10.3)$$

where ε_a represents the amplitude of strain, E is the modulus of elasticity, $\sigma_f^{'}$ is the fatigue strength coefficient, b is the fatigue strength exponent, $\varepsilon_f^{'}$ is the fatigue ductility coefficient, c is the fatigue ductility exponent, $\sigma_{max} = \sigma_m + \sigma_a$, σ_m is the mean stress and N_f is the number of cycles to failure. The material of the coil spring in this study is SAE 1045 carbon steel, and its material properties are shown in Table 10.1.

In real-life situation, many engineering applications are subjected to complex loading conditions which include variable amplitude loading (VAL). Previously, fatigue properties of metals had been tested with constant amplitude loading (CAL) to determine the stress-life or strain-life curve [6]. However, large discrepancy in the VAL and CAL fatigue test results had been reported in many investigations

Table 10.1 Material properties of SAE 1045 carbon steel

Material Property	Value
Ultimate tensile strength, σ_u (MPa)	857
Modulus of elasticity, E	207
Yield strength, σ_y	652
Fatigue strength coefficient, $\sigma_f^{'}$ (MPa)	1193
Fatigue strength exponent, b	−0.074
Fatigue ductility coefficient, $\varepsilon_f^{'}$	0.33
Fatigue ductility exponent, c	−0.49

[8]. This is mainly due to the effects of load interactions between alternating stress loading which is also known as load sequence effects. To determine the effects of load sequence in the loading histories, an effective strain damage (ESD) model was proposed by Abdullah et al. [9]. The ESD model assumed the effective strain range as the fatigue damage parameter, as defined in the following:

$$E.\Delta\varepsilon^{*} = A\left(N_f\right)^{B} \tag{10.4}$$

where $\Delta\varepsilon^{*}$ is the net effective strain range in a closed hysteresis loop, N_f is the fatigue cycle to failure, A and B are material constants. The $E.\Delta\varepsilon^{*}$ is a function of crack-opening stress (S_{op}) and can be defined as:

$$E.\Delta\varepsilon^{*} = E\left(\varepsilon_{\max} - \varepsilon_{\mathrm{op}}\right) - E\varepsilon_{\mathrm{i}} \tag{10.5}$$

where $\varepsilon_{\max}$ and $\varepsilon_{\mathrm{op}}$ are the maximum and crack-opening strain of particular cycle. For a hysteresis loop, if $\varepsilon_{\mathrm{op}}$ is smaller than the minimum strain ($\varepsilon_{\min}$), then $\varepsilon_{\mathrm{op}}$ is equal to $\varepsilon_{\min}$. Then, ε_{i} is the intrinsic fatigue limit strain range of the material. Prior to the fatigue life calculation using the ESD model, the rainflow cycles in strain histories must be rearranged according to their starting time. This to ensure that load sequence effect is taken into account in the fatigue life assessment process.

As the fatigue damage of each cycle was determined using strain-life and the ESD models, Palmgren–Miner linear damage rule was applied to sum up all the fatigue damage of all cycle as expressed in the following:

$$D = \sum \frac{n}{N_f} = 1 \tag{10.6}$$

where n is the total cycle counted at specific strain range and mean, and N_f is the total cycle required to failure computed from strain-life and ESD models at specific range and mean. Fatigue failure happens as D reaches one. Palmgren–Miner linear damage rule was applied due to its simplicity and acceptable accuracy when used with rainflow cycle counting algorithm, especially when involving the VAL [10].

Next, the fatigue life obtained using strain-life models were compared against the fatigue life calculated by ESD models using correlation graph. This is to investigate the effects of load sequence in the VAL strain histories where strain-life models did not take into account.

10.3 Results and Discussion

Figure 10.3 shows the rainflow reconstruction process for three types of road used in this study. Based on the signals obtained, highway signal had least high amplitude

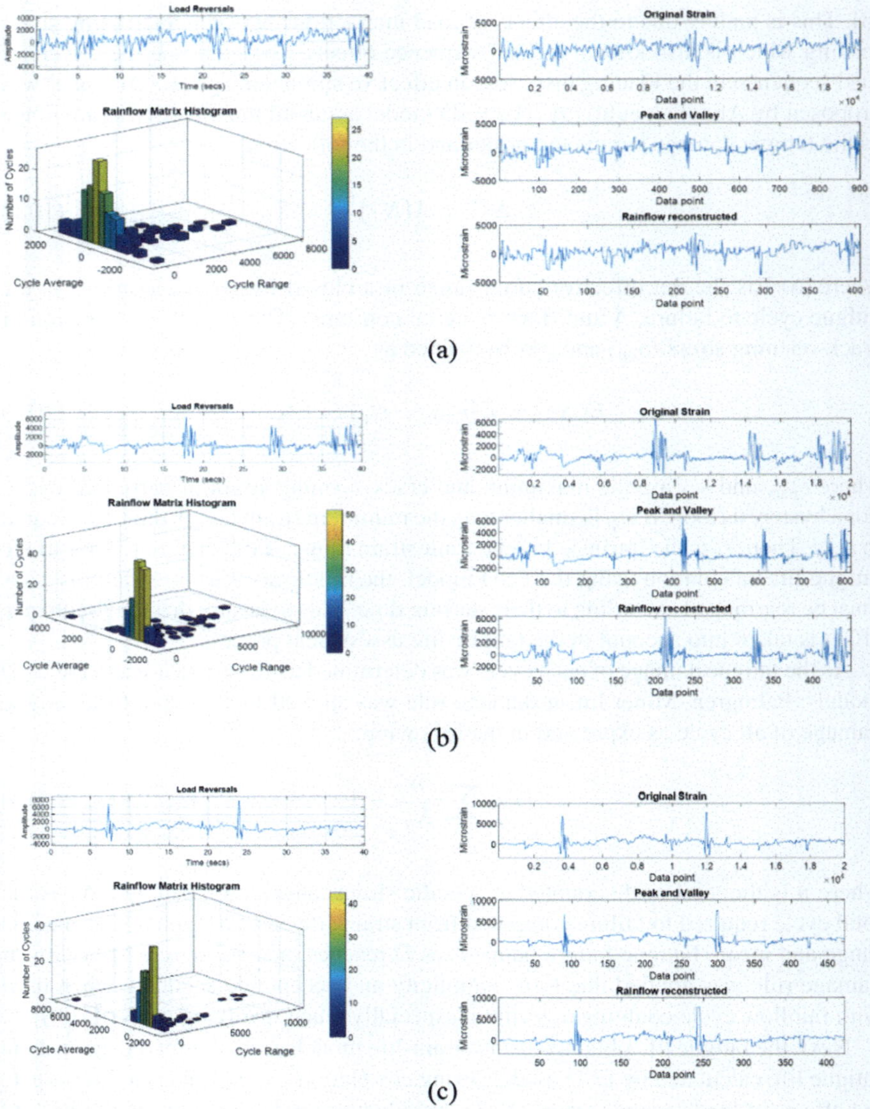

Fig. 10.3 Rainflow histogram and signal reconstruction of rainflow cycles; **a** highway, **b** campus road, **c** industrial road

Table 10.2 Fatigue life values for all type of road signals

	Fatigue life (block to failure)			
	coffin-Manson	Morrow	SWT	ESD
Highway	1.19×10^5	1.27×10^5	1.29×10^5	3.93×10^5
Campus	9.09×10^3	6.03×10^3	5.12×10^3	6.58×10^4
Industrial	2.93×10^3	2.19×10^3	1.91×10^3	3.93×10^4

events compared to other signals. This indicates less fatigue damage contribution to the coil spring for highway compared to campus and industrial road signals. The result for fatigue life values is shown in Table 10.2. The highway signal showed the longest fatigue life of around 10^5 blocks in loading. Meanwhile, the lowest fatigue life of between 1.19×10^3 and 2.93×10^3 blocks loading was found in the industrial road signal. This can be owing to the high amplitude events that frequently occurred on the industrial road that had contributed to the coil spring fatigue damage. This result is in line with the histogram shown in Fig. 10.4. It is found that the damage values for campus and industrial road are higher compared to highway with 10^{-4}. Based on the histogram shown, the high damage value involves large strain range amplitudes.

Figure 10.5 shows the fatigue life correlation between the fatigue life predicted by the ESD model and strain-life models. It can be found that the strain-life models had predicted a conservative fatigue life, especially in the campus and industrial road signals that contained many high amplitude events. The fatigue life of highway

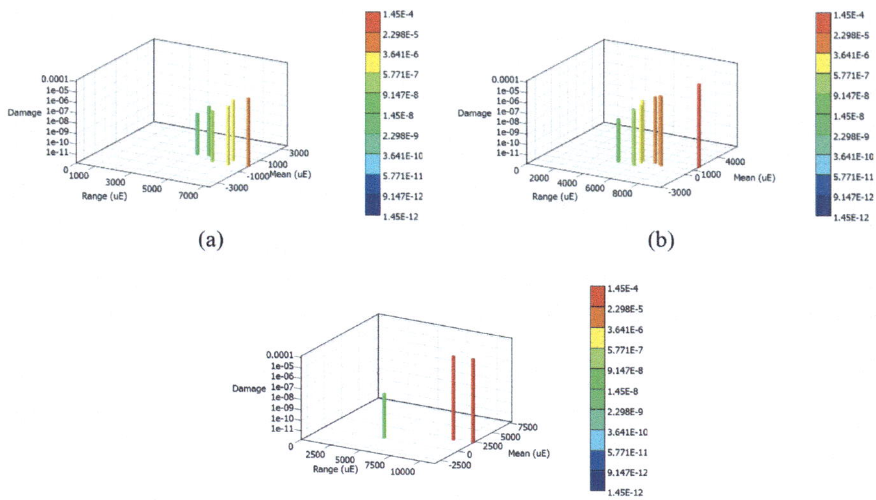

Fig. 10.4 Damage histogram for all type of signal; **a** highway, **b** campus road, **c** industrial road

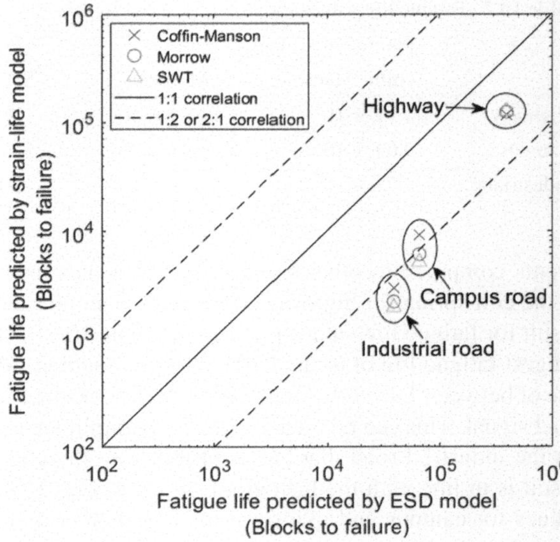

Fig. 10.5 Fatigue life correlation between ESD and strain-life models

signals predicted using the ESD model is close to the predicted value using strain-life models. This is mainly because the load sequence effect was not dominant in the durability prediction from the highway signal due to the absence of high amplitude event [11].

10.4 Conclusion

The load sequence effect was found to have more significant effects in strain signals that had many high amplitude events. Strain-life models were found to have predicted a more conservative fatigue life. For a strain signal that had smooth load profile such as the highway signal, the load sequence effect was not dominant in the durability prediction and hence the ESD model obtained consistent fatigue life results as the strain-life models. Therefore, it is important to consider the load sequence effect during fatigue life prediction to prevent any overly conservative prediction.

References

1. Nasir NNM, Abdullah S, Singh SSK, Haris SM (2020) Risk-based life assessment of prediction models on suspension system for various road profiles. Eng Failure Anal 114:104573
2. Kamaya M (2017) Fatigue assessment for seismic loads considering material degradation due to stress corrosion cracking. Nucl Eng Des 322:256–265

3. Manouchehrynia R, Abdullah S, Singh SSK (2020) Fatigue reliability assessment of an automobile coil spring under random strain loads using probabilistic technique. Metals 10(1):1–12
4. Ghahremani K, Walbridge S, Topper T (2016) A methodology for variable amplitude fatigue analysis of HFMI treated welds based on fracture mechanics and small-scale experiments. Eng Fract Mech 163:348–365
5. Diego F, José AC, Isidro AC, Soraya D, Estela R, María S, José Adolfo SA, Ana IC (2019) Experimental and finite element fatigue assessment of the spring clip of the SKL-1 railway fastening system. Eng Struct 188:553–563
6. Kadhim NA, Abdullah S, Ariffin AK (2012) Effective strain damage model associated with finite element modelling and experimental validation. Int J Fatigue 36:194–205
7. Haiba M, Barton DC, Brooks PC, Levesley MC (2003) The development of an optimization algorithm based on fatigue life. Int J Fatigue 25(4):299–310
8. Carvalho ALM, Martins JP, Voorlwad HJC (2010) Fatigue damage accumulation in aluminum 7050–T7451 alloy subjected to block programs loading under step-down sequence. Procedia Eng 2:2037–2043
9. Abdullah S, Choi JC, Giacomin JA, Yates JR (2006) Bump extraction algorithm for variable amplitude fatigue loading. Int J Fatigue 28:675–691
10. Putra TE, Abdullah S, Schramm D, Bruckmann T (2017) Reducing cyclic testing time for components of automotive suspension system utilising the wavelet transform and the Fuzzy C-Means. Mech Syst Signal Process 90:1–14
11. Baktheer A, Hegger J, Chudoba R (2019) Enhanced assessment rule for concrete fatigue under compression considering the nonlinear effect of loading sequence. Int J Fatigue 126:130–142

Chapter 11
Fatigue in Trapezoidal Leaf Springs of Suspensions in Two-Axle Wagons—An Overview and Simulation

V. M. G. Gomes, J. Correia, R. Calçada, R. S. Barbosa, and A. M. P. de Jesus

Abstract The Portuguese railway sector has suffered a significant shortage of resources, generating many issues associated with the maintenance of the rolling stock affecting its operational ability. Many of these problems are caused by the fatigue phenomenon in leaf springs of freight wagons. Thus, the study of fatigue becomes an important role in the structural integrity of leaf springs as well as for whole rolling stock. Fatigue analysis with cyclic parameters estimated from monotonic characteristics is a good option for mechanical design. Also, nominal stress obtained from classical beam theory for straight beams may be applied for trapezoidal leaf springs, even for loads close to the yielding limit of the master leaf. Despite high mean and alternating stresses for the most severe loadings, trapezoidal leaf springs have shown a fatigue-life range coincident with the high-cycle fatigue regime.

Keywords Railway · Suspension leaf spring · Spring steel · Fatigue failure

11.1 Introduction

11.1.1 Background of the Railway Sector

Over the years, the railway has been a method for transportation of goods and passengers for both small and long distances. Rail freight transport has been shown a great

V. M. G. Gomes (✉) · J. Correia · R. Calçada · A. M. P. de Jesus
Faculdade de Engenharia da Universidade Do Porto, s/n, R. Dr. Roberto Frias, 4200465 Porto, Portugal
e-mail: vtgomes@fe.up.pt

A. M. P. de Jesus
INEGI, Universidade Do Porto, Campus da FEUP, R. Dr. Roberto Frias 400, 4200465 Porto, Portugal

R. S. Barbosa
Escola Politécnica da USP, Cidade Universitária CEP, Av. Prof. Mello Moraes 2231, São Paulo 05508-030, Brazil

G. Lesiuk et al. (eds.), *Structural Integrity and Fatigue Failure Analysis*, Structural Integrity 25, https://doi.org/10.1007/978-3-030-91847-7_11

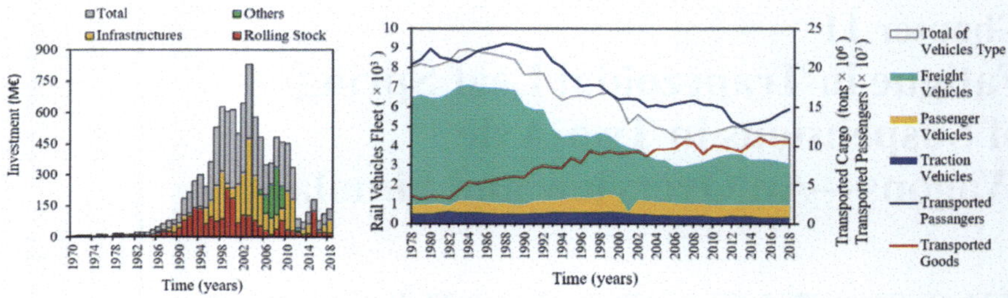

Fig. 11.1 Left—The investment made: By Portuguese government and railway companies in railway infrastructures, rolling stock, and other subsectors [3, 4]; Right—Variation over the years of rail vehicles fleet in Portugal and the number of goods and passengers transported [6, 9]

impact on the world economy since 2000, reaching values ranging between 4600 and 5600 million tons per km between years 2007 and 2013 [1, 2].

Despite the strong socio-economic and environmental responsibility that the freight rail sector has shown at the European and Worldwide, in Portugal, a great devaluation has been observed while the goods transportation trend is growing (see Fig. 11.1—right) by gray color. The disinvestment in the sector has begun in the second decade of 2000 by both the Portuguese Government and the railway companies in the generality of subsectors, including the rolling stock (see Fig. 11.1—left) [3, 4].

According to national and ministerial entities, the merchandise exports have been increasing totalizing around 10,200 million transported tons in 2018 in Portugal, in which around 1870 (18.3%) tons correspond to the international market of good transportation [5–8].

11.1.2 UIC Double Link Suspension

Freight wagons are unpowered rail vehicles and have been developed over the years. Freight wagons are constituted by different important parts such as their body, wagon frame or underframe, bogies, wheel sets, suspension components, couplings, draft gear, and braking system. Depending on the type of cargo and goods to carry, they are classified as heavy haul or mixed freight wagons [10].

In Europe, freight wagon running gear construction has been standardized by the International Union of Railways (UIC), aiming at interchangeability, and interoperability of the freight wagons traveling within the European field. Some of the most used running gears are composed of single-axle link suspension or link bogie. They may operate up to a maximum payload of 20 e 22.5 tons until 100 km/h [11–13].

Suspensions in freight wagons are typically single-stage with primary (axle box type) or secondary type (center type) suspension [10–14]. They are often constituted by dampers, linkages, connecting components, roller bearings of the wheel set, and a leaf or helicoidal springs [10]. The suspension system has an important role in vehicle dynamics once permits to compensate the vertical displacement among wheels and reduce dynamic forces and accelerations due to track irregularities. It also ensures a safer vehicle motion on the track due to oscillatory lateral motions and loss of steering control by speed above its critical value [14].

In Europe, UIC has normalized three types of freight wagon running gears, the link suspension bogie with leaf springs, the bogie with coil springs, and the single-axle running gear with the link suspension system.

The link suspension has been used for more than 100 years and has been applied mainly to two-axle freight wagons. The car body is connected to the leaf spring by the double-link suspension system, which is on the axle box. The link suspension is based on the physical principle of a pendulum, in which the axle box moves relative to the car body in the lateral and longitudinal directions. The relative motion is possible due to the existence of the axle box game at the axle horns, and the gap between the axle and car body structure [15].

This type of suspension is still attractive due to a low percentage of the total height and weight, becoming less expensive [13].

11.1.3 UIC Leaf Springs

Leaf springs have been applied in the suspension of two-axle freight wagons with single-axle running gear. UIC has standardized the spring leaves as single- or double-stage, in case of constant or progressive stiffness, respectively. They have been designed to work mainly in flexure conditions; however, when installed in rail vehicle suspension, they are subjected to adverse conditions that cause more complex load states such as circulation stability, cargo distribution on the wagon, and geometric irregularities of the track. As a result of this, suspensions of two-axle wagons work under longitudinal, lateral, and vertical dynamic loadings [13, 15, 16].

Regarding its geometry, they are constituted mainly by the spring buckle, the taper key, and leaves. According to UIC, leaf springs in the freight rail sector are classified as trapezoidal and parabolic geometry [12].

Trapezoidal leaves have a constant thickness and a slight curvature, remaining together among each other (Fig. 11.2—left). On the other hand, the parabolic geometry has a parabolic thickness in the middle of the leaf while, at the ends or at the central position, the thickness remains constant (Fig. 11.2—right). The configuration of the parabolic leaf has developed in order to become more efficient and stronger than trapezoidal geometry. This geometry provides a reduction of friction effect, reducing fretting fatigue issues, and increases specific strength because of reduction of the number of leaves and optimization of its shape [13, 15].

Leaf springs are made of alloy steels with alloying elements such manganese (Mn); silicon (Si); molybdenum (Mo); chromium (Cr); and vanadium (V). The spring steel is quenched and tempered to achieve high-strength, hardness, a few ductility, and toughness, and improved fatigue resistance [17, 18]. In accordance with ISO 683–14

Fig. 11.2 Double-link suspension with single-stage trapezoidal leaf spring (left), and double-stage parabolic leaf spring (right)

standard [19], spring steels reach a tensile strength between 1450 and 1600 MPa, after quenching and tempering processes. Once leaf springs are quenched and tempered, they are also shot-peened under pre-stressing only on tensile surfaces.

11.1.4 Fatigue Failure

Although the fatigue strength of leaf springs is a topic constantly investigated, it is still possible to verify that most of these researches are focused on road vehicles instead of rail ones [20–24].

Although there is no significant information about fatigue studies in leaf springs for rail vehicles, the dominant causes of delays and accidents on rail vehicles, especially in freight wagons, are due to the failure of the suspension members and supporting framework. Most breakages of suspension elements occur in leaves, where 70% are associated with master leaves in the middle that is closer to spring buckle, while 30% may occur on the other half of the spring (Fig. 11.4, row B) [16]. Also, fractures are observed in the middle of the leaf in the clamping zone (Fig. 11.4, row A), where the notch effects take greater importance for nucleation cracks, and at ends or close (Fig. 11.4, row C). The red points in Fig. 11.4 show the crack initiation sites, once the crack initiation is quite notable in these fractured samples.

Therefore, an extensive investigation in relation to the most likely sites for crack nucleation is mandatory, in order to prevent the failure of the leaf, avoiding occurrences of derailments that in addition to economic losses, threaten human life [25, 26].

This paper approaches the fatigue initiation in leaf springs considering the cyclic properties of the material estimated from monotonic properties available in the literature and standards. The numerical model of trapezoidal leaf spring is described and used to evaluate different load conditions due to the vertical oscillations and the transported cargo on the wagon frame. Numerical stresses are decomposed for posterior comparison between linearized numerical bending stress and the bending stress obtained from classic Bernoulli's theory. Lastly, the initiation fatigue-life is determined using the strain-life approach, considering the mean stress effect.

11.2 Simulation Details

11.2.1 Fatigue-Life Evaluation

This section aims at providing a design approach to assess the fatigue life at the surface in leaf springs in double-link suspension under vertical loadings. Fatigue life is determined by using the strain-life method, in which the coefficients and exponents of Coffin-Manson and Basquin (Eq. (11.1)) are estimated from previous knowledge of monotonic properties [27].

$$\varepsilon_a = \frac{\sigma_f^{'}}{E}(2N_f)^b + \varepsilon_f^{'}(2N_f)^c \tag{11.1}$$

There are a lot of different models to estimate the coefficients, $\sigma_f^{'}$ and $\varepsilon_f^{'}$, and the exponents, b and c. The most applicable methods for steels are the modified four-point method, the modified universal slope method, and the hardness method [28–31]. In this work, the hardness method is considered, such that the coefficients and exponents are given by equations from (11.2) until (11.5), as a function of Brinell's hardness, HB.

$$\sigma_f^{'} = 4.25\text{HB} + 225 \tag{11.2}$$

$$\varepsilon_f^{'} = \frac{0.32\text{HB}^2 - 487\text{HB} + 191000}{E} \tag{11.3}$$

$$b = -0.09 \tag{11.4}$$

$$c = -0.56 \tag{11.5}$$

Once leaf springs are previously loaded due to the weight of the wagon frame in empty or loaded, the mean stress has an important role in the fatigue assessment. Therefore, the Smith Watson and Topper model (SWT) (Eq. (11.6)) adapted for the strain-life method is considered to evaluate the critical sites. The Smith Watson and Topper model is considered because the critical plane according to Fig. 11.3 is perpendicular to the bending stress.

$$\sigma_{max}\varepsilon_a = \frac{\left(\sigma_f^{'}\right)^2}{E}(2N_f)^{2b} + \sigma_f^{'}\varepsilon_f^{'}(2N_f)^{b+c} \tag{11.6}$$

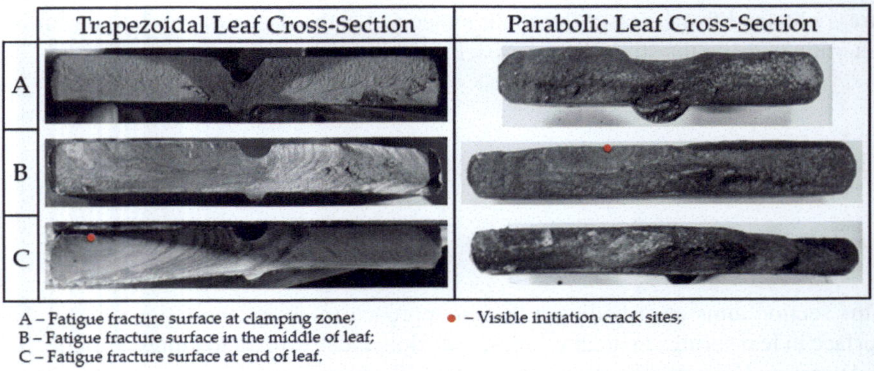

Fig. 11.3 Fatigue fracture surfaces at different positions along the length of leaf springs

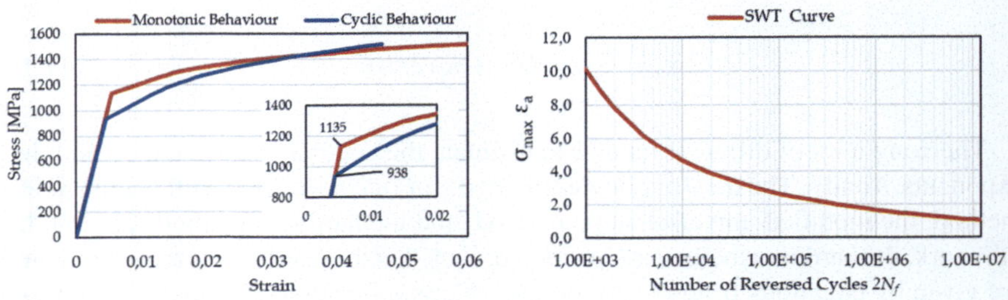

Fig. 11.4 Left—Monotonic and cyclic stress–strain curve; Right—Smith Watson and Topper model estimated by monotonic mechanical properties

11.2.2 Material

The material of leaf spring is a high-strength alloyed steel grade standardized designed as DIN 51Si7, with high silicon, and manganese contents (Table 11.1). DIN 51Si7 steel grade shows a high yielding point, σ_Y, and tensile strength, σ_{UTS}. In terms of ductility, ε_f, DIN 51Si7 has a low ductility, with values presented in Table 11.1.

The strain-life parameters are estimated from the hardness method, assuming a relationship between ultimate strength and Brinell's hardness equals $\sigma_{UTS}/3.45$ (Table 11.2). After determining the strain-life parameters, the SWT model may be built in

Table 11.1 Chemical and mechanical properties of DIN 51Si7 steel grade

Chemical composition (weight, %)				Monotonic mechanical properties			
C	Si	Mn	Others	σ_Y (MPa)	E (GPa)	σ_{UTS} (MPa)	ε_f (%)
0.51	1.65	0.65	Max. N = 0.007	1130	210	1320–1570	6

Table 11.2 Strain-life and cyclic parameters estimated from hardness model

σ_{UTS}(MPa)	HB	$\sigma_f^{'}$(MPa)	b	$\varepsilon_f^{'}$	c	K'(MPa)	n'
1445	418*	2005	– 0.09	0.21	– 0.56	2576	0.16

* within range $380 \leq \text{HB} \leq 420$ in accordance with ISO 683–14 [19]

order to predict the fatigue life under mean stress and maximum stress effects, as illustrated in Fig. 11.4—right.

The cyclic hardening stress–strain curve by Ramberg–Osgood's model is determined by Eqs. (11.7) and (11.8), relating the strain-life parameters and the hardening coefficient, K', and hardening exponent, n'. For DIN 51Si7 steel grade results in $K' = 2576$ MPa, and $n' = 0.16$ (Table 11.2).

$$n' = \frac{b}{c} \tag{11.7}$$

$$K' = \frac{\sigma_f'}{\varepsilon_f'^{\,n'}} \tag{11.8}$$

The cyclic curve by Ramberg–Osgood is illustrated in Fig. 11.4—left and is compared to the monotonic true stress–strain curve. It is noted the existence of a softening in the plastic hardening when the material is under cyclic loading as has been proven for high-strength steels [27]. The cyclic yield stress, σ_Y', assumes the value of 938 MPa.

11.2.3 Geometric Model

The trapezoidal leaf spring, illustrated in Fig. 11.5, is a UIC spring [12] and has a non-progressive behavior under static loading. Its leaves have a constant width and thickness along the length and a central notch at the clamping zone and along its length.

Table 11.3 shows the nominal dimensions such as length between support points, thickness and width of leaves, and the number of leaves of the UIC trapezoidal leaf spring. Also, the spring buckle's dimensions are presented.

Fig. 11.5 Geometry of single-stage UIC trapezoidal leaf spring: Left—Tridimensional model; Right—Technical drawing

Table 11.3 Nominal dimensions of single-stage UIC trapezoidal leaf spring (spring leaves and spring buckle)

Spring Leaves	Dimension (mm)		Value	Spring Buckle	Dimension (mm)		Value
	Length (between supports)	L mm	1190		Length	e mm	100
	Thickness	t mm	16				
	Width	b mm	120		Height	f mm	179
	Number of leaves	n_{leaves}	8				
	Support diameter	d mm	36		Width	g mm	150

11.2.4 Material Definition and Discretization of Domain

The numerical model of the trapezoidal leaf spring was built in ANSYS FEM software (R20.1). Spring leaves were modeled with isotropic elasticity, $E = 210$ GPa, and $\nu = 0.3$, and multilinear kinematic hardening given by Ramberg–Osgood equation with cyclic parameters, $K' = 2576$ MPa, $n' = 0.16$, and $\sigma_Y' = 938$ MPa. On the other hand, the spring buckle and the taper key were only considered as isotropic linear elastic components, $E = 210$ GPa, and $\nu = 0.3$. No yield is assumed for these components.

Due to physical and geometrical symmetries, a quarter of the model was considered for analysis. Regarding the finite element mesh, quadratic hexahedral elements with 20 nodes and quadratic tetrahedral elements with 10 nodes were applied. These two elements are both 3D solid elements with three degrees of freedom at node (x, y, z). They are available in the ANSYS library under the designations SOLID186 and SOLID187, respectively.

For the spring buckle and the taper key, the SOLID186 element was used with an average size equals to 8 mm. On the other hand, SOLID186 and SOLID187 elements were used along the length of each leaf. At central notches, tetrahedral elements were used, whereas the rest of the clamping zone (covered by spring buckle) prismatic shape is also considered with an average size equals to 5 and 8 mm, respectively. The numerical model is composed of 126,975 nodes and 32,863 elements.

The finite element model is presented in Fig. 11.6. Note that the mesh of the master leaf spring has refined along the length closer to the spring buckle because of the higher bending stresses and high likelihood to fracture by fatigue.

11.2.5 Definition of Contact Model

No separation contact models permit to converge to a solution faster than frictional models. They may be found in contact models of leaf spring [32]; however, they can cause convergence issues for large rotations due to high distortions of the finite

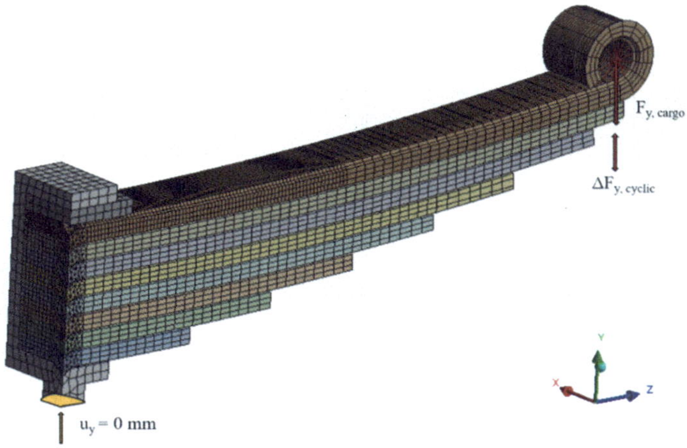

Fig. 11.6 Representation of the numerical model (¼ of the model) and respective boundary conditions—Mean force and force amplitude applied in the eye of the master leaf

element. Therefore, the contact model was modeled by a frictional model according Coulomb's law.

The contact elements used are surface-to-surface CONTA174 and TARGE170, which are suitable for 3D solid elements in contact. The friction model is isotropic with a static friction coefficient assumed to be equal to the dynamic friction coefficient, μ_F. In accordance with [33], for a clean and dry surface, μ_F, equals 0.151.

The contact algorithm is the Augmented Lagrangian method [34], with updating of stiffness in each iteration. The contact detection method is based on Gauss points with a penetration factor tolerance equals to 0.1 for lateral surfaces and taper-key surface with the spring buckle. On the other hand, surfaces in contact with leave surfaces have a factor penetration tolerance equals 0.01. Figure 11.7 illustrates the contact surfaces considered for the master leaf spring. Note that for the rest of springs, superior surfaces are totally meshed along their length.

11.2.6 Loading Conditions

The cyclic loading analysis was conducted in two phases (Fig. 11.6). The first phase consists of increasing the loading until the mean loading of cycle, $F_{y,\text{cargo}}$, and then, a second phase, the loading is made to vary cyclically counting the alternating force, $\Delta F_{y,\text{ cyclic}}/2$. The mean force, $F_{y,\text{ cargo}}$, corresponds to the weight of the wagon in empty or with cargo at the eye ends. The transmission of loading is along the vertical direction, generating bending stresses on leaves, which will play as mean stress during

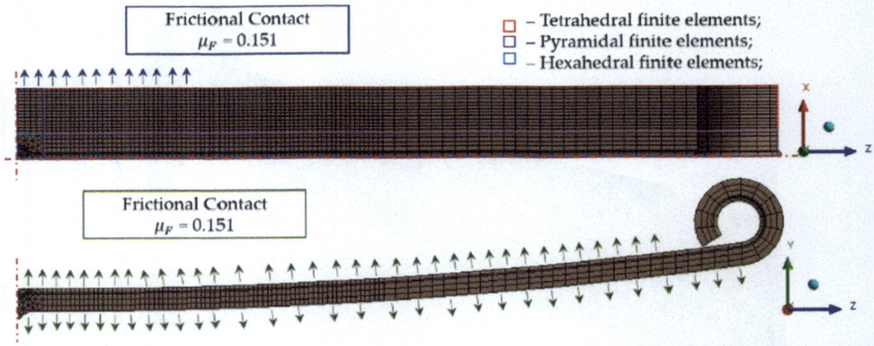

Fig. 11.7 Finite element mesh used in the master leaf spring and the respective contact conditions along its length

cyclic loading. The oscillatory loading, $\Delta F_{y,\text{cyclic}}$ incremented to the vertical component of $F_{y,\text{cargo}}$ is related to the cyclic loading caused for example from irregularities of tracks.

According to the UIC standard for suspension elements [12], the minimum mean loading applied on leaf springs will be 20 kN, corresponding to the force caused by the empty wagon frame. This value is considered as a reference.

The selection process of the simulation matrix has conducted considering the minimum mean loading, the loading corresponding to a detection of localized plastic strain in any leaf, and the yielding of master leaf spring. In addition, medium levels have considered, as well as different stress ratios (*R*), 0.1, 0.3, and 0.5. Table 11.4 shows the simulation matrix.

The solution process was conducted iteratively in a static analysis. The iterative solver was the Preconditioned Conjugate-Gradient, PCG, with a tolerance equal to

Table 11.4 Loading levels (kN) considered in simulations taking into account different stress ratios

Level of load	$F_{y,\text{ cargo}}$	R	$\Delta F_{y,\text{cyclic}}/2$	$F_{y,\text{ cargo max}}$	$F_{y,\text{ cargo min}}$
Level I	20.0	0.1	16.4	36.4	3.64
Level I	20.0	0.3	10.8	30.8	9.23
Level I	20.0	0.5	6.7	26.7	13.33
Level II	39.7	0.1	32.5	72.2	7.22
Level II	39.7	0.3	21.4	61.1	18.32
Level II	39.7	0.5	13.2	52.9	26.46
Level III	83.0	0.1	67.9	150.9	15.09
Level III	83.0	0.3	44.7	127.7	38.32
Level III	83.0	0.5	27.7	110.7	55.35
Level IV	120.0	0.1	98.2	218.2	21.82
Level V	160.0	0.5	53.3	213.3	106.67

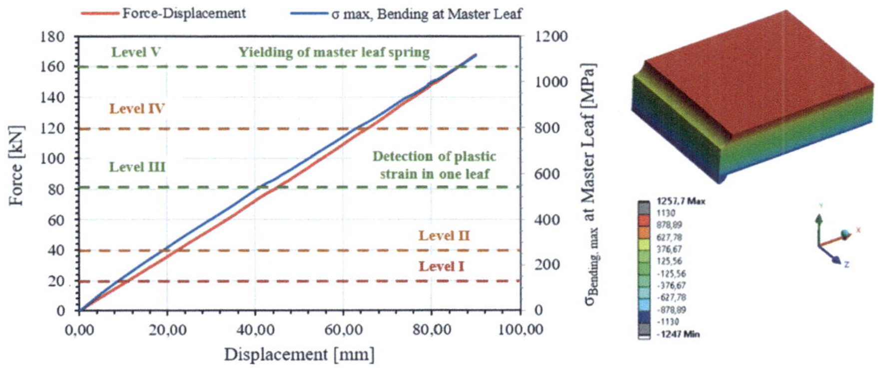

Fig. 11.8 Force–displacement global monotonic curve of the single-stage trapezoidal leaf spring under bending loading conditions

1.00000E-08. Once this bending problem requires high deflections of leaves and plasticity, the formulations of large rotation, the large strain, and the stress stiffening are considered [35–38].

11.2.7 Stress–strain State Under Cyclic Conditions

In order to define the load levels presented in the simulation matrix (Table 11.4), the numerical model was evaluated in terms of its monotonic behavior. Therefore, the elastoplastic monotonic curve was considered (red color in Fig. 11.5—left). The bending loading was increased until the reference values presented in UIC standard [12], which corresponds to the beginning of localized plastic deformation on the master leaf spring. Figure 11.8 illustrates displacement against applied force and also the bending stress at the surface of the master leaf spring. The stiffness of leaf spring is constant with displacement as was expected for a single-stage leaf spring.

After determining the loading levels based on the initial elastoplastic curve, the monotonic elastoplastic curve in the numerical model is changed to a cyclic elastoplastic curve to simulate the fatigue life under stabilized cyclic conditions.

11.3 Results and Discussion

11.3.1 Categorization of Stresses

In bodies with reduced thickness in comparison with the rest of its dimensions, models based on thin, thick plates and shells are more suitable and advisable due to

Table 11.5 Numerical bending, membrane, and peak stress and Bernoulli's bending stress for each loading

	Finite element method				Bernoulli's method
F [kN]	σ_{bd}	σ_{mb}/σ_{bd}	σ_{pk}/σ_{bd}	σ_{tot}	$\sigma_{bd,B}$
26.7	1.71E+02	2.82E−02	2.02E−05	1.75E+02	1.75E+02
30.8	1.97E+02	2.89E−02	−2.98E−04	2.03E+02	2.01E+02
36.4	2.32E+02	3.05E−02	−7.52E−04	2.39E+02	2.38E+02
53.0	3.37E+02	3.12E−02	−2.20E−03	3.47E+02	3.46E+02
61.1	3.88E+02	3.20E−02	−2.95E−03	3.99E+02	4.00E+02
72.2	4.57E+02	3.41E−02	−3.99E−03	4.70E+02	4.72E+02
110.7	6.93E+02	3.55E−02	−6.82E−03	7.13E+02	7.25E+02
127.7	7.95E+02	3.88E−02	−7.73E−03	8.19E+02	8.36E+02
151.0	9.33E+02	4.13E−02	−8.60E−03	9.63E+02	9.88E+02

low computational resources and its low stiffness in flexure conditions as occurs for solid element models.

Along the thickness, the distribution of stress state of thin plates, which may be nonlinear, may be decomposed as a sum of three components of stress, membrane, σ_{mb}, bending, σ_{bd}, and peak, σ_{pk}, such that [39]:

$$\sigma_{mb} = \frac{1}{t}\int_{-t/2}^{t/2} \sigma_{zz}\mathrm{d}y \tag{11.9}$$

$$\sigma_{bd} = -\frac{6}{t}\int_{-t/2}^{t/2} \sigma_{zz}y\,\mathrm{d}y \tag{11.10}$$

$$\sigma_{pk} = \sigma_{tot} - \sigma_{mb} - \sigma_{bd} \tag{11.11}$$

where σ_{tot} represents the total value of stress calculated from the finite element method. Each component was determined and is presented as a ratio in relation to the linearized bending stress in Table 11.5. From Table 11.5 is verified as the loading is increased, the membrane component has a higher increase than the bending stress. In addition, within the analyzed loading range, one verifies the peak stress is practically inexistent.

11.3.2 Normal Bending Stress for Straight Beams

The determination of bending stress from Bernoulli's theory is very common in design's codes and books, even for leaf springs with a slight curvature. The Bernoulli's bending stress, $\sigma^{i}_{bd,B}$, may be adapted for leaf springs with n_{leaves} leaves, considering that the total force is equally distributed along the length of each leaf. Thus, the bending stress is computed for a total force applied at the ends by [40]:

$$\sigma^{i}_{\mathrm{bd,B}} = \frac{F}{2n_{\mathrm{leaves}}} \frac{(L/2 - z_g)}{I^{i}_{zz}} y^{i} \tag{11.12}$$

where I^{i}_{zz} represents the area moment inertia and $y^{i}_{\max}$ the distance from the neutral line to fiber, referred to leave i. F is the applied total force that corresponds to $F/2$ at the ends.

The results for different values of the total force, F, are shown in Table 11.5. A comparison between Eq. (11.12) and the numerical values for linearized bending stress was performed. Figure 11.9 suggests that the deviation between the beam model and numerical model increases while the force increases. However, this increase seems to make the Bernoulli's model more conservative, with evaluated maximum deviation in relation to finite element model approximately 6%.

Figure 11.10 illustrates the resultant displacement for load case of $F_{y,\text{ cargo, max}} =$ 36.4kN and $F_{y,\text{ cargo max}} = 218.2$ kN, that is extreme cases. The maximum resultant displacement is 23.56 and 150.0 mm, respectively.

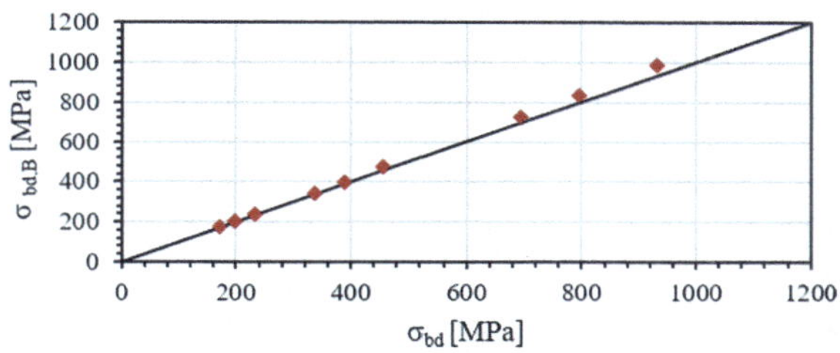

Fig. 11.9 Deviation between numerical bending stress results and Bernoulli's bending stress

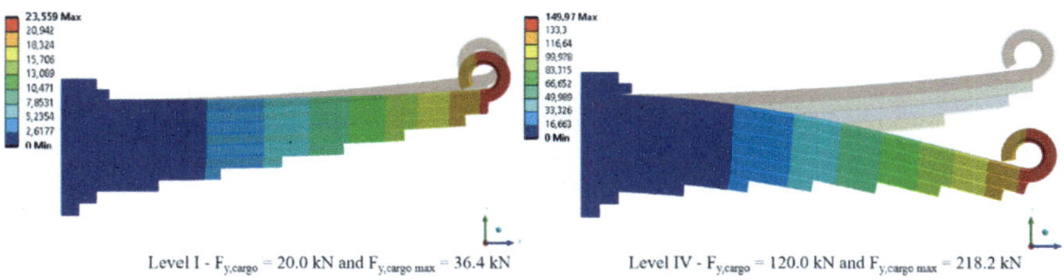

Fig. 11.10 Resultant displacement for: level I (left) and level IV (right)

11.3.3 Determination of the Fatigue Crack Initiation Life

The total fatigue life is a sum of nucleation, propagation of micro-cracks, and propagation of macro-cracks until the collapse of the structure. In engineering practices, nucleation and micro-propagation are both considered together. Thus, the total fatigue life is given by:

$$N_{\mathrm{f}} = N_{\mathrm{f}}^{i} + N_{\mathrm{f}}^{p} \tag{11.13}$$

where N_{f}^{i} and N_{f}^{p} are respectively the components of life for initiation and propagation.

From analyzing the maximum longitudinal strain and maximum longitudinal stress at surfaces, Fig. 11.11 illustrates the highest values occurring at sites closing to spring buckle (15 mm proximity) and at the sharp edge.

Combining the results of strain amplitude and maximum stress on longitudinal direction, its product, that is, the SWT, reveals infinite life until loading level III (see Table 11.6). Also, for the analyzed load range, the fatigue life falls in the high-cycle fatigue regime (Fig. 11.12).

The last step consists in determining an alternating and mean nominal stress, based on Bernoulli's model, $\sigma_{\mathrm{a,B}}$, and $\sigma_{\mathrm{m,B}}$, respectively, replacing F for $\Delta F_{y,\,\mathrm{cyclic}}/2$ and $F_{y,\,\mathrm{cargo}}$ in Eq. (11.12). The nominal stress is usually applied in an SN approach when a cyclic behavior of the structure is not known.

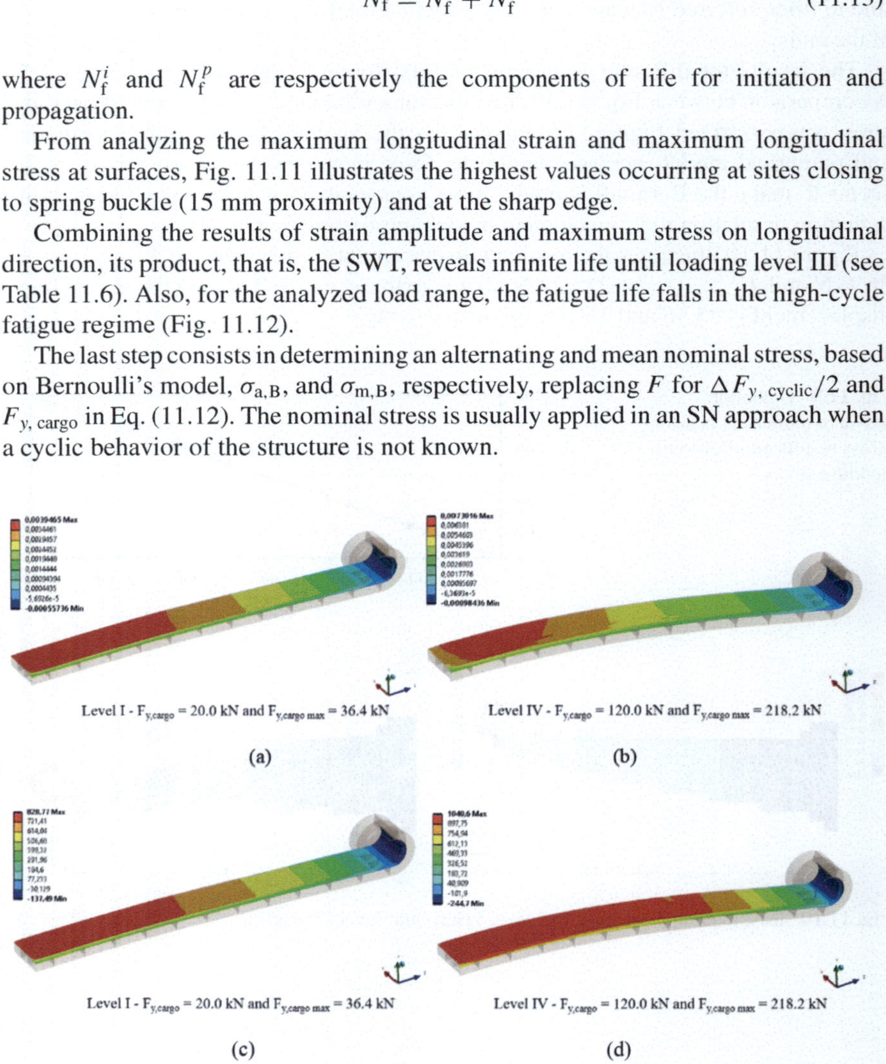

Fig. 11.11 **a** and **b**—Maximum bending strain; **c** and **d**—Maximum bending stress

Table 11.6 Assessment of fatigue life for master leaf spring evaluating SWT parameter

Load	R	ε_a	σ_{max}	SWT	N_f^i	$\sigma_{a,B}$	$\sigma_{m,B}$
Level I	0.1	5.17E−04	2.39E+02	1.24E−01	7.30E+11*	1.07E+02	1.31E+02
Level I	0.3	3.35E−04	2.03E+02	6.80E−02	2.00E+13*	7.05E+01	1.31E+02
Level I	0.5	2.02E−04	1.77E+03	3.55E−02	7.00E+14*	4.36E+01	1.31E+02
Level II	0.1	1.01E−04	4.73E+02	4.80E−01	3.60E+08*	2.13E+02	2.60E + 02
Level II	0.3	6.57E−04	4.01E+02	2.63E−01	1.00E+10*	1.40E+02	2.60E+02
Level II	0.5	3.95E−04	3.48E+02	1.38E−01	4.00E+11*	8.66E+01	2.60E+02
Level III	0.1	2.07E−03	9.74E+02	2.01E+00	1.83E+05	4.45E+02	5.44E+02
Level III	0.3	2.07E−03	8.30E+02	1.71E+00	3.95E+05	2.93E+02	5.44E+02
Level III	0.5	8.03E−04	7.20E+02	5.79E−01	1.40E+08*	1.81E+02	5.44E+02
Level IV	0.1	2.57E−03	1.01E+03	2.59E+00	5.54E+04	6.43E+02	7.86E+02
Level V	0.5	1.29E−03	1.00E+03	1.29E+00	1.81E+06	3.49E+02	1.05E+03

*Number of cycles to failure greater than superior limit to consider infinite life (5.00E+06 cycles)

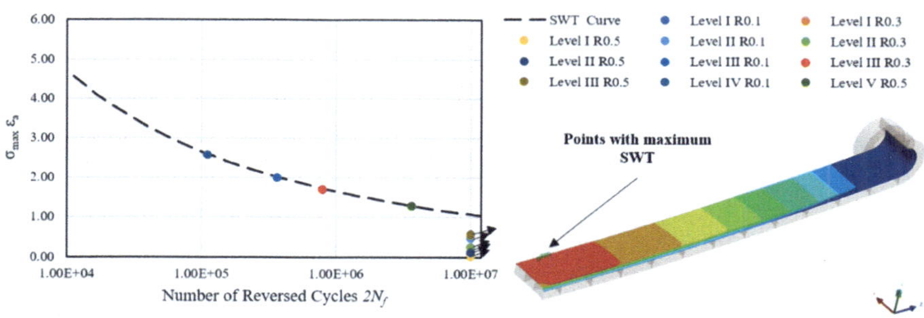

Fig. 11.12 Left—Analysis of fatigue life for initiation based on SWT parameter for different cyclic loading conditions; Right—Plot of SWT parameter along the surface of master leaf and the points with maximum SWT

11.4 Conclusions

A review about causes of fatigue phenomenon in leaf springs commonly applied in freight wagons and its impact in the rail sector has mentioned initially. Three different sites were identified as critical for crack initiation. Thus, the study of fatigue presents an important role in the structural integrity of the leaf springs.

The overall study consists of a design perspective together with the finite element method. The material cyclic behavior and the properties of strain-life fatigue are estimated from the monotonic properties.

Nominal stress amplitude and nominal mean stress were computed based on the classical Bernoulli's theory and posteriorly compared to the linearized bending stress, revealing a good agreement within the analyzed loading range.

The linearized bending stresses were obtained by total stress decomposition into its membrane, bending, and peak components. These were compared to linearized bending stress, revealing that by increasing the loading, the membrane stress increases more that linearized. On the other hand, the peak stress remains relatively inexistent.

The SWT model adapted to strain-life fatigue permitted to evaluate of the fatigue life for crack initiation. This analysis reveals that the crack initiates closer to the width ends, for a distance to spring buckle equal to 15 mm. For mean stress below 39.7 kN and R greater than 0.5 (inclusive) originates a fatigue life greater than 5E6 cycles. Therefore, trapezoidal leaf springs operate in the high-cycle fatigue regime.

Acknowledgements The authors thank to its industrial partner, MEDWAY (Maintenance and Repair). Also, the Doctoral Programme iRail- Innovation in Railway Systems and Technologies funding by the Portuguese Foundation for Science and Technology, IP (FCT) through the PhD grant (PD/BD/143141/2019).

References

1. Knoema (2020) World—passengers carried by railways. https://knoema.com/atlas/World/topics/Transportation/Rail-transport/Passengerscarried-by-railways. Accessed 2 Nov 2020
2. Knoema (2020) World—goods transported by railways. https://knoema.com/atlas/World/topics/Transportation/Rail-transport/Goodstransported-by-railways. Accessed 2 Nov 2020
3. PorData (2020) Investimentos efectuados no transporte ferroviário: total, pelo estado e pelas empresas ferroviárias - continente. https://www.pordata.pt/Portugal/Investimentos+efectuados+no+transporte+ferroviário+total++pelo+Estado+e+pelas+empresas+ferroviário+++Continente-3113. Accessed 2 Nov 2020
4. PorData (2020) Investimentos efectuados no transporte ferroviário: total, pelo estado e pelas empresas ferroviárias - continente. https://www.pordata.pt/Portugal/Investimentos+efectuados+no+transporte+ferroviário+total+e+por+área+++Continente-3114. Accessed 2 Nov 2020
5. Knoema (2020) Portugal—passengers carried by railways. https://knoema.com/atlas/Portugal/topics/Transportation/Rail-transport/Passengerscarried-by-railways. Accessed 2 Nov 2020
6. PorData (2020) Passageiros transportados no sistema ferroviário: total e por tipo de tráfego - continente. https://www.pordata.pt/Portugal/Passageiros+transportados+no+sistema+ferroviário+total+e+por+tipo+de+tráfego+++Continente-3116. Accessed 2 Nov 2020
7. Knoema (2020) Portugal—goods transported by railways. https://knoema.com/atlas/Portugal/topics/Transportation/Rail-transport/Goodstransported-by-railways. Accessed 2 Nov 2020
8. PorData (2020) Mercadorias transportadas por ferrovia: total, transporte doméstico e transporte internacional. https://www.pordata.pt/Europa/Mercadorias+transportadas+por+ferrovia+total++transporte+doméstico+e+transport+international-2721-315004. Accessed 2 Nov 2020
9. PorData (2020) Veículos do transporte ferroviário - continente. https://www.pordata.pt/Portugal/Veículos+do+transporte+ferroviário+++Continente-3111. Accessed 2 Nov 2020
10. Spiryagin M, Cole C, McClanachan SYQM, Spiryagin V, McSweeney T (2014) Design and simulation of rail vehicles. CRC Press, Taylor and Francis Group
11. I. U. of Railways (1995) Wagons—running gear—normalization.
12. I. U. of Railways (2007) Wagons—suspension gear—standardisation

13. Jönsson P (2004) Modelling and laboratory investigations on freight wagon link suspensions with respect to vehicle-track dynamic interaction European freight vehicle running gear: today's position and future demands, technical report, SE-100 44 Stockholm
14. Iwnicki S, Spiryagin M, Cole C, McSweeney T (2020) Handbook of railway vehicle dynamics, 2nd edn. CRC Press, Taylor and Francis Group
15. Hecht M (2001) European freight vehicle running gear: today's position and future demands. Proc Inst Mech Engrs Part F J Rail Rapid 215:1–11
16. Petrovi D, Gai BM, Savkovi M (2012) Increasing the efficiency of railway transport by improvement of suspension of freight wagons. Promet-Traff Transp 24(6):487–493. https://doi.org/10.7307/ptt.v24i6.1202
17. Smith WF (1999) Principles of materials science and engineering
18. Yamada Y (2007) Materials for springs. Springer, Berlin. https://doi.org/10.1007/978-3-540-73812-1
19. ISO 683–14 (2004) Heat-treatable steels, alloy steels and free-cutting steels- part 14: hot-rolled steels for quenched and tempered springs, 3rd edn. International Standard Organization
20. Husaini NA. Riantoni R, Putra TE, Husin H (2019) Study of leaf spring fracture behavior used in the suspension systems in the diesel truck vehicles. In: IOP Conf Ser Mater Sci Eng 541. https://doi.org/10.1088/1757-899X/541/1/012046
21. Noronha B, Yesudasan S, Chacko S (2020) Static and dynamic analysis of automotive leaf spring: a comparative study of various materials using ANSYS. J Fail Anal Prev 20(3):804–818. https://doi.org/10.1007/s11668-020-00877y
22. Ceyhanli UT, Bozca M (2020) Experimental and numerical analysis of the static strength and fatigue life reliability of parabolic leaf springs in heavy commercial trucks. Adv Mech Eng 12(7):804–818. https://doi.org/10.1177/1687814020941956
23. Kumar K, Aggarwal ML (2013) Computer aided fea simulation of EN45A parabolic leaf spring. Int J Ind Eng Comput 4(2):297–304. https://doi.org/10.5267/j.ijiec.2013.01.005
24. Fragoudakis R, Saigal A, Savaidis G, Malikoutsakis M, Bazios I, Savaidis A, Pappas G, Karditsas S (2013) Fatigue assessment and failure analysis of shot-peened leaf springs. Fatigue Fract Eng Mater Struct 36(2):92–101. https://doi.org/10.1111/j.1460-2695.2011.01661.x
25. Atig A, Sghaier RB, Seddik R, Fathallah R (2017) Reliability-based high cycle fatigue design approach of parabolic leaf spring. Fatigue Fract Eng Mater Struct. https://doi.org/10.1177/1464420716680499
26. Federal Department of the Environment, Transport, Energy and Communications DETEC, Federal Office for Transport BAV, Derailment in Daillens VD, report according to RID 1.8.5.4, 2016. Available at http://otif.org/en/
27. Kumbhalkar MA, Bhope DV, Vanalkar AV, Chaoji PP (2018) Failure analysis of primary suspension spring of rail road vehicle. J Fail Anal Preven. https://doi.org/10.1007/s11668-018-0542-1
28. Stephens RI, Fatemi A, Stephens RR, Fuchs HO (2001) Metal fatigue in engineering, 2nd edn
29. Hartman D (2013) Robust model for fatigue life estimation from monotonic properties data for steels, Master thesis, University of Waterloo, Waterloo, Ontario, Canada
30. Ong JH (1993) An improved technique for the prediction of axial fatigue life from tensile data. Int J Fatigue 15:213–219
31. Muralidharan U, Manson SS (1988) A modified universal slopes equation for estimation of fatigue characteristics of metals. J Eng Mater Technol 110:55–58
32. Roessle ML, Fatemi A (2000) Strain-controlled fatigue properties of steels and some simple approximations. Int J Fatigue 20:495–511
33. Arora VL, Bhushan G, Aggarwal ML (2014) Fatigue life assessment of 65Si7 leaf springs: a comparative study. Int Sch Res Not 2014. Article ID 607272. https://doi.org/10.1155/2014/607272
34. Hryciów Z, Krasoń W, Wysocki J (2018) The experimental tests on the friction coefficient between the leaves of the multi-leaf spring considering a condition of the friction surfaces. Maintenance Reliab 20(4). https://doi.org/10.17531/ein.2018.4.19

35. Simo JC, Laursen T (1992) An augmented Lagrangian treatment of contact problems involving friction. Comput Struct 42:97–116. https://doi.org/10.1016/0045-7949(92)90540-G
36. Hughes TJR (1984) Numerical Implementation of constitutive models: rate-independent deviatoric plasticity. In: Nemat-Nasser S, Asaro RJ, Hegemier GA (eds) Theoretical foundation for large-scale computations for nonlinear material behavior. Mechanics of elastic and inelastic solids 6, vol 6. Springer, Dordrecht. https://doi.org/10.1007/978-94-009-6213-2_3
37. Rankin CC, Brogan FA (1986) An element independent corotational procedure for the treatment of large rotations. ASME J Press Vessel Technol 108(2):165–174. https://doi.org/10.1115/1.3264765
38. Argyris J (1982) An excursion into large rotations. Comput Methods Appl Mech Eng 32(1–3):85–155. https://doi.org/10.1016/0045-7825(82)90069-X
39. Malvern LE (1969) Introduction to the mechanics of a continuous medium. Prenticle-Hall, Inc., Englewood Cliffs
40. Gordon JL (1976) Outcur: an automated evaluation of two dimensional finite element stresses according to ASME. ASME, paper 76-WA/PVP-16
41. Schmid SR, Hamrock BJ, Jacobson BO (2013) Fundamentals of machine elements, 2nd edn. McGraw-Hill International Edition

Part III
Additive Manufacturing Metals

Chapter 12
FE Modelling of the Fatigue Behaviour of L-PBF Inconel 718 with As-Built Surfaces

Federico Uriati, Gianni Nicoletto, and Radomila Konečná

Abstract The evaluation of fatigue properties of additively manufactured metal parts with surfaces at the as-built state is matter of great interest since it has been demonstrated that surface roughness has a key influence on high-cycle fatigue properties of components. As-built surfaces contain notch-like features that work as stress concentration sites where fatigue crack nucleation may occur. These typical morphologies due to Laser-Powder Bed Fusion (L-PBF) processing need to be considered in evaluating the fatigue performance; therefore, parameters related to as-built surface quality should be investigated and considered in part design. An innovative fatigue test method using miniature specimens is adopted to evaluate fatigue strength of the material; different specimen orientations are produced and tested. The fatigue results show the directional nature of the as-built surface and the surface differences introduced by the specimen orientation. In this work, as-built surfaces are investigated with two methodologies: microscopical observation of specimen cross section and optical interferometer observation of specimen surface. The irregular profiles are digitally reconstructed and evaluated via finite element analysis and the stress distributions interpreted. A correlation between stress-based parameters of surface models and fatigue strength is attempted.

Keywords Powder bed fusion · Inconel 718 · Fatigue · FEA

12.1 Introduction

Additive manufacturing (AM) has opened a new design approach in relation with shape complexity of mechanical components. Parts are manufactured by adding material layer by layer, allowing to obtain part geometries previously impossible

F. Uriati (✉) · G. Nicoletto
Università Degli Studi Di Parma, 43124 Parma, Italy
e-mail: federico.uriati@unipr.it

R. Konečná
University of Žilina, Žilina 01026, Slovakia

G. Lesiuk et al. (eds.), *Structural Integrity and Fatigue Failure Analysis*,
Structural Integrity 25, https://doi.org/10.1007/978-3-030-91847-7_12

to realize with traditional manufacturing processes. Concerning metal parts, laser-powder bed fusion (L-PBF) is an additive manufacturing technology which allows the accurate production of parts with internal cavities and features not producible using subtractive methods.

One of the main challenges when qualifying AM components is that the material properties are dependent on the process itself. Porosity, internal defects and surface roughness are dependent of the manufacturing process and define the usability of the parts. The main aim of this work is the evaluation of the fatigue strength of parts with surfaces in the as-built state because surface morphology has a key influence on fatigue properties. Since surface irregularities are recognized to be highly detrimental on fatigue properties [1], the as-built surface condition needs to be investigated and considered in the design workflow for structural integrity. A methodology for fatigue life prediction of as-built surface effects under cyclical loading is investigated here. It is based on (i) the experimental evaluation of the HCF behaviour of specimens with as-built surfaces of different quality according to different directions with respect to the build direction, and (ii) structural FEM simulations of the material surface having profiles obtained with two different techniques and reconstructed. Surface roughness appears to be the predominant parameter affecting fatigue life and the present work focuses on the modelling of the effect of this parameter on fatigue strength.

12.2 As-Build Fatigue Behaviour

12.2.1 Material and Specimen Fabrication

The material considered in this study is gas-atomized Inconel 718, a nickel–chromium alloy extensively used in gas turbines, rocket motors and space crafts since it is characterized by excellent mechanical properties at high temperatures. The chemical composition listed in Table 12.1.

Fatigue specimens were manufactured by service provider BEAM-IT (Fornovo Taro, Italy) with a SLM Solutions 280HL system (SLM Solutions, Germany). A layer thickness of 50 μm was used together with a laser power $P = 250$ W, hatch spacing $h = 0.12$ mm and scan speed $v = 760$ mm/s.

After L-PBF fabrication, the post-processing steps were heat treatment before specimen removal from the base plate and a two-step heat treatment after removal given by stress relief and age hardening by double ageing. No surface post-processing was performed on the specimens.

Table 12.1 Chemical composition of IN 718 powder

Element	Ni	Cr	Fe	Nb	Mo	Co	Ti	Al	Cu
Wt. %	51.56	17.9	18.2	5.23	3.21	0.15	1.14	2.19	0.05

12.2.2 Fatigue Testing and Results

The fatigue test method adopted is based on the miniature specimen procedure [2] characterized by an original 22-mm-long specimen geometry shown in Fig. 12.1a. The advantages of the methodology are the significant reduction of material use, reduction of production costs and freedom in the orientation of specimen on the build plate. According to this method, two different specimen orientations with respect to the build direction are fabricated as shown in Fig. 12.1b. The specimens were tested under cyclic plane bending generating a pulsating tensile stress ($R = 0$) on the flat surface opposite to the notch. Fatigue testing was performed at a frequency of 25 Hz and test run-out was fixed at 2×10^6 cycles.

The as-built surface quality of the two specimen orientations shown in Fig. 12.1b is different: the top surface of type A is influenced by the hatching and contour strategy (Fig. 12.2a); the flat test surface of type C is influenced by the layer thickness and the contour parameters (Fig. 12.2b).

The fatigue results of Fig. 12.3 show the directional nature of the fatigue behaviour due to the different as-built surface quality as it is affected by process parameters and surface orientation with respect to fabrication direction. Specifically, fatigue strengths are estimated in 410 MPa for type C and in 315 MPa for type A.

Thus, the role of surface morphology on the fatigue performance is demonstrated to be very relevant. The origin of the difference is attributed to the irregularities of the as-built surfaces that contain micro-notches and partially welded particles that work as local stress concentrators.

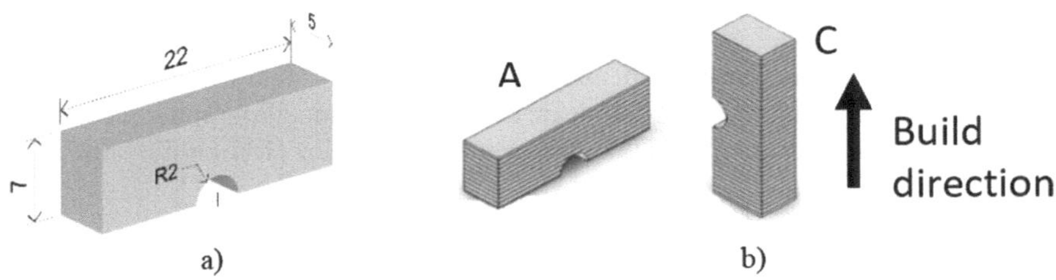

Fig. 12.1 **a** Miniature specimen dimension, **b** denomination of miniature specimens and layer direction with respect to build direction

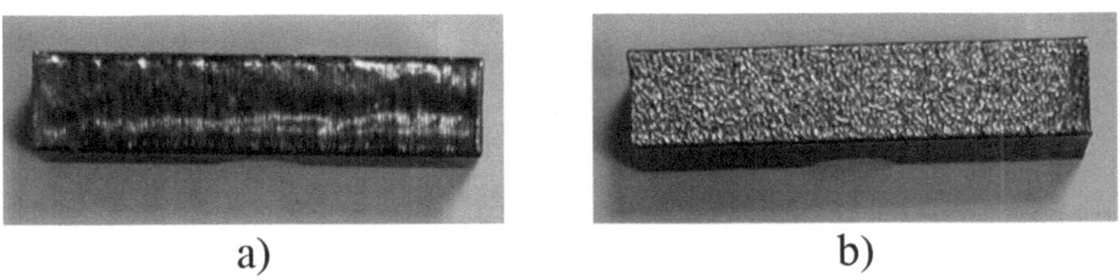

Fig. 12.2 **a** Type A specimen flat surface; **b** type C specimen flat surface

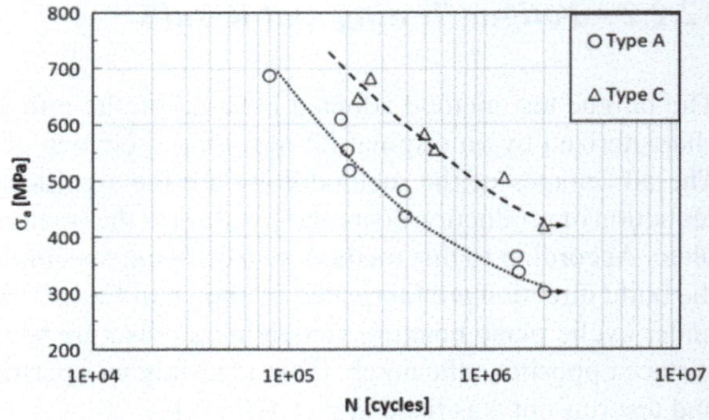

Fig. 12.3 Directional fatigue data for as-built Inconel 718 produced by SLM 280HL

12.3 FE Analysis of As-Built Surfaces

Previous works such as [4] already investigated experimentally and by finite element modelling the influence of surface morphology on the fatigue behaviour of as-built L-PBF Ti6Al4V. This study applies that approach to investigate the stress distribution at rough surfaces of L-PBF Inconel 718.

12.3.1 Characterization of As-Build Surface Profiles

The proposed approach, following [4], requires the observation of the real surface morphology to be transformed into a structural FE model to investigate predictive methods of fatigue performance of metal AM parts. Here, two different methods of profile extraction and reconstruction are considered: i) metallographic sectioning and light optical microscopy (LOM) observation; ii) Taylor Hobson Talysurf CCI optical surface mapping.

Optical metallographic cross-sectioning. The first method is based on the optical observation of specimen cross-sections, Fig. 12.4a. Metallographic specimens were prepared according to standard techniques and observed using the Neophot 32 light microscope. The profiles of the flat surfaces of the specimens were observed.

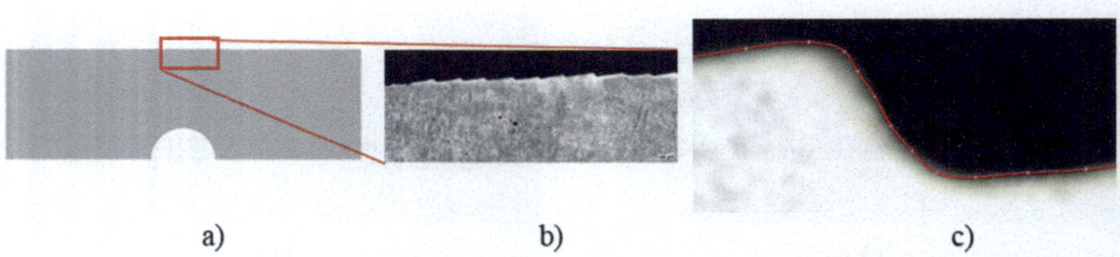

Fig. 12.4 **a** Miniature specimen cross section representation and, **b** type A specimen magnified profile detail and, **c** digital reconstruction

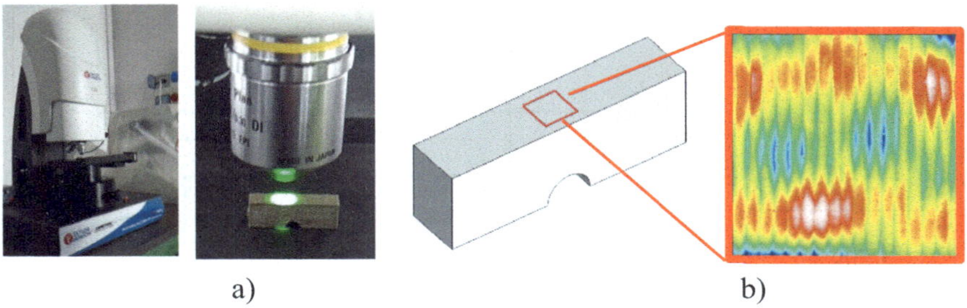

Fig. 12.5 **a** Taylor Hobson Talysurf CCI white light interferometer; **b** example of type A miniature specimen surface mapping using MountainLab software

Figure 12.4b shows a detail of type A specimen profile where the characteristic features due to the laser scanning strategy and the hatching spacing are readily recognized as a saw-tooth pattern.

The magnified images of the as-built specimens were used to digitally reconstruct the profile and a detailed 2D FE model of the surface layer. The cross-section images, obtained with optical observation, were used as a reference to manually reconstruct the real surface profile using the software Rhinoceros (Fig. 12.4c).

Interferometric surface mapping. The second method is a non-destructive procedure useful to investigate the specimens' surface topography. The area of interest is measured with a Taylor Hobson CCI optical profiler belonging to the class of white light interferometer, Fig. 12.5a; the profiler was capable of scanning with a horizontal resolution of 1.3 μm and a vertical resolution of 1 nm. The total measured area is equal to 3×3 mm^2, the procedure and the areal map results are shown in Fig. 12.5b.

As-built three-dimensional surface, extracted with this methodology, is analysed using the software MountainLab (Digital Surf), which allows to adjust measurement errors and perform data filtering. The software computation is based on the standard ISO 25178 (Geometric Product Specifications (GPS)—Surface texture: areal). To avoid or at least reduce errors, the default values defined in ISO 25178 part 3 for areal surfaces are applied to the surface investigation performed for this study. The surfaces are filtered using a low pass filter $\lambda_s = 50$ μm. Figures 12.5b and 12.6a show typical maps for the two specimen orientations. Profiles for each 3D map for the two types of specimens were extracted (see Fig. 12.6b); the profile is reconstructed from point cloud distribution with third-order polynomial interpolation. The software Rhinoceros is used for the profile modelling and surface preparation for FE analysis.

12.3.2 FE Analysis of As-Built Surfaces

Stress analysis of near-surface material layer delimited by the profiles was performed by the finite element (FE) method. Surface profiles obtained by the two experimental

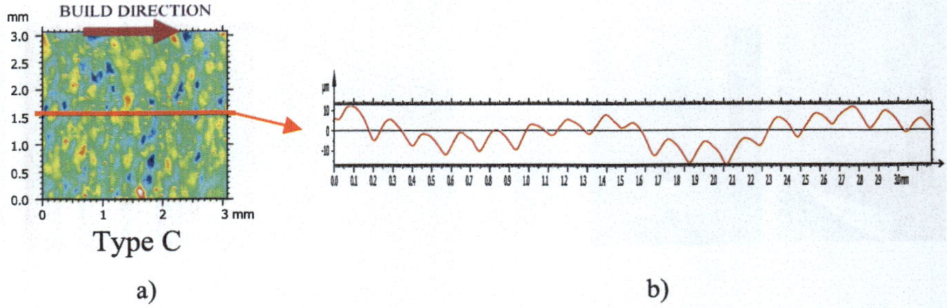

Fig. 12.6 **a** Surface topography obtained with white light interferometer and, **b** profile extraction

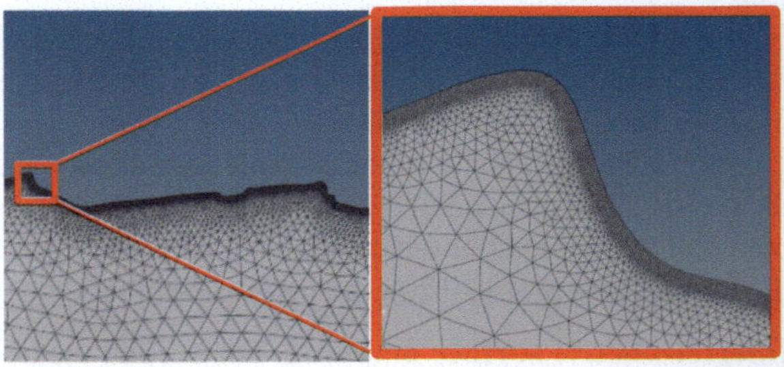

Fig. 12.7 Reference surface profile and detailed mesh distribution at details

methods were used to generate a suitable FE model meshing and application of boundary conditions. Detailed near-surface elastic stress distributions were obtained by linear solution of finely meshed FEA models.

12.3.2.1 Surface Profile Meshing and Analysis

The FE analysis is developed in Altair HyperMesh using a graded triangular meshing method. A very fine finite element mesh is needed to resolve the geometrical features shown in Fig. 12.4 and provide detailed stresses and their gradients at the notch roots. After a sensitivity analysis, the minimum element size was set to 0.1 μm, Fig. 12.7. This minimum element size guarantees the determination of local stress peaks at sharp notches typical of as-built profiles. Element size is then gradually increased from this size to element size of 50 μm to mesh subsurface material.

12.3.2.2 As-Built Surface Stresses

Irregular profiles of as-built surfaces induce a high number of micro-notches that locally generate high stress gradients. Theoretical stress concentration factors K_t of surface notches are calculated as the ratio between the local peak stress at the notch root and the nominal tensile stress. The definition of $K_t = \frac{\sigma_{eff}}{\sigma_{nom}}$ where σ_{eff} is the stress

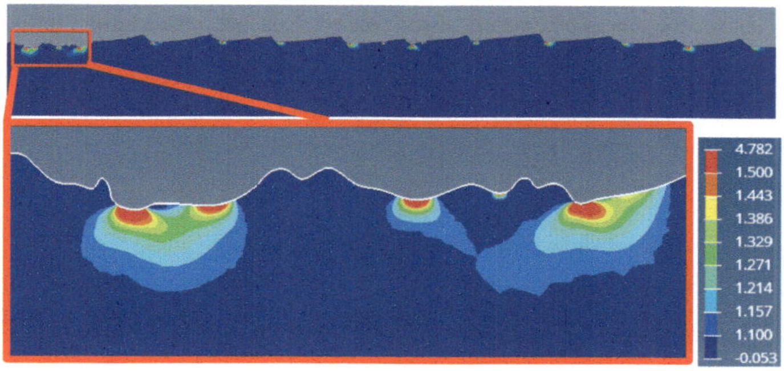

Fig. 12.8 Macro-FE contour plot and local stress concentration factor distribution on the irregular reconstructed profile

in the notch root and σ_{nom} is the nominal stress is locally computed, readily used and directly mapped, Fig. 12.8.

In addition to this straightforward approach, previous works [3] proposed non-local approaches to correlate experimental fatigue strength of L-PBF Inconel 718 with FEA results of as-build profiles. Besides the local parameter (i.e. highest K_t), the extension of a highly stressed area (HSA) is proposed in this work. It is defined as the ratio between the area covered by elements whose nodes have $K_t > 1.5$ and the length of the profile

$$\text{HSA} = \frac{\text{area where } K_t > 1.5}{\text{profile length}}$$

12.4 Results and Discussion

This section refers to the fatigue strengths determined experimentally on directional specimens and compares to (i) roughness measurements of the specimen surfaces according to the optical interferometric method and (ii) FEA results obtained using the experimental profiles determined by two methods.

Areal roughness parameters were calculated in compliance with ISO 25178 for the two specimen surfaces. The more representative values are: i) Sq (Root mean square height - RMS); ii) area surface roughness (Sa); iii) maximum peak height (Sp); iv) maximum valley depth (Sv). A preliminary qualitative correlation is reported in Table12.2, experimental fatigue strength is inversely correlated with measured surface roughness.

Table 12.3 shows the FEA results in terms of peak stresses and highly stressed areas in dependence of the two profile reconstruction methods for the two specimen orientations. Peak stress data strongly depend on the reconstruction method and do

Table 12.2 Surface roughness parameters of specimen flat surface versus fatigue strength

Specimen type	Sq (μm)	Sa (μm)	Sp (μm)	Sv (μm)	Exp. fatigue strength (MPa)
A	9.61	7.61	27.95	35.13	300
C	6.47	4.989	37.5	27.2	420

Table 12.3 FEA analysis of surface profiles versus fatigue strength

Profile reconstruction technique	Specimen type	Local peak stress (FEA)	Highly stressed area (FEA) HSA	Exp. Fatigue strength
Optical cross section	A	4.86	$1.87\ 10^{-3}$	300 MPa
	C	6.56	$1.46\ 10^{-3}$	420 MPa
Surface topography	A	2.00	$1.26\ 10^{-4}$	300 MPa
	C	1.98	$1.21\ 10^{-4}$	420 MPa

not correlate with fatigue data. The highly stressed area data also depend on the reconstruction method but, at least, inversely correlate with fatigue data.

This contribution represents a preliminary assessment of the usefulness of the FEA approach to predict fatigue behaviour of complex rough surfaces such as those associated with selective laser melting [4]. It is pointed out that 3D effects are neglected in this 2D FEA approach. Additional work is ongoing, specifically aimed at different ways of interpreting the FEA results and the identification of equivalent notch models.

12.5 Conclusions

In this work, as-built specimens of L-PBF Inconel 718 were experimentally demonstrated to have fatigue strength that depends on their orientation with respect to the build direction. The influence is very significant, and it is attributed to the different surface morphology which was visible also to the naked eye. Simple surface roughness parameters, such as Sa, were found to correlate with the directional fatigue strengths.

The morphology on the as-build surfaces was determined using two reconstruction methods, namely metallographic cross-sectioning and interferometric surface mapping. The first method is experimentally laborious but gives more detailed information and requires a manual intervention; the second is experimentally quicker and allows to observe a larger area but is more approximated.

Reconstructed profiles were then used to develop FE models of the rough surface layers and to determine near-surface stress distributions. Computed stresses depend on the surface reconstruction method. An initial correlation of peak stresses and non-local measures with experimental fatigue strengths is presented. However, more work needs to be performed to demonstrate that the FE modelling of rough surface is viable to predict fatigue strength.

Acknowledgements The authors wish to acknowledge the company BEAM-IT srl., Fornovo Taro, Italy, for skilfully fabricating the Inconel 718 specimens.

References

1. Murakami Y (2019) Effect of surface roughness on fatigue strength of Ti-6Al-4V alloy manufactured by additive manufacturing. Procedia Struct Integr 294–301. https://doi.org/10.1016/b978-0-12-813876-2.00016-9
2. Nicoletto G (2017) Anisotropic high cycle fatigue behavior of Ti–6Al–4V obtained by powder bed laser fusion. Int J Fatigue 94:255–262. https://doi.org/10.1016/j.ijfatigue.2016.04.032
3. Solberg K, Berto F (2020) The effect of defects and notches in quasi-static and fatigue loading of Inconel 718 specimens produced by selective laser melting. Int J Fatigue 137. https://doi.org/10.1016/j.ijfatigue.2020.105637
4. Vayssette B, Saintier N, Brugger C, El May M, Pessard E (2019) Numerical modelling of surface roughness effect on the fatigue behavior of Ti-6Al-4V obtained by additive manufacturing. Int J Fatigue 123:180–195. https://doi.org/10.1016/j.ijfatigue.2019.02.014

Chapter 13
Three-Dimensional Fractography for Conventional and Additive Manufactured Steels After Bending-Torsion Fatigue

Wojciech Macek, Ricardo Branco, Jarosław Trembacz, José Domingos Costa, J. A. M. Ferreira, and Carlos Capela

Abstract In this study, fracture surface topography parameters were measured to investigate the effects of multiaxial loading. In order to assess the metrological aspects of fracture for notched specimens made of high-strength steels processed by both conventional and additively manufacturing (AM) techniques, an optical surface profilometer was used. Three bending moment to torsion moment ratios (*B*/*T*) were studied, i.e. 2, 1 and 2/3. The geometries were solid round bars with lateral notches made of conventional steel and hollow round bars with transversal holes for AM specimens. The investigations indicate that arithmetical mean height, Sa, decreased and fractal dimension, Df, increased with higher *B*/*T* ratios and higher fatigue lives.

Keywords Multiaxial Fatigue · Bending-Torsion · Fatigue Fracture · Surface Topography

13.1 Introduction

Multiaxial fatigue is a common mode of failure in critical components made of high-strength steels using both subtractive and additive techniques [1, 2]. Utilitarian features play a significant role for elements containing notches and other

W. Macek (✉)
Faculty of Mechanical Engineering and Ship Technology, Gdańsk University of Technology, Narutowicza 11/12, 80-233 Gdańsk, Poland
e-mail: wojciech.macek@pg.edu.pl

R. Branco · J. D. Costa · J. A. M. Ferreira · C. Capela
Department of Mechanical Engineering, University of Coimbra, CEMMPRE, 3030-788 Coimbra, Portugal

J. Trembacz
Opole University of Technology, Prószkowska 76, 45-758 Opole, Poland

G. Lesiuk et al. (eds.), *Structural Integrity and Fatigue Failure Analysis*,
Structural Integrity 25, https://doi.org/10.1007/978-3-030-91847-7_13

stress concentrators. In this way, fracture mechanics and fractography, which investigate the mechanical behaviour of structural elements, have been utilized to characterize the failure mechanisms [3–7]. One efficient method of fracture surface topography analysis is FRASTA, developed by SRI International [8], and tuned by Macek et al. and Cao et al., respectively [9, 10]. Singh et al. [11] and Susmel with Petrone [12] also analysed this type of loading for fatigue crack initiation and propagation behaviour. Pawliczek and Prażmowski [13] tested S355 steel specimens with mean block bending loadings. Samples after fatigue tests were subjected to metallographic examinations. These papers showed how important it is to look at the surface and morphology of the material in different perspectives, namely from nano- to macro-scales.

The morphology of the material can be examined by both in the depth of the material [14–16] and on its surface [17–19]. In some works, scientists have tried to combine all possible methods [20–22] and link them to fracture mechanisms and fatigue life, especially using surface roughness for utilizing specimen areas or their fractures [23–25]. Therefore, in this paper, the authors attempt to link the fracture surface topography parameters with the multiaxial fatigue scenarios, especially under proportional bending-torsion loading. Surface analysis was taken out on the whole fracture area without partition into regions of initiation and propagation.

13.2 Materials and Methods

The materials considered for this study were the 18Ni300 maraging steel produced by selective laser melting and the 34CrNiMo6 martensitic steel produced by conventional manufacturing [26, 27]. The chemical compositions of the investigated materials in weight percentage can be viewed in Table 13.1. Detailed information on the steels used and their material properties are presented in the paper [2].

The specimen geometries are schematized in Fig. 13.1. We have used a hollow round bar with a transversal hole and a solid round bar with a lateral *U* notch. Fatigue tests were conducted using a 100 kN DARTEC servo-hydraulic machine connected to a custom-made gripping system under proportional bending-torsion with a stress ratio (R) equal to 0. The fatigue testing programme, summarized in Table 13.2, encompassed three bending moment to torsion moment (B/T) ratios, i.e. B/T equal to 2, 1 and 2/3.

Fracture surfaces were examined via an optical three-dimensional non-contact focus-variation microscope. The InfiniteFocus G4 made by Alicona connects the

Table 13.1 Nominal chemical composition of **a** AISI 18Ni300 and **b** DIN 34CrNiMo6 (wt%)

	C	Ni	Mn	Co	Mo	Ti	Al	Cr	P	Si	Mn	Fe
(a)	0.01	18.2	0.65	9.0	5.0	0.6	0.05	0.3	0.01	0.1	0.04	Bal
(b)	0.34	1.5	0.65		0.22			1.5		0.4		

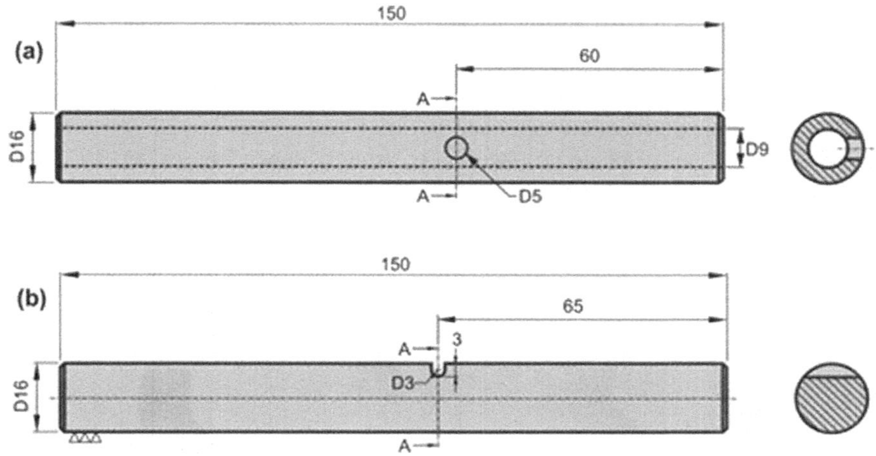

Fig. 13.1 Specimen geometries used in the test: **a** Hollow bar with transversal hole (18Ni300); and **b** solid bar with lateral notch (34CrNiMo6). Dimensions in millimetres

Table 13.2 Fatigue testing programme for specimens: **a** AISI 18Ni300 and **b** DIN 34CrNiMo6

	B/*T*	σ_a (MPa)	σ_m (MPa)	τ_a (MPa)	τ_m (MPa)	*R*	N_f (cycle)
(a)	2/3	79.6	87.6	59.7	65.7	0.04	85,361
	1	79.6	87.6	39.8	43.8	0.04	106,883
	2	79.6	87.6	19.9	21.9	0.04	159,714
(b)	2/3	179.1	194.0	134.3	145.5	0.04	50,025
	1	179.1	194.0	89.6	97.0	0.04	81,995
	2	179.1	194.0	44.8	48.5	0.04	253,501

small depth of focus of an optical microscope with vertical scanning to receive topographical and colour information [28, 29]. Due to the restricted field of view, 19 rows × 13 columns were stitched together to map the entire fracture area. Each individual micrograph had a vertical resolution 57.3 nm with a lateral resolution 3.91 μm level. Generated real 3D surface images were also processed with MountaisMap® software produced by the company Digital Surf. Alicona (*.al3d) files were resampled into height maps at a resolution automatically determined by software. Surfaces were analysed in relative coordinates (*X*, *Y* and *Z* axes) with the *Z* axis in heights from the lowest point by default. No additional filters were used. The total area of the fracture was chosen for imaging. All 3D parameters were calculated on the whole studied surfaces. Fracture surface measurements were conducted using surface parameters defined in accordance with ISO 25178 standard [30].

13.3 Results

The extracted areas (pseudo-colour views) and original measured areas (grey scale objects) of fracture surfaces are exhibited in Fig. 13.2.

The degree of out-of-plane propagation is strongly dependent on the *B*/*T* ratio. The analysis of both kind of specimens AISI 18Ni300 (Fig. 13.2a–c) and 34CrNiMo6

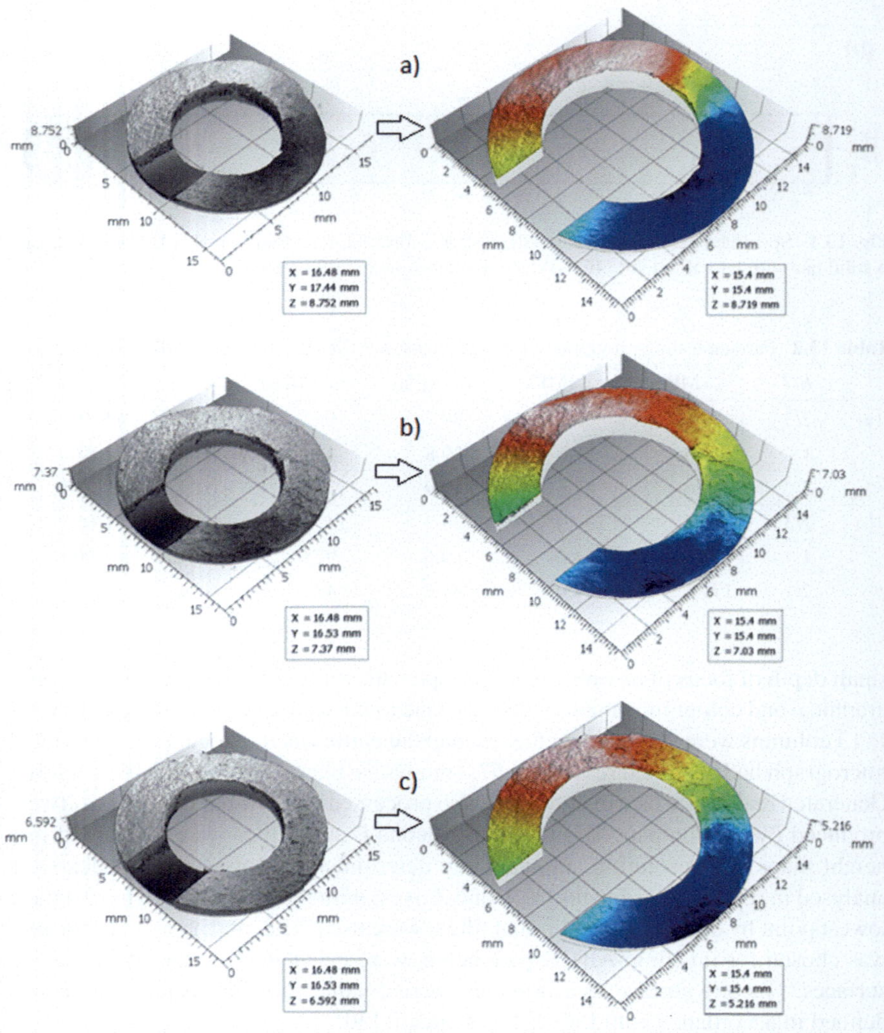

Fig. 13.2 Fracture surface transform from original to extracted: hollow bar with transversal hole 18Ni300: **a** *B*/*T* = 2/3; **b** *B*/*T* = 1; **c** *B*/*T* = 2; and solid bar with lateral notch 34CrNiMo6: **d** *B*/*T* = 2/3; **e** *B*/*T* = 1; **f** *B*/*T* = 2. Dimensions in millimetres

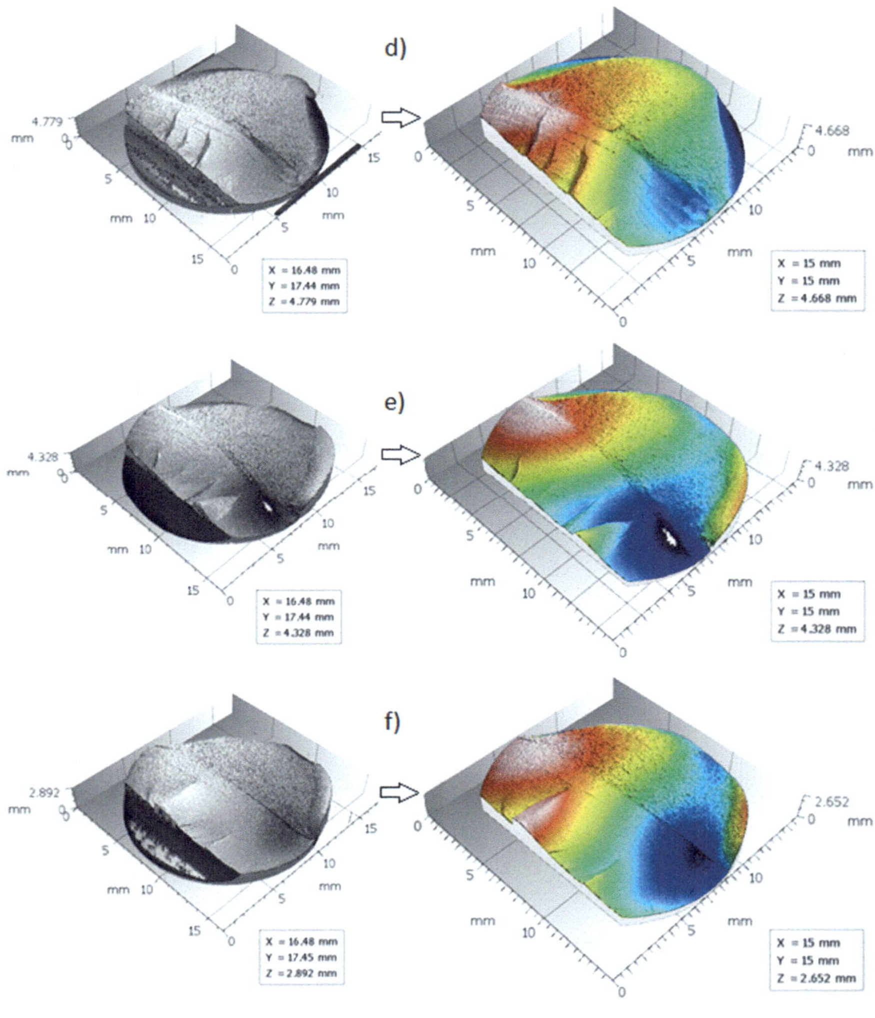

Fig. 13.2 (continued)

(Fig. 13.2d, e) shows that the decrease of shear stress level, i.e. the increase of *B*/*T* ratio, leads to flatter surfaces.

The analysis of the fracture surfaces based on the entire total area method using fractal dimension Df [31] and arithmetical mean height Sa [32] confirmed the previous observation (see Fig. 13.3). In fact, AM steel specimens have higher Sa and smaller Df values than those made of conventional steel. Investigated surface topography parameters were also correlated with fatigue life (Nf). In general, as shown in Fig. 13.3, we can observe that arithmetical mean height Sa decreased and fractal dimension Df increased for higher fatigue lives and higher *B*/*T* ratios. Fatigue

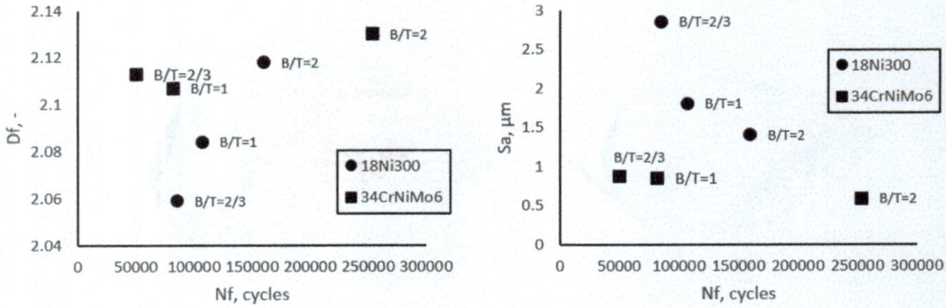

Fig. 13.3 Surface parameters Df and Sa versus fatigue life Nf

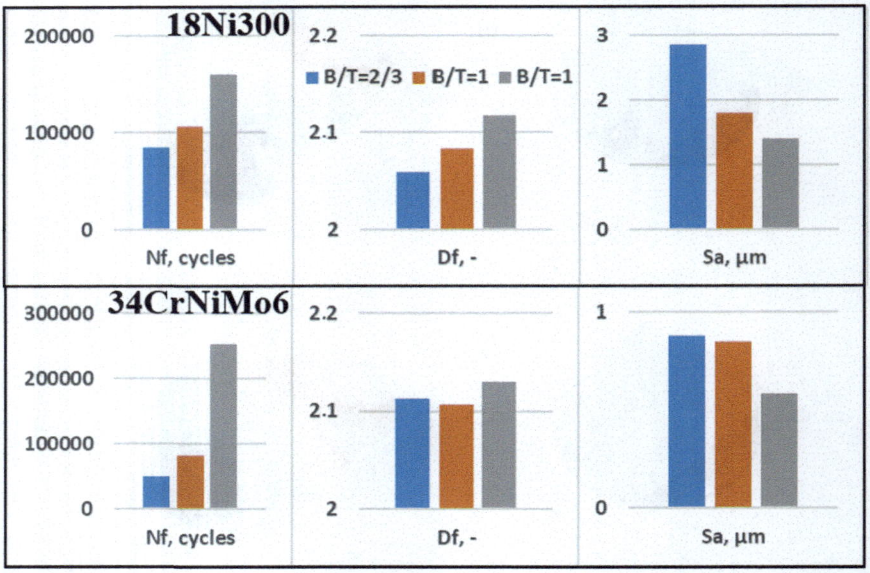

Fig. 13.4 Fatigue life Nf and surface parameters Df, Sa versus *B*/*T* ratio

life Nf and surface parameters Df, Sa versus *B*/*T* ratio, shown in Fig. 13.4, confirm the above relationship in detail in bar charts.

13.4 Conclusions

The effect of loading scenario on surface topography in notched samples made of high-strength steels processed by both conventional and additively manufactured techniques under bending-torsion fatigue loading was experimentally investigated.

Fracture surfaces were examined with fractal dimension Df and arithmetical mean height Sa using the entire total area method. Experimentally, it was proven that:

- Fracture topographies are strongly affected by the *B*/*T* ratio, irrespective of the notch geometry and the processed material;
- Similarly, the complexity of fracture topographies, namely the degree of non-planar propagation, increased with the *B*/*T* ratio;
- Arithmetical mean height Sa decreased with higher *B*/*T* ratios and higher fatigue lives. A reverse trend takes place for fractal dimension Df.

References

1. Tisza M, Czinege I (2018) Comparative study of the application of steels and aluminium in lightweight production of automotive parts. Int J Light Mater Manuf. https://doi.org/10.1016/j.ijlmm.2018.09.001
2. Macek W, Branco R, Trembacz J, Costa JD, Ferreira JA, Capela C (2020) Effect of multiaxial bending-torsion loading on fracture surface parameters in high-strength steels processed by conventional and additive manufacturing. Eng Fail Anal 118:104784. https://doi.org/10.1016/j.engfailanal.2020.104784
3. Aliakbari K, Nejad RM, Toroq SKP, Macek W, Branco R (2022) Assessment of unusual failure in crankshaft of heavy-duty truck engine. Eng Fail Anal 134:106085. https://doi.org/10.1016/j.engfailanal.2022.106085
4. Henry R, Blay T, Douillard T, Descamps-Mandine A, Zacharie-Aubrun I, Gatt J-M, Langlois C, Meille S (2019) Local fracture toughness measurements in polycrystalline cubic zirconia using micro-cantilever bending tests. Mech Mater 136:103086. https://doi.org/10.1016/J.MECHMAT.2019.103086
5. Rozumek D, Marciniak Z, Lesiuk G, Correia JA, de Jesus AMP (2018) Experimental and numerical investigation of mixed mode I + II and I + III fatigue crack growth in S355J0 steel. Int J Fatigue 113:160–170. https://doi.org/10.1016/J.IJFATIGUE.2018.04.005
6. Gosch A, Berer M, Hutař P, Slavik O, Vojtek T, Arbeiter FJ, Pinter G (2020) Pinter, mixed mode I/III fatigue fracture characterization of polyoxymethylene. Int J Fatigue 130:105269. https://doi.org/10.1016/J.IJFATIGUE.2019.105269
7. Branco R, Costa JD, Berto F, Antunes FV (2017) Effect of loading orientation on fatigue behaviour in severely notched round bars under non-zero mean stress bending-torsion. Theor Appl Fract Mech 92:185–197. https://doi.org/10.1016/J.TAFMEC.2017.07.015
8. Kobayashi T, Shockey DA (2010) Fracture surface topography analysis (FRASTA)—development, accomplishments, and future applications. Eng Fract Mech 77:2370–2384. https://doi.org/10.1016/J.ENGFRACMECH.2010.05.016
9. Macek W, Sampath D, Pejkowski Ł, Żak K (2022) A brief note on monotonic and fatigue fracture events investigation of thin-walled tubular austenitic steel specimens via fracture surface topography analysis (FRASTA). Eng Fail Anal 134:106048. https://doi.org/10.1016/j.engfailanal.2022.106048
10. Cao Y, Nie W, Yu J, Tanaka K (2014) A novel method for failure analysis based on three-dimensional analysis of fracture surfaces. Eng Fail Anal 44:74–84. https://doi.org/10.1016/J.ENGFAILANAL.2014.04.032
11. Singh AK, Datta S, Chattopadhyay A, Riddick JC, Hall AJ (2019) Fatigue crack initiation and propagation behavior in Al—7075 alloy under in-phase bending-torsion loading. Int J Fatigue 126:346–356. https://doi.org/10.1016/J.IJFATIGUE.2019.05.024

12. Susmel L, Petrone N (2003) Multiaxial fatigue life estimations for 6082–T6 cylindrical specimens under in-phase and out-of-phase biaxial loadings. Eur Struct Integr Soc 31:83–104. https://doi.org/10.1016/S1566-1369(03)80006-7
13. Pawliczek R, Prażmowski M (2015) Study on material property changes of mild steel S355 caused by block loads with varying mean stress. Int J Fatigue. https://doi.org/10.1016/j.ijfatigue.2015.05.019
14. Macek W, Martins RF, Branco R, Marciniak Z, Szala M, Wronski S (2022) Fatigue fracture morphology of AISI H13 steel obtained by additive manufacturing. Int J Fract 1–20. https://doi.org/10.1007/s10704-022-00615-5
15. Fisher K, Marquis E (2016) Comparing plasma-FIB and Ga-FIB preparation of atom probe tomography samples. MicroscMicroanal. https://doi.org/10.1017/S1431927616004311
16. Wang Z, Wu W, Qian G, Sun L, Li X, Correia JAFO (2019) In-situ SEM investigation on fatigue behaviors of additive manufactured Al-Si10-Mg alloy at elevated temperature. Eng Fract Mech 214:149–163. https://doi.org/10.1016/J.ENGFRACMECH.2019.03.040
17. Macek W, Rozumek D, Krolczyk GM (2020) Surface topography analysis based on fatigue fractures obtained with bending of the 2017A-T4 alloy. Measurement. 152:107347. https://doi.org/10.1016/J.MEASUREMENT.2019.107347
18. Kida K, Ishida M, Mizobe K, Yoshida I, Tamura A, Nakane K (2017) Fatigue of low carbon alloy steel (JIS S45C) and a new method of fracture surface analysis. https://doi.org/10.4028/www.scientific.net/MSF.893.181
19. Lauschmann H, Šiška F (2012) The reference texture: A proposal of a physical explanation. Int J Fatigue 43:120–127. https://doi.org/10.1016/J.IJFATIGUE.2012.03.002
20. Pomberger S, Stoschka M, Aigner R, Leitner M, Ehart R (2020) Areal fatigue strength assessment of cast aluminium surface layers. Int J Fatigue 133:105423. https://doi.org/10.1016/J.IJFATIGUE.2019.105423
21. K. Slámečka, J. Pokluda, M. Kianicová, S. Major, I. Dvořák, Quantitative fractography of fish-eye crack formation under bending-torsion fatigue, Int. J. Fatigue. 32 (2010). doi:https://doi.org/10.1016/j.ijfatigue.2009.07.009.
22. Sampath D, Akid R, Morana R (2018) Estimation of crack initiation stress and local fracture toughness of Ni-alloys 945X (UNS N09946) and 718 (UNS N07718) under hydrogen environment via fracture surface topography analysis. Eng Fract Mech 191:324–343. https://doi.org/10.1016/j.engfracmech.2017.12.010
23. Macek W (2021) Fracture surface formation of notched 2017A-T4 aluminium alloy under bending fatigue. Int J Fract 1–17. https://doi.org/10.1007/s10704-021-00579-y
24. Macek W, Branco R, Korpyś M, Łagoda T (2021) Fractal dimension for bending–torsion fatigue fracture characterisation. Measurement 184:109910. https://doi.org/10.1016/j.measurement.2021.109910
25. Macek W, Branco R, Costa JD, Pereira C (2021) Strain sequence effect on fatigue life and fracture surface topography of 7075-T651 aluminium alloy. Mech Mater 160:103972. https://doi.org/10.1016/j.mechmat.2021.103972
26. Branco R, Costa JDM, Berto F, Razavi SMJ, Ferreira JAM, Capela C, Santos L, Antunes F (2018) Low-cycle fatigue behaviour of AISI 18Ni300 maraging steel produced by selective laser melting. Metals (Basel). https://doi.org/10.3390/met8010032
27. Branco R, Costa JDM, Antunes FV, Perdigão S (2016) Monotonic and cyclic behavior of DIN 34CrNiMo6 tempered alloy steel. Metals (Basel). https://doi.org/10.3390/met6050098
28. Macek W (2021) Correlation between fractal dimension and areal surface parameters for fracture analysis after bending-torsion fatigue. Metals 11:1790. https://doi.org/10.3390/met11111790
29. Podulka P (2022) Selection of methods of surface texture characterisation for reduction of the frequency-based errors in the measurement and data analysis processes. Sensors 22:791. https://doi.org/10.3390/s22030791
30. International Organisation of Standardization. ISO 25178, Geom. Prod. Specif.—Surf. Texture Areal (2010)

31. Molent L, Spagnoli A, Carpinteri A, Jones R (2017) Using the lead crack concept and fractal geometry for fatigue lifing of metallic structural components. Int J Fatigue 102:214–220. https://doi.org/10.1016/J.IJFATIGUE.2017.04.001
32. Todhunter LD, Leach RK, Lawes SDA, Blateyron F (2017) Industrial survey of ISO surface texture parameters. CIRP J Manuf Sci Technol 19:84–92. https://doi.org/10.1016/J.CIRPJ.2017.06.001

Chapter 14
Finite Element Analysis of Distortions, Residual Stresses and Residual Strains in Laser Powder Bed Fusion-Produced Components

Jorge Gil, Felipe Fiorentin, J. C. R. Pereira, A. M. P. de Jesus, and Ana Reis

Abstract Metallic Additive Manufacturing (AM) has experienced increasing levels of demand due to its flexibility in producing complex geometries, bolstered by the growing availability of topology optimization methods. However, the multiple thermal cycles associated with these processes lead to the appearance of residual stresses that can worsen the dimensional accuracy and mechanical behavior of components. Consequently, efforts have been made to propose numerical models capable of distortion and residual stress prediction of parts manufactured through Laser Powder Bed Fusion (LPBF). This article introduces the mechanisms associated with residual stress generation and reviews known Finite Element Method (FEM) formulations of the additive manufacturing problem. Simulations are proposed and conduced that aim at comparing two software packages against each other and against experimental data; results are discussed, and lastly, conclusions are achieved.

Keywords Laser Powder Bed Fusion · Finite Element Method · Metallic Additive Manufacturing

14.1 Introduction

Additive manufacturing, 3D printing or rapid prototyping are all terms that describe the same process through which a component is produced via the successive overlap of two-dimensional layers. Initially reserved to the production of components with mainly aesthetic purposes that aided the product development workflow [1] with emphasis on polymeric materials, AM has expanded its scope to the production of

J. Gil (✉) · F. Fiorentin · A. M. P. de Jesus · A. Reis
Faculdade de Engenharia da Universidade Do Porto, s/n, R. Dr. Roberto Frias, 4200-465 Porto, Portugal
e-mail: jgil@inegi.up.pt

J. Gil · F. Fiorentin · J. C. R. Pereira · A. M. P. de Jesus · A. Reis
INEGI, Universidade Do Porto, Campus da FEUP, R. Dr. Roberto Frias 400, 4200-465 Porto, Portugal

G. Lesiuk et al. (eds.), *Structural Integrity and Fatigue Failure Analysis*, Structural Integrity 25, https://doi.org/10.1007/978-3-030-91847-7_14

end use parts, as well as encompassing metallic alloys [2]. In fact, metallic AM has emerged as an alternative to conventional processes [3], with metallic AM evolving more rapidly than any other segment within the field, experiencing an 80% increase in metal systems in the 2017/2018 years alone [4]. There are four predominant metallic AM processes: (i) direct energy deposition (DED), (ii) sheet lamination, (iii) binder jetting and (iv) powder bed fusion. The latter, in which this present work focuses, is further subdivided into two differing methods: Electron Beam Melting (EBM) and Laser Powder Bed Fusion (LPBF), also named Selective Laser Melting (SLM). The main difference between them lies in the heat source, as the former uses an electron beam and the latter a laser [5], a feature that carries several implications in the process. A schematic representation of a typical LPBF machine setup is shown in Fig. 14.1. Despite the several advantages associated with SLM, such as the geometric flexibility and relatively unrestricted components it can produce [3, 5–7], its mostly autonomous procedure and capacity for production of parts with good mechanical properties [8, 9], it also features a plethora of defects that greatly hinder the mechanical behavior of the produced components.

Thermal defects, such as keyhole porosity (rapid vaporization of material that is trapped in the melt pool) [11], balling (appearance of spherical formations of molten material due to surface tension) [12] and lack-of-fusion (incomplete penetration of the laser's energy into the powder) [13] are all problematic phenomena associated with additive manufacturing serving as crack initiation zones, negatively affecting mechanical performance. Additionally, residual stresses, characterized by being stresses that remain present in the body in the absence of external loads, thermal or otherwise, are common in AM parts. In LPBF, they develop in two different stages: an initial phase where the deposited layer is heated, and its thermal expansion is constrained by the cooler, bottom layers, inducing compressive strains in the top layer and tensile strains in the lower layers. Later, however, the upper layer begins to cool down, a process in which the component's core restricts the layer's contraction and tensile strains are originated in the top layer, while the core is in compression [14, 15]. This process is schematically displayed in Fig. 14.2. Residual

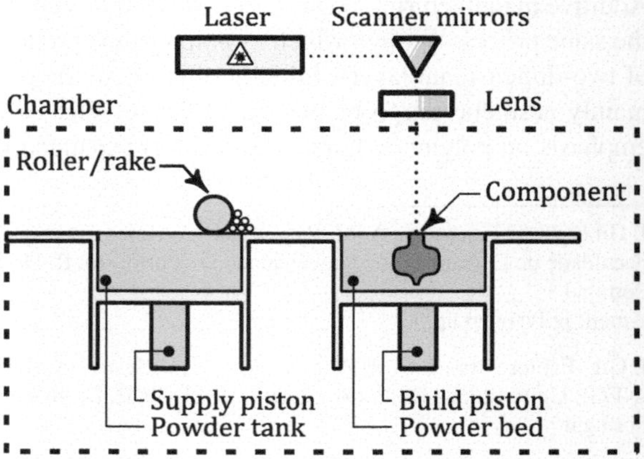

Fig. 14.1 Schematic representation of common laser powder bed fusion machine setup (adapted from [10])

Fig. 14.2 Schematic representation highlighting basic mechanisms of residual stress and residual strain evolution throughout AM process. Adapted from [15]

stresses have several negative influences on a component's performance, other than the aforementioned fatigue inadequacy when tensile residual stresses are generated at potential crack initiation sites, appearing through the component's distortion with poor geometric tolerances and possible delaminations [15, 16].

This article focuses on the simulation of residual stress development in LPBF-produced components, providing an overall review of the existing finite element formulations of the additive manufacturing process, complementing this theoretical review using existing commercial FEA software packages and establishing comparisons via benchmark geometries within two numerical approaches and experimental data.

14.2 FEM Modeling of LPBF

14.2.1 Simulation Approaches

14.2.1.1 Inherent Strain

The Inherent Strain (IS) method was first introduced by Ueda et al. [17] in the calculation of strains in welded joints. IS solves the problem by computing the sum of all permanent strains induced by inelastic deformation, temperature gradients and other processes that lead to residual stresses at equilibrium [18]; the sum of these strains define the inherent strain vector, which corresponds to the strain present in a component after it undergoes the cooling down stage [19]. There are two ways of determining the inherent strain vector: (i) experimentally, in which a small workpiece, usually made out of two or three layers [17–19], is physically printed and used to measure coefficients that determine the inherent strain in the three normal directions (shear inherent strain is often disregarded due to its lower magnitudes [20]), as given by the Eq. (14.1):

$$\varepsilon^* = \mathrm{ASC} \cdot \mathrm{SSF} \cdot \frac{\sigma_y}{E} \tag{14.1}$$

where ε^* is the inherent strain vector, ASC is the anisotropic strain coefficient, SSF is the strain scaling factor, σ_y is the material's yield strength and E is the Young's modulus; (ii) numerically, in which a small-scale thermomechanical simulation is conducted, and the directional strains are determined through the finite element method. The nodal variables are applied as loads through regular FEM.

14.2.1.2 Weakly Coupled Thermomechanical Analysis

In weakly coupled thermomechanical analysis, the thermal simulation is entirely conducted before any structural consideration is account for [14, 21], implying that the thermal history is independent from the mechanical response. This solution is the usual method adopted in commercial software packages due to its lower computational costs while simultaneously delivering good approximations [22]. The generalized thermal and mechanical differential equations that define the AM problem are given in Eqs. (14.2)–(14.6):

$$\rho c_p \frac{\partial T}{\partial t} = \nabla(k \cdot \nabla T) + Q \tag{14.2}$$

$$\nabla \sigma = 0 \tag{14.3}$$

$$\sigma = D\varepsilon^{el} \tag{14.4}$$

$$\varepsilon^{tot} = \varepsilon^{el} + \varepsilon^{pl} + \varepsilon^{th} \tag{14.5}$$

$$\varepsilon^{th} = \alpha(T - T_0) \tag{14.6}$$

where ρ is the material's density, c_p is the specific heat, T is the temperature at a point, k is the thermal conductivity, Q is the body heat balance that may encompass the heat delivered through the laser beam and thermal losses through radiation, convection and/or conduction; ε_{th} is the thermal strain, α is the thermal expansion coefficient and T_0 is the initial temperature at a point.

14.2.1.3 Fully Coupled Thermomechanical Analysis

In a fully coupled thermomechanical analysis, the thermal history computation of a deposited layer is accompanied by a structural assessment at each time step, meaning that the thermal equilibrium is dependent of the component's mechanical response. The equations associated with this method are the same as the weakly coupled case.

14.2.2 Selected Simulation Approaches

Within the background of commercially available software packages, two different approaches to the problem were selected for this study:

- In approach # 1, the initial thermal problem is solved by setting the deposited layer's temperature to the material's melting temperature. The previous layer to the one being deposited is at chamber temperature, a user defined temperature that is maintained as a thermal boundary condition throughout the build process. The software then applies the governing differential equations to the deposited layer, until it cools to the chamber temperature and subsequent distortions are calculated, and the process is repeated for the following layer of material until the geometry is fulfilled. In this approach, temperature-dependent properties were not considered, hindering the precision of the results;
- Approach #2 is based on a weakly coupled thermomechanical formulation and therefore allows layer re-heating after its deposition. Layers cool down from the melting temperature, until the following deposition takes place in which heat bleeds through. When the total build is achieved, the component cools down to a pre-defined cool down temperature.

14.3 Simulations and Results

14.3.1 Convergence Study

The objective of the convergence study was to measure the influence of h-refinement, that is, the increase of the number of elements, in the residual stress field of an AM-produced part. The geometry in which the simulation was performed in is a simple prism, whose dimensions are shown in Fig. 14.3, as it is a computationally expedient geometry to mesh, and requires no support structures.

The chosen approach for this convergence study is the Approach #1.

The employed mesh sizes are found in Table 14.1: the in-plane (xoy) mesh size was independently varied from the build direction (z) mesh fineness, with the goal

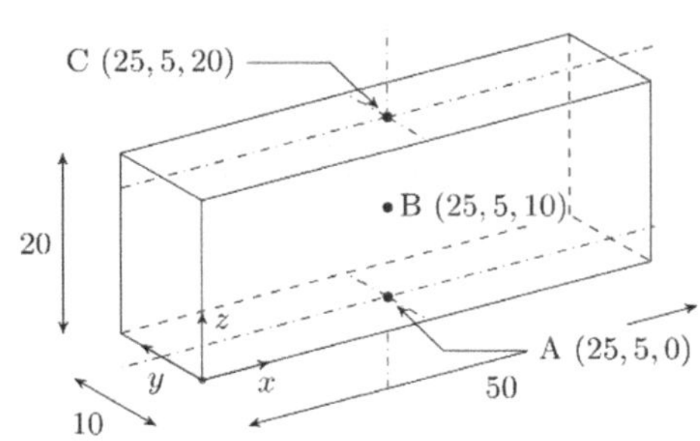

Fig. 14.3 Prism dimensions and coordinate system (dimensions in millimeters). Build direction coincides with z-axis. Nodes used to probe results appear as A, B and C

Table 14.1 Mesh sizes of conducted convergence simulations (the mesh size pertains to the in-plane element size)

Mesh size (mm)	Number of simulated layers
5.0	4, 10, 20, 50, 100, 250, 500
2.5	4, 10, 20, 50, 100, 250, 500
1.0	4, 10, 20, 50, 100, 250, 500
0.5	4, 10, 20, 50, 100, 250

Table 14.2 316L stainless steel thermal and mechanical properties, adopted in the convergence study

Melting temperature T_m	Thermal expansion coefficient α	Young's modulus E	Poisson ratio ν
1673.15 K	$1.990 \times 10^{-5} K^{-1}$	195 GPa	0.29

of determining whether one is predominant over the other. The material of the prism was chosen to be 316L stainless steel, whose properties are shown in Table 14.2. The baseplate is also composed of 316L steel, and its dimensions were chosen to be 100 × 40 × 20 mm. The obtained results can be found represented in three-dimensional, semi-logarithmic plots in Fig. 14.4. It was found that results obtained in node *(C)* converged to a 2% difference to the value relative to the finer mesh for each simulation group, when reducing the in-plane mesh size and maintaining the lumping factor; furthermore, it can be said that increasing the amount of simulated layers proves to be the most determining factor in obtaining a converging solution.

Additionally, an "overshoot" can be observed in the vertical residual stress σ_{zz} measured in node (A), for simulated lumped layer values of 10, 20 and 50. This fact is attributed to an adaptive remeshing that automatically occurred in the used software that performed a local h-refinement in the interface between the baseplate and the component, which introduced numerical noise for coarser in-plane meshes.

14.3.2 Comparison Between Approaches

An effort to establish a comparison between approaches was undergone using a simple simulation problem. The chosen geometry was a unitary layer, meaning that the component requires only one deposition to be built. This geometry is schematically shown in Fig. 14.5.

The results for the simulation were extracted from point (D), in coordinates (25, 5, 1).

The results are shown in Table 14.3. A distinction is made between two extracted results: (i) build values, corresponding to residual stresses after the layer has been

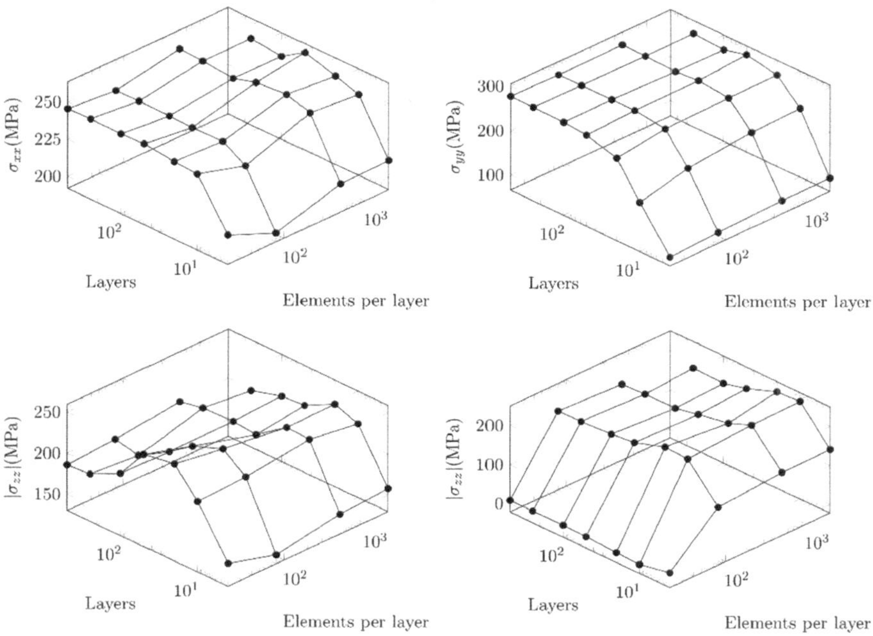

Fig. 14.4 Three-dimensional semi-logarithmic (*x*- and *y*-axis are logarithmic; *z*-axis is linear) convergence curves. Top left curve is the longitudinal (x-axis) residual stress σ_{xx} measured at node (C); top right curve is the transverse (y-axis) residual stress σ_{yy} at node (C); bottom left curve is the vertical (z-axis) residual stress σ_{zz} measured in node (A); bottom right curve is the vertical (z-axis) residual stress σ_{zz} measured in node (B)

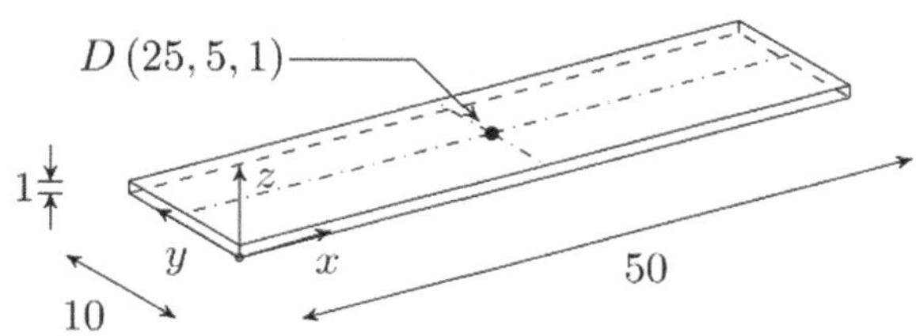

Fig. 14.5 Unitary layer dimensions and coordinate system (dimensions in millimeters). Build direction coincides with z-axis. Node D used to probe results

Table 14.3 Stress values for the unitary layer case, for both simulation approaches

Variable	#1 (MPa)	#2's Build (MPa)	Δ (%)	#2 Final (MPa)	Δ Final (%)
σ_{xx}	277.75	279.60	0.66	307.75	9.75
σ_{yy}	273.55	263.18	3.94	227.70	20.13

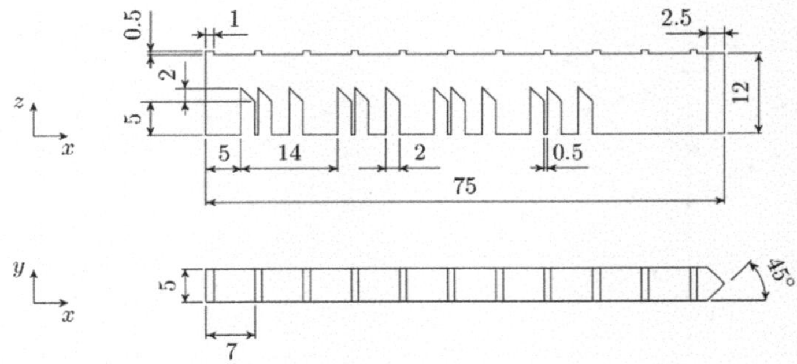

Fig. 14.6 NIST's benchmark bridge dimensions (in millimetres) and geometric configuration

built, and its temperature has not yet achieved thermal equilibrium with the environment temperature; (ii) final values, in which the entire system has undergone full cooling down reaching thermal equilibrium with the surrounding environment. The simulation's main objective—establishing an analogous case between both methods, with as close as possible mesh configuration, equal component and baseplate dimensions, and material definition—proved to be successful in retrieving similar results, with the difference being 0.66% and 3.94% for the longitudinal and transverse stresses, respectively.

14.3.3 Comparison with Experimental Data

Due to its extensive documentation [23], both from the manufacturing and simulation side, the chosen component to simulate consists of the National Institute of Standards and Technology (NIST) benchmark bridge, specially designed to challenge the additive process due to narrow details and overhang areas. Its geometry is shown in Fig. 14.6

14.3.3.1 Residual Strain Analysis

The residual strain analysis was solely conducted through approach #2, although the deflection analysis was performed by both approaches. The same manufacturing properties as the ones disclosed by NIST [24] were employed and are shown in Table 14.4.

The results regarding the residual strains are presented in Fig. 14.7. In general, the values obtained by the finite element method displayed greater amplitudes, when compared to the experimental residual strain, which may be explained by the approximated nature of the re-heating cycles the layers are subjected to, accompanied by an overestimation of the contact stiffness between the baseplate and the component.

Table 14.4 Materials and input variables used in residual strain experimental simulation

Component	Baseplate	Powder thickness	Laser speed	Hatch spacing	Pre-heat temperature
Inconel 625	Inconel 625	0.02 mm	0.95 ms^{-1}	0.10 mm	100 °C

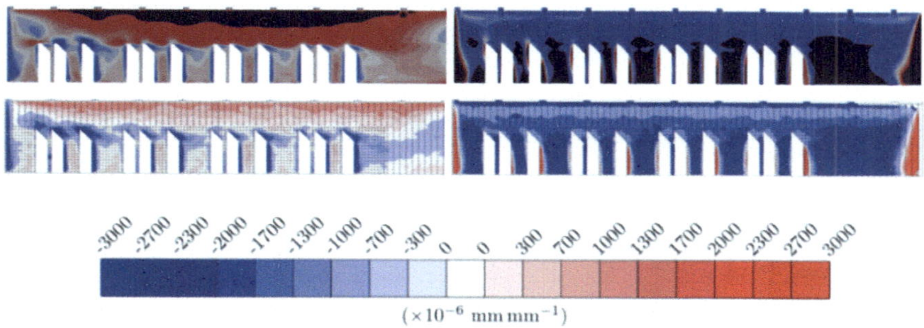

Fig. 14.7 Numerical (top) and experimental (bottom) comparison between longitudinal residual strain ε_{xx} (left) and vertical residual strain ε_{zz} (right) contour plots, extracted from $y = 2.5$ mm plane in Inconel 625 benchmark bridges

14.3.3.2 Deflection Analysis

An additional analysis was conducted using the benchmark bridge geometry, studying the cantilever effect the workpiece experiences when it is physically separated from the baseplate. This procedure was conducted through Wire Electrical Discharge Machining (WEDM), which separated the bridge legs one at a time (from left to right when looking at Fig. 14.7), introducing vertical displacements due to the internal residual stresses present in the component.

In terms of numerical simulation, this was done through relaxing the boundary conditions between the baseplate, defining new free surfaces that were initially restrained. Ideally, this process is made at different timesteps, although this pre-processing procedure is tedious, and the used software may not allow it altogether.

The obtained result is shown in Fig. 14.8, which highlights the maximum error of 16.06% experienced by approach #2 and 28.41% of approach #1, at the left-most point. The positive values for displacements are expected, when considering the residual strains shown in Fig. 14.7: the bulk structure above of the bridge legs is overwhelmingly tensile, which forces the structure to bend upwards to accommodate the originated residual stress field.

This error seems in line with the conclusion taken in the computation of the residual strains, in which the numerical model overestimates the contact stiffness between the component and the baseplate.

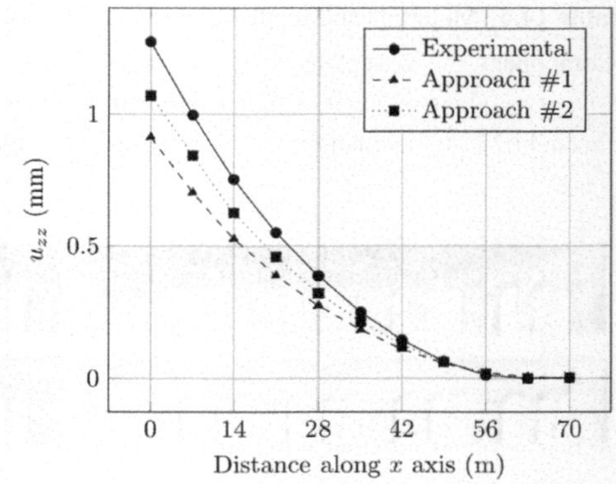

Fig. 14.8 Vertical displacement u_{zz} experienced by the benchmark bridge, along its longitudinal direction, after its legs are cut from the baseplate

14.4 Conclusions

Metallic additive manufacturing methods were discussed and introduced; the most common finite element method formulations of the problem were assessed, and two different approaches presented. Convergence analysis was conducted, in which the mesh h-refinement in the build direction proved to be predominant in result improvement; the unitary layer simulation was successful, establishing an analogous case between the two differing simulation approaches by using the same mesh size and material, which result in comparable residual stress values; similar residual strain fields to the experimental NIST bridges were achieved in approach #2, although displaying larger strain amplitudes, attributable to the possibility of an overestimation of the contact stiffness between the baseplate and the component, idea that is reinforced through the displacement analysis.

Acknowledgements The project Add.Strength entitled "Enhanced Mechanical Properties in Additive Manufactured Components" funded by the Programa Operacional Competitividade e Internacionalização, and Programa Operacional Regional de Lisboa funded by FEDER and National Funds (FCT) is acknowledged (Reference PTDC/EME-EME/31307/2017). The BIC scholarship funded by the Add.Strength project is also acknowledged. The project "MAMTool—Machinability of Additive Manufactured Parts for Tooling Industry" funded by the Programa Operacional Competitividade e Internacionalização and the Programa Operacional Regional de Lisboa funded by FEDER and National Funds (FCT) is acknowledged (Reference PTDC/EME-EME/31895/2017). Finally, LAETA—Laboratório Associado de Energia, Transportes e Aeronáutica is acknowledged through the contract UIDP/50022/2020 supported by the FCT/MCTES involving PIDDAC national funds.

References

1. Yang et al L (2017) Additive manufacturing of metals: the technology, materials, design and production. Springer
2. A. N. S. Institute (2017) Standardization roadmap for additive manufacturing

3. Zhang Y et al (2018) Additive manufacturing of metallic materials: a review. J Mater Eng Perform 27(1):1–13
4. AMFG (2019) The additive manufacturing landscape
5. Debroy T et al (2018) Progress in materials Science Additive manufacturing of metallic components—process, structure and properties 92:112–224
6. Saboori A, Piscopo G, Lai M, Salmi A, Biamino S (2020) Materials science and engineering a an investigation on the effect of deposition pattern on the microstructure, mechanical properties and residual stress of 316L produced by directed energy deposition. Mater Sci Eng A 780(2):139179
7. Murr LE et al (2012) Invited review metal fabrication by additive manufacturing using laser and electron beam melting technologies. J Mater Sci Technol 28(1):1–14
8. Hanzl P, Zetek M, Bakša T, Kroupa T (2015) The influence of processing parameters on the mechanical properties of SLM parts. Procedia Eng. 100:1405–1413
9. Wycisk E, Emmelmann C, Siddique S, Walther F (2013) High Cycle fatigue (HCF) performance of Ti-6Al-4V alloy processed by selective laser melting. Adv Mater Res 816–817:134–139
10. Frazier WE (2014) Metal additive manufacturing : a review 23(6):1917–1928
11. du Plessis A, Yadroitsava I, Yadroitsev I (2020) Effects of defects on mechanical properties in metal additive manufacturing: a review focusing on X-ray tomography insights. Mater Des 187:108385
12. Sanaei N, Fatemi A (2020) Defects in additive manufactured metals and their effect on fatigue performance: a state-of-the-art review. Prog Mater Sci 100724
13. Mukherjee T, Zuback J, De A, Debroy T (2016) Printability of alloys for additive manufacturing. Sci Rep 6:1–6
14. Bartlett JL, Li X (2019) An overview of residual stresses in metal powder bed fusion. Addit Manuf 27(March):131–149
15. Mercelis P, Kruth J (2006) Residual stresses in selective laser sintering and selective laser melting 5(3):254–265
16. Lewis GK, Schlienger E (2000) Practical considerations and capabilities for laser assisted direct metal deposition. Mater Des 21(4):417–423
17. Ueda Y, Fukuda K, Nakacho K, Endo S (1975) A new measuring method of residual stresses with the aid of finite element method and reliability of estimated values. J Soc Nav Archit Japan 1975(138):499–507
18. Bugatti M, Semeraro Q (2018) Limitations of the inherent strain method in simulating powder bed fusion processes. Addit Manuf 23(6)2017:329–346
19. Liang X, Cheng L, Chen Q, Yang Q, To AC (2018) A modified method for estimating inherent strains from detailed process simulation for fast residual distortion prediction of single-walled structures fabricated by directed energy deposition. Addit Manuf 23:471–486
20. Liang X et al (2020) Inherent strain homogenization for fast residual deformation simulation of thin-walled lattice support structures built by laser powder bed fusion additive manufacturing. Addit Manuf 32:101091
21. Hajializadeh F, Ince A (2020) Short review on modeling approaches for metal additive manufacturing process 2019:1–7
22. Yang Q, Zhang P, Cheng L, Min Z, Chyu M, To AC (2016) Finite element modeling and validation of thermomechanical behavior of Ti-6Al-4V in directed energy deposition additive manufacturing. Addit Manuf 12:169–177
23. Phan TQ et al (2019) Elastic residual strain and stress measurements and corresponding part deflections of 3D additive manufacturing builds of IN625 AM-bench artifacts using neutron diffraction, synchrotron X-ray diffraction, and contour method. Integr Mater Manuf Innov 8(3):318–334
24. Mayer T, Brändle G, Schönenberger A, Eberlein R (2020) Simulation and validation of residual deformations in additive manufacturing of metal parts. Heliyon 6(5):e03987

Chapter 15
Metal Additive Manufacturing of End-Use Components and Parts: A Practical Overview

Milan Sljivic, Stefan Wagner, Ana Pavlovic, and Dragan Marinkovic

Abstract This paper offers a practical overview of the most common procedures and materials currently available in the production of end-use metal components and parts by additive manufacturing (AM). The latest achievements and news are also discussed with special attention on defining the real production capability of each technological AM option under investigation. As general conclusion, by the present technical review, it was possible to confirm the proper capabilities of AM in accelerating the development of ready-to-market metal products supporting the overall industrial investment strategy. This is especially true in the case of geometrically complex parts or small/variable production batches. Furthermore, it also emerges that a direct metal 3D printing opens to new possibilities with relevant impacts on the contemporary industry, especially if driven by with appropriate changes in the information system and technology.

Keywords Metals production technology · Additive manufacturing · Rapid prototyping

15.1 Introduction

Additive manufacturing (AM) is an advanced manufacturing technology in which the final object is made by adding materials, layer by layer. It all starts with three-dimensional object geometry, expressed in a CAD file, which is exported to an

M. Sljivic (✉)
Centre for Virtual Technology, University of Banja Luka, Banja Luka, Bosnia and Herzegovina
e-mail: milan.sljivic@mf.unibl.org

S. Wagner
Faculty of Mechanical Engineering, University of Applied Sciences Esslingen, Esslingen, Germany

A. Pavlovic
Department of Industrial Engineering, University of Bologna, Bologna, Italy
e-mail: ana.pavlovic@unibo.it

D. Marinkovic
Department of Structural Analysis, Technical University of Berlin, Berlin, Germany
e-mail: dragan.marinkovic@tu-berlin.de

G. Lesiuk et al. (eds.), *Structural Integrity and Fatigue Failure Analysis*,
Structural Integrity 25, https://doi.org/10.1007/978-3-030-91847-7_15

STL file or G-Code thanks to specific software tools (such as CATIA, Solidworks, Autodesk, or many others). This file is entered into Slicing software to process all relevant print settings such as material selection, build time, temperature, speed, and choice of supporting structures and more following the printer settings, which is the universal input of all 3D printers [1–8]. Printing the finished piece follows after the settings of the slicing tool based on the print parameters that are set up in such a way that they better meet many different engineering and market needs. The capabilities of AM technology in the production of metal end parts are crucially dependent on the metal input material. Materials suitable for AM are steel (mostly stainless steel and tool steel), super-alloys, titanium, copper and aluminum. These materials are made from the fine metal powder used for most AM processes. Metal powder is also the basis for making filaments for some additive manufacturing processes. Therefore, AM technology is gradually replacing other consolidated techniques in the production of end-use items by conquering different industrial sectors. It mainly happens in the case of functional prototypes, custom tools and components, sports items, toy figures, surgery and medical equipment. Respecting other 3D printing processes, AM in the case of metals needs state-of-the-art solutions, sophisticated and intelligent machines, capable of producing high precision components at the low overall cost. At the same time, the significant factors that affect 3D metal printing, such as software and hardware limitations, energy used, material optimization, real-time process monitoring, and the like, are almost the same as any other 3D printing technology. Here, we give critical practical overviews of the current state and ability to create typical applications using 3D metal printing.

15.2 Classification of Metal Additive Manufacturing Technologies

Additive manufacturing for metal is a significantly more advanced process than traditional printing. It has matured from its inception in the research phase to the development of a wide range of industrial and commercial functional applications. Different methods of metal AM process are available that differ in their terminology from the companies and institutes that developed, patented and marketed them. However, the classification is mainly done by type of energy, material (mostly fine powder, wire, metal paste) and type of metal printing equipment. The following four procedures probably represent the current standard in metal AM production of end-use parts [9–29]: (a) Metal Powder Bed Fusion, (b) Direct Energy Deposition, (c) Binder Jetting, and (d) Metal Extrusion, schematically shown in Fig. 15.1.

It can be noticed how each type of printer has a unique technological process with own significant strengths and weaknesses. Differences are also reflected in the way machines convert the material into reliable metal parts. It is noted, e.g., that most AM processes such as Metal Powder Bed Fusion and Binder Jetting use metal powder, while other methods use mostly filament. It should be also noted that the materials

(a) Metal Powder Bed Fusion	(b) Direct Energy Deposition	(c) Binder Jetting	(d) Metal Extrusion
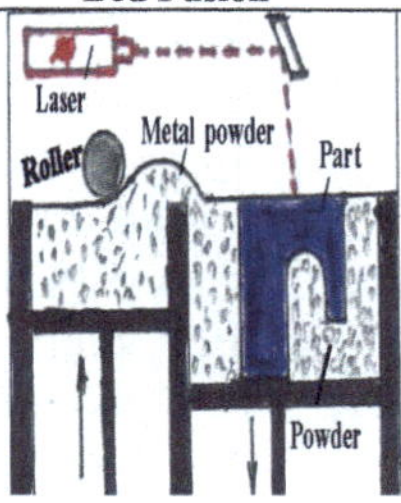	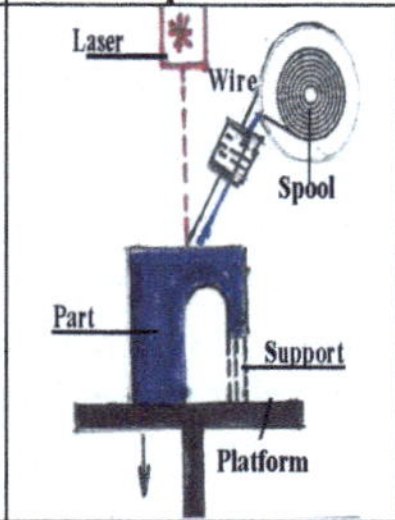	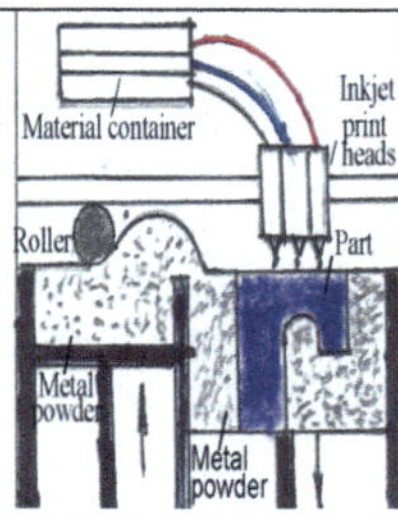	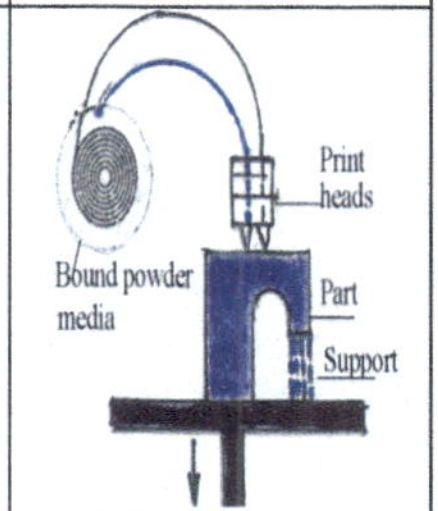
Post-processing ventilation, post-processing: surface finishing, CNC, EDM, HIP, heat treatment, etc.	**Post-processing** ventilation, post-processing: CNC, surface finishing, heat treatment, etc.	**Post-processing** batch sintering solution, surface finishing, heat treatment, etc.	**Post-processing** wach and sinter, for some parts heat treatment, CNC, surface finishing, etc.
Processes - Selective Laser Melting/ /Sintering (SLM) /(SLS) - Direct Metal Laser Sintering (DMLS). - Electron Beam Melting (EBM)	**Processes** - Powder Direct Energy Deposition (DED). - Wire Direct Energy Deposition (DED).	**Processes** - Material Jetting. - Binder Jetting. - Powder Bed Fusion	**Processes** -Metal Fused Filament Fabrication (FFF/FDM). - Atomic Diffusion Additive Manufacturing (ADAM).
Corresponding Materials Metal powder: stainless steel, nickel, titanium, ceramic powder, polyamide powder, etc.	**Corresponding Materials** Bound metal printing, molten metal powder, etc.	**Corresponding Materials** Metal powder, ceramic powder, polymer powder, wax, photopolymer, etc.	**Corresponding Materials** Bound metal printing, thermoplastics, ceramic slurries, metal pastes, composites, etc.

Fig. 15.1 Classification of metal additive manufacturing (AM) technologies by type of energy, post-processing required, type of process metal printing and corresponding material

used are generally consistent with the type of machine and the manufacturer, which also develop the appropriate materials for those machines.

15.3 Classification of Metal Additive Manufacturing Technologies

Thanks to the accelerated development of AM technology, metal 3D printing has now been successfully trans-formed into a more affordable production of many key finishing components over conventional techniques such as metal processing by deformation and cutting. Unlike traditional technological processes, it can produce very intricate geometric and complex parts; in general, all designed structures up to optimized processes successfully print. An up-to-date practical overview of the state and capabilities of creating typical applications is here reported.

15.3.1 Metal Powder Bed Fusion

Powder Bed Fusion is a very successful type of 3D printing from which standard technologies have evolved: Selective Laser Melting (SLM), Direct Metal Laser Sintering (DMLS) and Electron Beam Melting (EBM), Fig. 15.1a. The basic feature of all these machines is that they use a high power laser power source. This enables the selective dissolution of loose metal powder to form a precise and intricate geometric figure. There is a greater choice of machines and materials that can produce the most precision high-quality repeatable metal end-use components. The choice of metal powder materials with outstanding mechanical properties includes stainless steel, aluminum, nickel, titanium and others are currently large [9, 27–31]. EOS, a German manufacturer in the field of industrial 3D printing of metals and polymers, is one of the world's leading manufacturers and has developed four new metallic materials: EOS Stainless Steel CX, EOS Aluminum AlF357, EOS Titanium Ti64 Grade 5 and EOS Titanium Ti64 Level 23 [31]. These materials are adapted to almost all types of metal 3D printing and can be used for a wide range of applications in all areas of metal production. The disadvantage is that parts require post-processing such as surface finishing, CNC, EDM and heat treatment. Otherwise, these machines are recommended for industrial users where it is necessary to fabricate metal parts of high complexity and precision. The leading manufacturers of Metal Powder Bed Fusion 3D printers in today's market are EOS GmbH, SLM Solution, 3D SYSTEMS, RENISHAW, TRUMPF and others [27–31].

Direct Metal Laser Sintering (DMLS). The DMLS process is a typical representative of the Metal Powder Bed Fusion technology, which is based on the gradual melting of wonderful metal powder by laser beam, Fig. 15.1a. The laser beam energy dissolves the metal powder in layers with precise geometry defined by the 3D CAD model of the manufacturing component. Upon completion of the automatic mode of the production process, which lasts from several hours to a week, the printed product must be separated from the workspace and platform. The following is a post-processing finishing system such as surface finishing, CNC, heat treatment if necessary and possibly other refinement processes. DMLS is a method for the rapid and accurate production of highly durable components of functional prototype or final products. The scope is broad and is used for many industrial goods such as aviation, automotive, electronics, architecture, medicine, consumer goods and more. A full field of application is the production of molds and tools for the production of plastic or metal products [9–13, 31]. Figure 15.2 shows: (a) a used metal powder with one end-use metal part (source: EOS GmbH), (b) lightweight turbine cover door hinge for Airbus A380, Material: EOS Titanium Ti64 (source: EADS/EOS GmbH), (c) one tooling insert for child cup, Material: EOS Maraging Steel MS1 (source: Polymold) made using the process DMLS [9–13, 31].

The main advantages of the DMLS method are the direct fabrication of parts from a 3D CAD file, high geometric flexibility for the components design, high strength and toughness of the manufactured parts, shortening production time, low material consumption with the possibility of reusing molten powder, etc. The disadvantages of

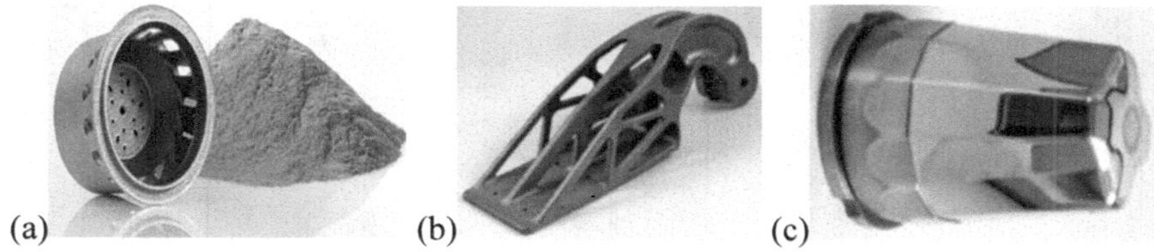

Fig. 15.2 **a** Metal powder whit one end-use metal part (*Source* EOS GmbH), **b** lightweight turbine cover door hinge (*Source* EADS/EOS GmbH), **c** tooling insert for child cup (*Source* Polymold/EOS GmbH) [31]

the DMLS method are the higher costs of purchasing production equipment and more labor-intensive work to operate, the relatively smaller overall dimensions of the work parts (recommended part size: 2–350 mm), demanding post-processing finishing of printed parts, removal of supporting structures and possibly other requirements which are required [31].

15.3.2 Direct Energy Deposition

Direct Energy Deposition (DED) are types of 3D printers that use metal powder or wire to produce parts, shown in Fig. 15.1b. It is carried out by bringing metallic materials (wire or powder) into the print nozzle and pointing to the object with a simultaneous source of energy using a laser, electron beam or plasma arc to melt the material. The process is stratified until a given CAD model is formed or an item is repaired. DED machines that use a metal wire (typically 1–3 mm diameter wire is used) are the largest and fastest types of 3D printers and are significantly quicker than powder printers (powder particle sizes are between 50 and 150 microns) [14, 32]. With this technology, it is possible to produce metal components of larger dimensions, recommended part size: 50–2000 mm. It is very cost-effective to repair damaged components that cannot be corrected by traditional methods such as turbine blades or other industrial facilities. It should be noted that DED printers can be used for component production with polymers, ceramics and composites. The leading manufacturer of DED printers that use lasers as heat sources is BeAM. Other manufacturers of laser DED machines include Trumpf, Optomec, FormAlloy, DMG Mori, Relativity, etc. We give several cases of BeAM manufacturer applications that have been made with DED technology. Figure 15.3a shows the implementation of this technology to repair the worn shaft seal, thereby extending the life of this vital component [15, 27, 32].

The other two cases of using DED technology are the optimum fabrication of Near Net Shape building of high-value parts, shown in Fig. 15.3b, c parts of complex geometries that have been made using traditional forming technology [15, 32].

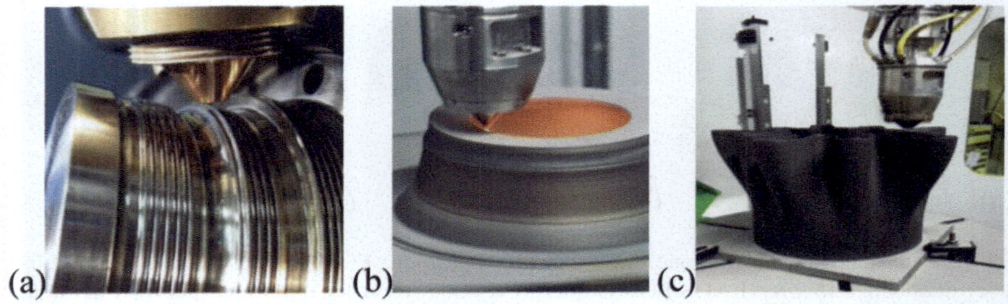

Fig. 15.3 **a** Repair of worn shaft seal (*Source* BeAM), **b** end-use part (*Source* BeAM), **c** end-use component of complex geometries (*Source* BeAM) [32]

15.3.3 Binder Jetting

Binder Jetting is a method of printing using metallic powder and a binder jet of complex geometric forms of high precision, very quickly and it is possible for batch production, shown in Fig. 15.1c [16, 17]. It uses an inkjet-style print head to promptly apply a liquid binder to a fine layer of metal powder, and it can also use other powders (ceramics, sand or composites, etc.). The process is repeated in layers until the object is complete. Unlike the Powder Bed Fusion procedure, lasers are not used here. The printed parts are quite brittle, so further sintering is required [16, 17, 33]. Otherwise, the process involves complex powder management, mandatory sintering and post-processing recommended such as surface finishing and heat treatment. Binder Jetting is a slightly cheaper process than Powder Bed Fusion. Still, it can print smaller part sizes (recommended part sizes 1–150 mm) while delivering precision, density and repeatability at significantly lower levels. The ExOne Company is a global leader in the industrial production of metal 3D printers using Binder Jetting technology and includes Company: Stratasys, DIGITAL METAL, 3DEO, etc. [16, 17, 27, 33]. Representative case studies of finished metal use components manufactured by ExOne Company with Binder Jetting 3D printing are shown in Fig. 15.4a

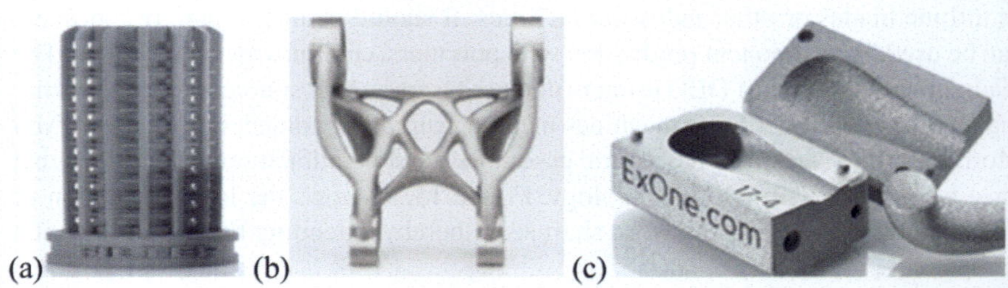

Fig. 15.4 Case studies finished of metal end-use components manufactured by ExOne Company with Binder Jetting 3D Printing, **a** metal filter components for strainers and metallic filter application (*Source* ExOne), **b** case study one optimized of part (*Source* ExOne) and **c** binder jetting printed casting mold (*Source* ExOne) [33]

metal filter components for strainers and metallic filter applications, (b) case study one optimized of the part and (c) one new Binder Jetting printed casting mold [33].

15.3.4 Metal Additive Manufacturing Extrusion

Metal extrusion-based additive manufacturing is a type of 3D printing that has been developed under the name Fused Deposition Modeling (FDM) or Fused Filament Fabrication (FFF) and is mainly used for plastics, but 3D printers have been designed using the same principle that has been used successfully for metals and ceramics [6, 18, 20, 21]. For this purpose, a suitable metal Bound powder filament has been developed where the metal powder integrates with the wax polymers into the wire filament and thus enables AM extrusion. For the printed parts of this filament, the sintering process should be used subsequently to completely transform the printed parts into metal [19, 29, 34]. Another technology developed on the principle of metal AM extrusion is Atomic Diffusion Additive Manufacturing (ADAM), which has a simple and inexpensive fabrication and is suitable for a wide variety of manufacturing applications [25, 29]. These technologies require requirements for additional facilities, such as ventilation for wash and sinter, post-processing for some parts such as heats treatment, CNC, surface finishing and, if necessary, others. The recommended size for printing parts is: 10–250 mm [29, 34].

Metal Fused Filament Fabrication (FFF). Metal FFF is an advanced new process in additive manufacturing based on plastic Fused Deposition Modeling (FDM), where the heated metal filament is deposited through the nozzles layer by layer. The printed metal part needs to be sintered in the next phase and thus created into the final metal part. Post-processing Metal FFF parts are required for both wash and sintering, while heat treatment and surface finishing treatments depend on the complexity of the parts. This processing accounts for almost 50% of the cost of total metal 3D printing. Even so, the process is far more favorable than traditional technologies [18, 20–24, 34]. The world's leading producer of metal FFF processes is 3DEO Company. Figure 15.5a shows an FFF 3D printer from 3DEO capable of producing

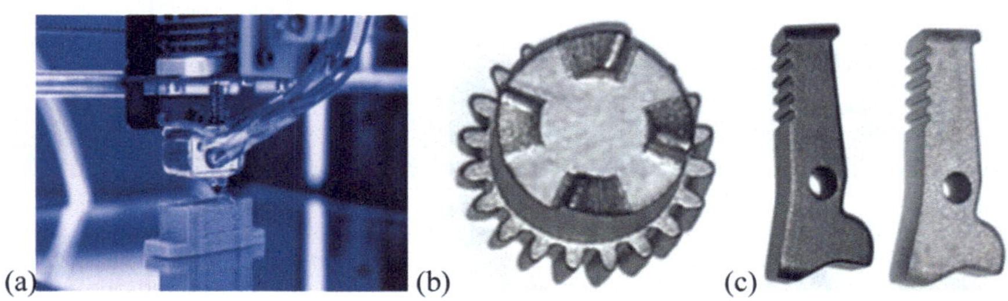

Fig. 15.5 **a** Metal FFF 3D printer from 3DEO (*Source* 3DEO), **b** one metal end-use component (*Source* 3DEO) and **c** one optimized rack-wheel part (*Source* 3DEO)

high-quality metal components of small- to medium-size max up to 250 mm [34]. Two case studies of metal 3D printing with design freedom and manufacturing flexibility manufactured in this company are shown in Fig. 15.5b, c [34].

Atomic Diffusion Additive Manufacturing (ADAM). The Atomic Diffusion Additive Manufacturing (ADAM) process is based on the Bound Powder Extrusion patented by Markforged for metal printing using its own developed Markforged Metal X printer [25, 29]. It is a very advanced 3D printer that prints a filament of metal powder enclosed in a plastic binder shown in Fig. 15.6a. In a short time, the Metal X system has proven to be affordable, reliable and easy to use, which is up to 10 × cheaper than alternative metal additive manufacturing technologies as well as up to 100 × less than traditional machining or casting [25, 29].

The workflow of the ADAM process includes: (i) CAD Design of the part, (ii) upload the STL file in the Slice tool Eiger with defining all the working parameters of the metal printing components and generating all supports as needed, (iii) printing a part from metal powder bound in plastic to a Markforged Metal X printer having a build volume of 300 × 220 × 180 mm Fig. 15.6a, (iv) the printed parts are placed in a Wash-1 debinding system that uses a special fluid for dissolving plastic binders Fig. 15.6b, (v) the washed parts now go into the Sintering system where heat treatment is carried out to burn the remaining binder and solidify the metal powder into the final metal component. Figure 15.6c, d shows a Markforged Sinter-2 sintering system where the components are cemented to a high accuracy comparable to production by traditional methods. We present several case studies of end-use components that have been optimized and make using the Markforged Metal X printing system.

Case study of Actuator housing, Fig. 15.7a is constructed from four parts into one joint piece without supporting material and is printed on the principle of metallic AM extrusion using a Markforged 3D printing system. This reduces the complexity of the part and also reduces the volume so that by producing a single actuator housing, it saves costs by 92% and thus saves time by 95% [29].

The Wheel shaft is the second case study redesigned to be a three-part assembly with a ~90% savings of basic steel material, shown in Fig. 15.7b [29]. This complex

Fig. 15.6 **a** Markforged Metal X printer (*Source* Markforged), **b** the Wash-1 debinding system is a specialized fluid dispenser that dissolves primary bonding material and prepares for sintering (*Source* Markforged), **c** view of the Markforged Sinter-2 sintering system that converts the washed parts into their high-quality thick metal form (*Source* Markforged), **d** view of Sinter-2 sintering system in the completed phase of the end-use component sintering process (*Source* Mark-forged) [29]

Fig. 15.7 **a** One optimized end-use part (*Source* Markforged), **b** the Wheel shaft redesigned to be a three-part assembly (*Source* Markforged), **c** the Thermoset Mold tool for thermoset plastic (*Source* Markforged) [29]

flange was printed on the Markforged Metal X system and successfully passed the usability *and durability testing.*

The significant advancement of AM technology has created sustainable solutions in 3D printing injection molds, thermoset molds and other tools and parts. The development time and total manufacturing costs of designing, prototyping, repetition and production tools were significantly reduced using 3D printing than traditional processes. Here is the case of the Thermoset Mold tool for thermoset plastic, which was printed using Markforged technology, shown in Fig. 15.7c [29]. The mold met all the necessary standards such as strength and precision for production, and the massive savings of mold printing was 4 $\times$ cheaper and 2.5 times faster than traditional technologies [29].

15.4 Conclusion

The presented metal AM processes in the production of metal components have shown that they have enormous potential capabilities over traditional technological methods of production. Concerning the classification of metal AM technology, Fig. 15.1a–d, the right choice of the AM process for the correct application is critical as it enables a faster, easier and more convenient transition from design to fabrication of complex metal parts. Each of the four AM processes shown is characterized by the manufacturer or scientific institute that developed it. Still, implementation of the market has shown which machines have better production capabilities and advantages. These metal AM printing techniques provide outstanding optimized application capabilities in almost all industrial fields as well as in medical applications, from the design, testing and rapid prototype phases to a wide range of commercial functional component printing. The many advantages of using metal 3D printing over conventional fabrication methods are reflected in the following significant metal AM procedures. (i) CAD design where complete geometric freedom can be applied to optimize processes and design the most complex structures. The CAD model provides metal slicer 3D software that generates and processes all operational parameters with a selection of print materials and launches on the appropriate 3D printer. (ii) The design and production of traditional tools required by many forms of metal component manufacturing such as casting, bending, deep drawing, extrusion, forging and more are not necessary, because the 3D printing system forms the material in its final form. (iii) Metal 3D printers automatically print components from design files using a minimum number of skilled operators far more advantageous than traditional processing requiring more skilled labor. The main disadvantages of the metal AM process are the requirements for post-processing equipment, which is present in all methods depending on whether the material is powder or filament. Thus, in the Metal Powder Bed Fusion, Direct Energy Deposition and Binder Jetting processes, the following facility requirements are required, namely powder management, ventilation, post-processing depending on the use case, heat treatment, CNC, EDM, HIP, surface finish and others if necessary. Metal AM extrusion has the least requirements for post-processing equipment, depending on the case of the printed part: necessary ventilation, CNC, heat treatment and surface finishing. But it has to be said that, although they significantly increase the cost of metal 3D printing, they are not limited, because the benefits of metal 3D printing are great. The analysis showed that metal AM systems in the production of metal components achieve predominantly more favorable design flexibility, shorter fabrication time, high precision of complex shapes and a significantly lower cost of production compared to traditional machining processes.

References

1. Buchanan C, Gardner L (2019) Metal 3D printing in construction: a review of methods, research, applications, opportunities and challenges. Eng Struct 180:332
2. Bajaj P, Hariharan A, Kini A, Kurnsteiner P, Raabe D, Jagle EA (2020) Steels in additive manufacturing: a review of their microstructure and properties. Mater Sci Eng A 772:138633
3. Haleem A, Javaid M (2019) Additive manufacturing applications in industry 4.0: a review. J Indus Integr Manage 4(4):1930001
4. Sljivic M, Pavlovic A, Ilic J, Stanojevic M, Todorovic S (2017) Comparing the accuracy of professional and consumer grade 3D printers in complex models production. FME Transactions 45(3):348–353
5. Sljivic M, Pavlovic A, Kraisnik M, Ilic J (2019) Comparing the accuracy of 3D slicer software in printed end-use parts. In: IOP conference series: materials science and engineering, vol. 659, no 1. IOP Publishing, pp 1–9
6. Fragassa C, Minak G, Poodts E (2010) Mechanical characterization of photopolymer resins for rapid prototyping. In: Proceedings of of the 27th Danubia-Adria symposium on advances in experimental mechanics, DAS 2010, 22–25 Sept Wroclaw. Code 125161, pp 55–56
7. Sljivic M, Pavlovic A, Stanojevic J, Fragassa C (2016) Combining additive manufacturing and vacuum casting for an efficient manufacturing of safety glasses. FME Transactions 44(4):393–397
8. Frazier WE (2014) Metal additive manufacturing: a review. J Mater Eng Perform 23(6):1917–1928
9. Bhavar V, Kattire P, Patil V, Khot S, Gujar K, Singh R (2017) A review on powder bed fusion technology of metal additive manufacturing. In: Additive manufacturing handbook. CRC Press, pp 251–253

10. Duplakova D, Hatala M, Duplak J, Radchenko S, Steranka J (2018) Process direct metal laser sintering—possibility of application in production. Sci Res J 1(4):123–127
11. Bahnini I, Rivette M, Rechia A, Siadat A, Elmesbahi A (2019) Additive manufacturing technology: the status, applications. Int J Adv Manuf Technol 97:147–161
12. Tagliaferri V, Trovalusci F, Guarino S, Venettacci S (2019) Environmental and economic analysis of FDM, SLS and MJF additive manufacturing technologies. Materials 12:4161
13. Bertol LS, Junior WK, Da Silva FP, Aumund-Kopp C (2010) Medical design: direct metal laser sintering of Ti-6Al-4V. Mater Des 31(8):3982–3988
14. Sreekanth S, Ghassemali E, Hurtig K, Joshi S, Andersson J (2020) Effect of direct energy deposition process parameters on single-track deposits of alloy 718. Metals 10(1):96
15. Dass A, Moridi A (2019) State of the art in directed energy deposition: From additive manufacturing to materials design. Coatings 9(7):418
16. Mirzababaei S, Pasebani S (2019) A review on binder jet additive manufacturing of 316L stainless steel. J Manuf Mater Process 3(3):82
17. Kumar P at al (2017) Design and fabrication of powder based binder jetting 3d printer. Ceram Int 3(9):142–150
18. Gibson MA, Mykulowycz NM, Shim J, Fontana R, Schmitt P, Roberts A, Myerberg JS (2018) 3D printing metals like thermoplastics: fused filament fabrication of metallic glasses. Mater Today 21(7):697–702
19. Kang JW, Ma QX (2017) The role and impact of 3D printing technologies in casting. China Foundry 14(3):157–168
20. Tsirogiannis EC, Vosniakos G-C (2019) Redesign and topology optimization of an industrial robot link for additive manufacturing. Facta Univ Ser Mech Eng 17(3):415–424
21. Rane K, Strano M (2019) A comprehensive review of extrusion-based additive manufacturing processes for rapid production of metallic and ceramic parts. Adv Manuf 7(2):155–173
22. Annoni M, Giberti H, Strano M (2016) Feasibility study of an extrusion-based direct metal additive manufacturing technique. Procedia Manuf 5:916–927
23. Jabbari A, Abrinia K (2018) A metal additive manufacturing method: semi-solid metal extrusion and deposition. Int J Adv Manuf Technol 94(9–12):3819–3828
24. Gonzalez-Gutierrez J, Kukla C, Schuschnigg S, Duretek I, Holzer C (2016) Fused filament fabrication for metallic parts. In: Leobener Kunststoff-Kolloquium: Kunststoffgerechte Bauteilentwicklung vom Werkstoff zum Produkt. Pinter, G. Pilz G (eds) Leoben, Austria, vol 25, pp 187–188
25. Galati M, Minetola P (2019) Analysis of density, roughness, and accuracy of the atomic diffusion additive manufacturing (ADAM) process for metal parts. Materials 12(24):4122
26. Dilberoglu UM, Gharehpapagh B, Yaman U, Dolen M (2017) The role of additive manufacturing in the era of industry 4.0. Procedia Manuf 11:545–554
27. Cherdo L (2020) The best metal 3D printers in 2020. https://www.aniwaa.com/buyers-guide/3d-printers/best-metal-3d-printer
28. 3D SYSTEMS: Design guide- direct metal printing. www.3dsystems.com, last accessed 16 June 2019
29. Markforged: Guide to metal 3D printing. https://all3dp.com/1/markforged-metal-x-review-3d-printer-specs/
30. Sandvik: Metal powder by Sandvik. https://www.metalpowder.sandvik/en/. Last Accessed 25 May 2020
31. EOS GmbH Presents: EOS parameter sets for direct metal laser sintering (DMLS) and new materials and processes for series additive manufacturing. https://www.eos.info/press/eos-presents. Last Accessed 26 Mar 2020
32. BeAM Machines: Directed energy deposition. https://www.beam-machines.com/applications-process-3d-printing

33. ExOne Company: The most researched binder jetting systems in 3D printing: ExOne. https://www.exone.com/en-US/News/The-Most-Researched-Binder-Jetting-Systems-in-3D-P. Last Accessed 29 Mar 2020
34. 3DEO Company: Metal 3D printing processes—metal extrusion FFF/FDM. https://news.3deo.co/author/3deo

Chapter 16
Multiaxial Fatigue Behaviour of SLM 18Ni300 Steel

R. Branco, J. D. Costa, J. A. M. Ferreira, C. Capela, F. Berto, and W. Macek

Abstract Fatigue behaviour of SLM 18Ni300 steel under proportional bending-torsion loading is studied. The fatigue tests are conducted under pulsating loading conditions using tubular specimens with a transversal circular hole. Three ratios of the normal stress to shear stress are considered, namely $\sigma/\tau = 4, \sigma/\tau = 2$ and $\sigma/\tau = 4/3$. Crack initiation sites were found for two diametrically opposite points around the hole, whose locations are governed by the loading scenario. Crack angles at the early stage of growth were successfully predicted from the first principal direction at the nodes with maximum values of the first principal stress at the hole. Fatigue crack initiation life was estimated based on the SWT parameter, defined from uniaxial low-cycle fatigue tests performed for smooth standard specimens under fully reversed strain-controlled conditions. Overall, predicted lives are in agreement with the experiments.

Keywords Multiaxial fatigue · Bending-torsion · Notch effect · SWT parameter · Fatigue life prediction

16.1 Introduction

Maraging steels are a special class of advanced high-strength steels, widely used in aircraft, aerospace, military, offshore, subsea, tooling and moulding industries [1, 2].

R. Branco (✉) · J. D. Costa · J. A. M. Ferreira · C. Capela
CEMMPRE, Department of Mechanical Engineering, University of Coimbra, Coimbra, Portugal
e-mail: ricardo.branco@dem.uc.pt

C. Capela
Department of Mechanical Engineering, Instituto Politécnico de Leiria, ESTG, Leiria, Portugal

F. Berto
Department of Mechanical and Industrial Engineering, Norwegian University of Science and Technology, 7491 Trondheim, Norway

W. Macek
Faculty of Mechanical Engineering and Ship Technology, Gdańsk University of Technology, Gabriela Narutowicza 11/12, 80-233 Gdańsk, Poland

G. Lesiuk et al. (eds.), *Structural Integrity and Fatigue Failure Analysis*,
Structural Integrity 25, https://doi.org/10.1007/978-3-030-91847-7_16

This special class of steels, whose microstructure is formed by a cubic martensitic matrix, is hardened by finely dispersed nanometer-sized precipitates of intermetallic compounds, resulting from a metallurgical reaction involving no carbon, which seriously hampers the movements of dislocations and, thereby, promotes strengthening by precipitation hardening processes [3, 4].

Because of their martensitic matrix, these materials require a rapid quench from the austenitic region to temperatures below the martensite start temperature, which makes them particularly suited for the selective laser melting technology [4, 5]. This fact can be explained by the small size of the melt pool that leads to very high cooling rates. On the other hand, the cyclic reheating experienced by the material may trigger the nucleation of hardening precipitates without the need of additional heat treatments [4]. Last but not least, the geometrically complex shapes and the relatively reduced quantities associated with the typical application areas are two other important requirements that can be met by this process [4–6].

Selective laser melting is a very popular but unequivocally complex technique for metal processing that creates fully functional components, directly from three-dimensional digital models, by successive deposition of thin layers of material [4, 7]. Due to the nature of this additive manufacturing process, products are prone to different types of anomalies (e.g. porosities, inclusions, voids, cavities, micro-cracks, shrinkage, lack of fusion, excessive roughness, etc.) increasing the uncertainty concerning the mechanical behaviour [8, 9]. Therefore, a good understanding of structural integrity under monotonic and cyclic loading is pivotal to extend the field of application as well as to develop more sophisticated products. Regarding the 18Ni300 maraging steel, there are no studies focused on multiaxial loading. Therefore, this paper aims at fulfilling this gap. The paper studies the fatigue behaviour of tubular geometries with a transversal hole subjected to proportional bending-torsion loading.

16.2 Materials and Methods

The material considered in this study was a 18Ni300 maraging steel produced by selective laser melting (SLM). Its chemical composition and its main mechanical properties are summarised in Tables 16.1 and 16.2 [9].

Table 16.1 Nominal chemical composition of the AISI 18Ni300 steel

C	Ni	Mn	Co	Mo	Ti	Al	Cr	P	Si	Mn	Fe
0.01	18.2	0.65	9.0	5.0	0.6	0.05	0.3	0.01	0.1	0.04	Bal

Table 16.2 Main mechanical properties of the AISI 18Ni300 steel

Density	Young's modulus	Tensile strength	Yield strength	Porosity
7.42 g/cm^3	168 GPa	1147 MPa	910 MPa	0.74%

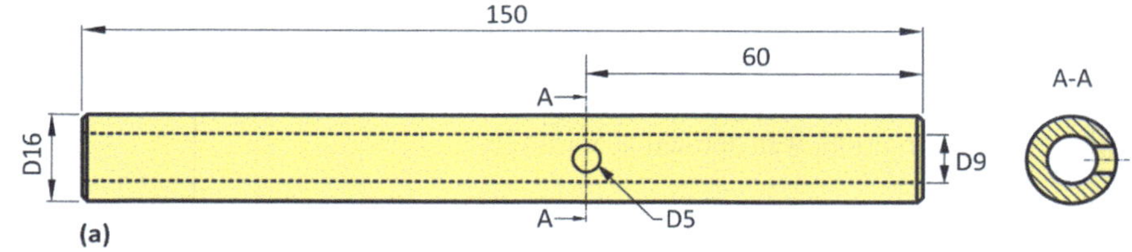

Fig. 16.1 Tubular specimen with a transversal circular hole fabricated via selective laser melting from 18Ni300 steel utilised in the multiaxial bending-torsion fatigue tests. Dimensions in millimetres

The specimen geometries, presented in Fig. 16.1, were fabricated in a vertical orientation, on the base plate, using a Concept Laser M3 linear printing system equipped with a Nd:YAG fibre laser. After the SLM process, specimens were machined at the external surface and polished to a scratch-free condition. The transversal hole was introduced by machining.

Fatigue tests were conducted using a DARTEC servo-hydraulic machine connected to a custom-made gripping system under proportional bending-torsion with a stress ratio (R) equal to 0. Three normal stress to shear stress (σ/τ) ratios were used, more precisely: $\sigma/\tau = 4$, $\sigma/\tau = 2$ and $\sigma/\tau = 4/3$. The nominal normal stress amplitudes varied between 78 and 96 MPa.

The evaluation of the stress and strain states at the notch vicinity was done through a finite-element model. Figure 16.2 shows the structured mesh developed in this study. It was developed in a parametric manner, with 8-node brick finite elements, assuming a homogeneous, linear-elastic and isotropic behaviour. At the notch region, an ultrarefined mesh was introduced to better describe the stress–strain gradients. The

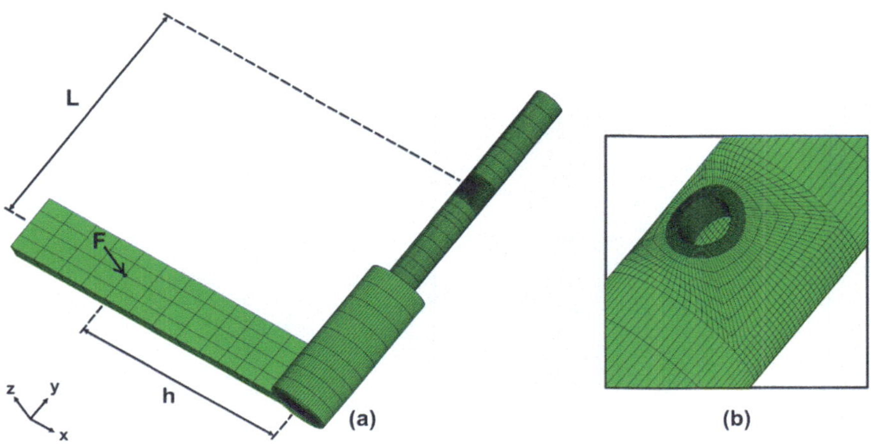

Fig. 16.2 Three-dimensional finite-element model developed in the present study: **a** Assembled mesh; and **b** hole region

assembled model contained 152,248 finite element and 163,138 nodes. The normal stress to shear stress ratios were defined using a single force (F) by adjusting its magnitude and its point of application. In this case, since L was fixed, the value of h, in the x-axis direction, was defined accordingly.

16.3 Results and Discussion

The fatigue process, in this geometry, is characterised by the initiation of two cracks at the hole in diametrically opposite points. Figure 16.3 shows an example for a test performed with $\sigma/\tau = 2$. In general, the experimental angles measured at the two sides, either α or β, are relatively similar, regardless of the normal stress to the shear stress ratio. The crack initiation angles, represented by β in Fig. 16.3, increased for lower σ/τ ratios, which is explained by the higher shear stress levels. Identical behaviour was found for the crack angles at the early stage of growth, represented in Fig. 16.4 by α. In this study, these angles were successfully predicted using the numerical model described above. Regarding the β angle, it can be reasonably estimated from the node with maximum value of the first principal stress. In relation to the α angle, it was determined by computing the first principal direction at the nodes with maximum value of the first principal stress. The good correlation between the experimental observations and the predicted values, either for the α angle or the β angle, is shown in Fig. 16.4. In order to simplify the analysis, scatter bands of $\pm 10°$ and $\pm 5°$ were added (dashed lines) in Fig. 16.4a, b, respectively. Overall, the average errors were 6.3° and 7.4° for β_1 and β_2; and 3.3° and 2.3° for α_1 and α_2, respectively.

The examination of the fracture surfaces by scanning electron microscopy showed that crack initiation occurred from un-melted regions, most probably due to a lack of fusion or inadequate penetration (see Fig. 16.5). These heterogeneities, in the

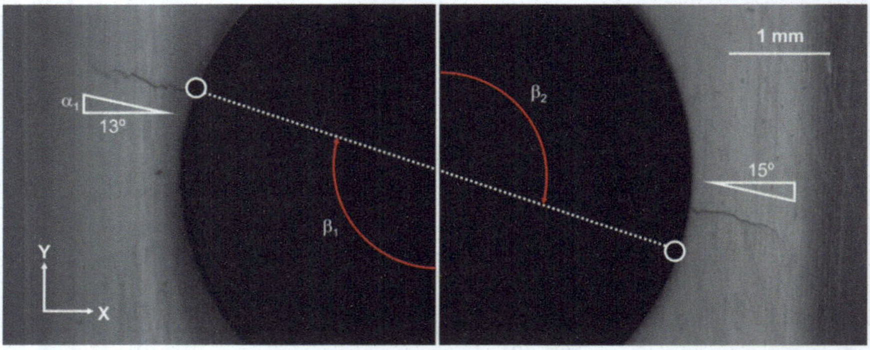

Fig. 16.3 Fatigue crack initiation sites observed experimentally for $\sigma/\tau = 2$ (α_1 and α_2 represent the angles on the left and right sides of the hole defined in accordance with the referential system of Fig. 16.2, and β_1 and β_2 represent the angles of the initiation sites measured in relation to the y-axis)

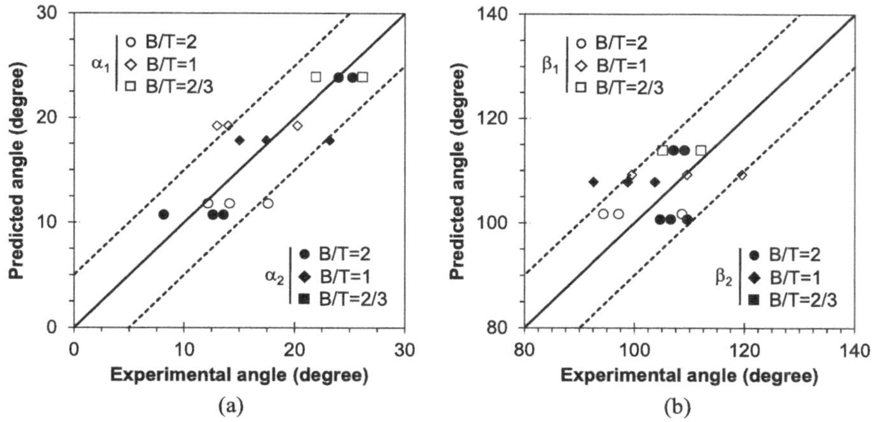

Fig. 16.4 Experimental observations versus predictions: **a** Crack initiation sites; and **b** crack angles at the early stage of growth

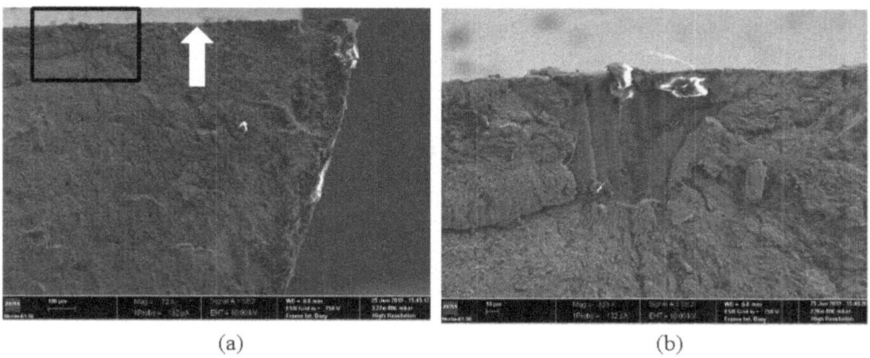

Fig. 16.5 Experimental observations versus predictions: **a** Crack initiation site; and **b** crack angle at the early stage of growth

presence of cyclic loads, created localised microscopic stresses, possibly larger than the yield stress, causing local plastic deformation and easing the fatigue crack initiation [10, 11]. Furthermore, it was also concluded that initiation occurred at the hole surface, near the outer diameter of the specimen (see Fig. 16.5a). This fact was corroborated by the numerical simulations. In Fig. 16.5a, the experimental initiation site, identified by the black box, and the predicted initiation site (based on the maximum value of the first principal stress) represented by the white arrow are compared. As can be seen, both locations are quite close, which is an interesting outcome.

Fatigue crack initiation life was predicted through a uniaxial model based on the SWT parameter [12]. It assumes two main premises [13]: (1) regardless of the geometric discontinuity, fatigue life is the same, if the stress–strain histories at the initiation sites are similar; and (2) fatigue failure occurs when a critical value of damage is reached at the initiation site. Firstly, we develop a uniaxial damage law relating the SWT parameter (product of strain amplitude to maximum stress, see Eq. (16.1)) with the number of cycles to crack initiation (N_i). For this specific material, this law assumed the form:

$$SWT = \varepsilon_a \sigma_{max} = 10^{1.5315} (N_f)^{-0.3567} \tag{16.1}$$

where ε_a is the strain amplitude, σ_{max} is the maximum stress, and N_f is the number of cycles to failure. This power function was established from a series of low-cycle fatigue tests performed using smooth standard specimens under fully reversed strain-controlled conditions for strain amplitudes in the range ±0.3 to ±1.0%. The gauge section of these specimens had a diameter of 6 mm and a length of 19 mm. In a second stage, it is computed the fatigue crack initiation life (N_i) for the notched geometries. The modus operandi consists of four main steps (see Fig. 16.6): the multiaxial stress state is reduced to a uniaxial stress state at the notch-controlled process zone using the equivalent von Mises stress range (Fig. 16.6a); the local stress distribution is averaged by applying the Theory of Critical Distances (Fig. 16.6b); a representative hysteresis loop is generated by applying the equivalent strain energy density concept and the effective von Mises stress range computed in the previous step (Fig. 16.6c); and, finally, the corresponding value of the SWT parameter is inserted into the fatigue damage law to estimate the fatigue life (Fig. 16.6d). Here, the size of the notch-controlled process zone was estimated through the El-Haddad parameter (a_0):

$$a_0 = \frac{1}{\pi} \left(\frac{\Delta K_{th}}{\Delta \sigma_0} \right)^2 \tag{16.2}$$

where ΔK_{th} is the stress intensity factor range threshold, and $\Delta \sigma_0$ is the fatigue limit stress range. Based on previous studies, for $R = 0$, the former value is equal to 5.2 MPa m$^{0.5}$ and the latter was equal to 266 MPa [11, 14]. Therefore, $a_0 =$ 121.7 MPa.

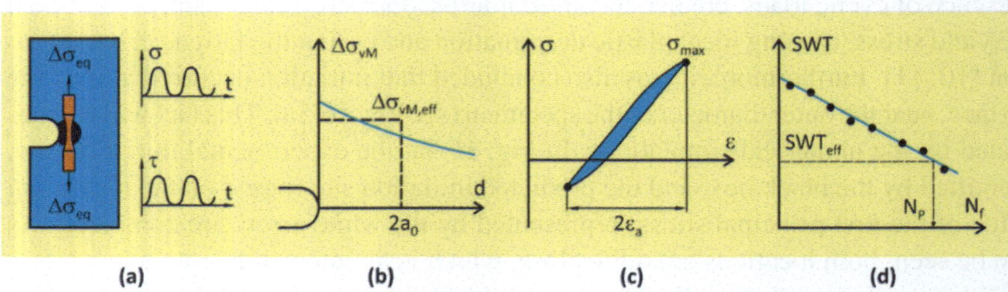

Fig. 16.6 Experimental observations versus predictions: **a** Crack initiation site; and **b** crack angle at the early stage of growth

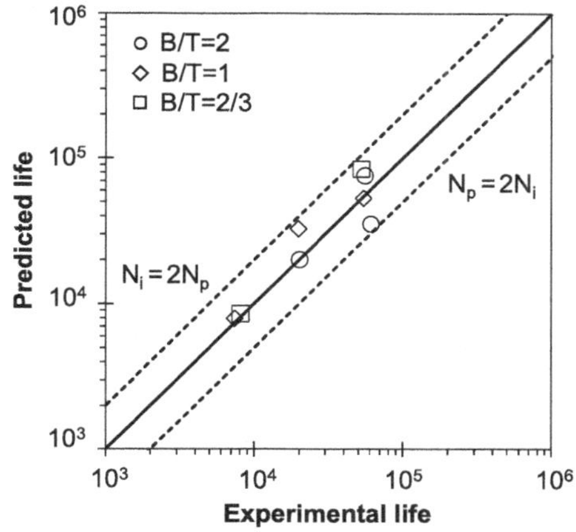

Fig. 16.7 Experimental fatigue life versus predicted fatigue life for the multiaxial fatigue testes performed in the present study

The value of a_0 was used to estimate the experimental fatigue life. During the tests, the hole was observed periodically (interval ranging from 2000 to 5000 cycles, depending on the applied stress level) with a high-resolution camera to correlate the crack length with the number of cycles. Based on this information, we develop the so-called a-N curves (i.e. curves relating the crack length with the number of cycles). Then, these curves were fitted by exponential functions. The fatigue life was estimated, using the a-N curves, for a crack length equal to a_0. Finally, the predicted fatigue lives (N_p) obtained by means of the procedure described above (see Fig. 16.6) were compared with those determined experimentally (N_i).

Figure 16.7 plots the experimental values against the predicted ones for the different loading scenarios studied in this paper. For the sake of clarity, scatter bands with factors of two were plotted (dashed lines), i.e. $2N_p = N_i$ and $2N_i = N_p$. Overall, the predictive capabilities of the proposed approaches are remarkable. In both cases, there is a strong correlation between the data, with all points within the delimited area. This clearly demonstrates the robustness of the proposed methodology. A close look at the figure shows a mixed behaviour; that is, predicted lives are either conservative or non-conservative. However, the data are very well correlated which is an interesting outcome.

16.4 Conclusions

This paper addressed the multiaxial fatigue behaviour of the 18Ni300 steel fabricated by selective laser melting. Tests were conducted under proportional bending-torsion loading for different normal stress to shear stress ratios using a tubular geometry with a transversal hole. A liner-elastic numerical model was developed to evaluate the stress and strain fields at the hole region. The following conclusions can be drawn:

- Fatigue failure was characterised by the initiation and growth of two cracks at diametrically opposite sites around the hole surface;
- Un-melted regions at the surface and sub-surface, resulting from insufficient fusion or inadequate penetration, were the cause of crack initiation;
- Fatigue crack initiation sites and fatigue crack angles at the early stage of growth were predicted from the principal stress field;
- The proposed fatigue life prediction model is suitable for SLM 18Ni300 components subjected to proportional bending-torsion loading.

Acknowledgment This research is sponsored by national funds through FCT - Fundação para a Ciência e a Tecnologia under the project UIDB/00285/2020.

References

1. Kempen K, Yasa E, Thijs L, Kruth JP, Van Humbeeck J (2011) Microstructure and mechanical properties of selective laser melted 18Ni-300 steel. In: Physics procedia
2. Kürnsteiner P, Wilms MB, Weisheit A, Barriobero-Vila P, Jägle EA, Raabe D (2017) Massive nanoprecipitation in an Fe-19Ni-xAl maraging steel triggered by the intrinsic heat treatment during laser metal deposition. Acta Mater
3. Mooney B, Kourousis KI, Raghavendra R (2019) Plastic anisotropy of additively manufactured maraging steel: Influence of the build orientation and heat treatments. Addit Manuf
4. Tan C, Zhou K, Ma W, Zhang P, Liu M, Kuang T (2017) Microstructural evolution, nanoprecipitation behavior and mechanical properties of selective laser melted high-performance grade 300 maraging steel. Mater Des
5. Jägle EA, Choi PP, Van Humbeeck J, Raabe D (2014) Precipitation and austenite reversion behavior of a maraging steel produced by selective laser melting. J Mater Res
6. Zhu SP, He JC, Liao D, Wang Q, Liu Y The effect of notch size on critical distance and fatigue life predictions. Mater Des 196:109095
7. Suzuki A, Nishida R, Takata N, Kobashi M, Kato M (2019) Design of laser parameters for selectively laser melted maraging steel based on deposited energy density. Addit Manuf
8. Meneghetti G, Rigon D, Gennari C (2019) An analysis of defects influence on axial fatigue strength of maraging steel specimens produced by additive manufacturing. Int J Fatigue
9. Branco R, Costa JDM, Berto F et al (2018) Low-cycle fatigue behaviour of AISI 18Ni300 maraging steel produced by selective laser melting. Metals (Basel) 8(1)
10. Bajaj P, Hariharan A, Kini A, Kürnsteiner P, Raabe D, Jägle EA (2020) Steels in additive manufacturing: a review of their microstructure and properties. Mater Sci Eng A
11. Branco R, Silva J, Ferreira JM et al (2019) Fatigue behaviour of maraging steel samples produced by SLM under constant and variable amplitude loading. Procedia Struct Integr
12. Correia J, Apetre N, Arcari A et al (2017) Generalized probabilistic model allowing for various fatigue damage variables. Int J Fatigue
13. Branco R, Prates PA, Costa JD, Borrego LP, Berto F, Kotousov A, Antunes FV (2019) Rapid assessment of multiaxial fatigue lifetime in notched components using an averaged strain energy density approach. Int J Fatigue (2019)124:89–98. https://doi.org/10.1016/j.ijfatigue.2019.02.005
14. Santos LMS, Ferreira JAM, Jesus JS, Costa JM, Capela C (2016) Fatigue behaviour of selective laser melting steel components. Theor Appl Fract Mech

Part IV
Composite Materials and Components

Chapter 17
Study of Viscoelastic Properties of Sandwich Composites for Manufacturing Loudspeaker Membranes

Natalia Raźny, Adrianna Filipiak-Kaczmarek, Wojciech Błażejewski, Romuald Bolejko, and Andrzej Dobrucki

Abstract The aim of the work was to find an innovative sandwich composite material that can be used as a material for loudspeaker membranes. Composite samples of a sandwich construction were made using different materials for the core: balsa wood, Poraver glass granulate, Coremat mat and polyurethane foam. Measurement of complex dynamic viscoelastic Young's modulus was carried out using the Oberst method, allowing to select a composite sample with the best parameters in comparison with other materials: possibly high Young's modulus and loss factor, low density and the high velocity of the longitudinal wave. The best properties are characterized by a composite whose core is made of balsa wood. The covers of all systems are made of the same material—carbon mat, known also as carbon paper. The experiment was run on selected materials for core and for outer layers.

Keywords Loudspeaker membranes · Sandwich composites · Carbon mat · Oberst method

N. Raźny (✉)
Faculty of Mechanical Engineering, Department of Lightweight Elements Engineering, Foundry and Automation, Wrocław University of Science and Technology, Wrocław, Poland
e-mail: natalia.razny@pwr.edu.pl

A. Filipiak-Kaczmarek · W. Błażejewski
Faculty of Mechanical Engineering, Department of Mechanics, Materials and Biomedical Engineering, Wrocław University of Science and Technology, Wrocław, Poland
e-mail: adrianna.filipiak@pwr.edu.pl

W. Błażejewski
e-mail: wojciech.blazejewski@pwr.edu.pl

R. Bolejko · A. Dobrucki
Faculty of Electronics, Photonics and Microsystems, Department of Acoustics, Multimedia and Signal Processing, Wrocław University of Science and Technology, Wrocław, Poland
e-mail: romuald.bolejko@pwr.edu.pl

A. Dobrucki
e-mail: andrzej.dobrucki@pwr.edu.pl

G. Lesiuk et al. (eds.), *Structural Integrity and Fatigue Failure Analysis*, Structural Integrity 25, https://doi.org/10.1007/978-3-030-91847-7_17

17.1 Introduction

The speaker is an electro-acoustic transducer whose task is to change the electrical signal into an acoustic signal. Usually, sound is radiated into space using a membrane. The task of membrane is to convert vibration into acoustical pressure. The basic parameter that defines energy conversion is efficiency—that is the ratio of the sound power of the emitted signal to the electric power which has been supplied from the signal source. Nowadays, the most commonly used speakers are open-ended magnetoelectric loudspeakers (Fig. 17.1). The most widespread structure is the conical construction, which allows to work in a wide range of frequencies. The electric signal is converted into vibration by the voice coil placed in permanent magnetic field. The coil is made of a thin, insulated wire wounded in two or four layers. Another element of speaker is a dust cap that protects the speaker mechanism against dust or moisture. The last element is a basket which is a skeleton of the entire construction. The basket allows the air to move on both sides of the membrane. The most important element of the speaker is the membrane, which is responsible for transmitting the acoustic wave to the medium surrounding the device—the membrane sets in motion the vibrating particles of air adjacent to it [1].

The loudspeaker membrane should be as light and rigid as possible. The efficiency of a loudspeaker in middle frequency range is given by the formula:

$$\eta_0 = \frac{\rho_0}{2\pi c} \cdot \frac{(Bl)^2 S_D^2}{M_{MS}^2 R_E} \tag{17.1}$$

where ρ_0—air density = 1205 kg/m^3, c—speed of sound = 344 m/s at 20 °C, Bl—force factor, S_D—surface of the diaphragm, R_E—electrical resistance of the voice coil, M_{MS}—mass of the moving system.

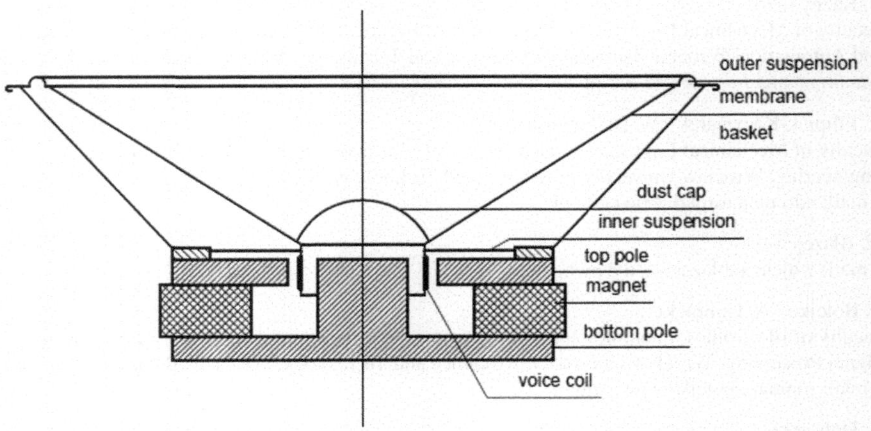

Fig. 17.1 Schematic construction of the cone speaker [1, 2]

Dominant part of the mass M_{MS} is mass of the membrane. If it increases, the efficiency of the loudspeaker decreases rapidly. To make this mass low, the density of the diaphragm material should be possibly low, and thickness should be also very small. The elasticity of the membrane causes resonances, which result in peaks and dips on the frequency response of the loudspeaker. What is more, the membrane should be as rigid as possible. There are two factors that influence the membrane stiffness: geometrical and material. Geometrical stiffness depends on the shape of membrane. The simplest shape, a flat plate is the worst solution. In flat plate, only flexural stresses appear which are very small in comparison with longitudinal (stretching) stresses. Then, the resonances of flat plate occur for very low frequencies in comparison with the resonances of non-flat shape of the diaphragm. This is the reason that the loudspeaker membranes have conical or dome form. For such shapes, the stretching stresses dominate over the flexural ones and the stiffness is high. The material stiffness is characterized by the Young's modulus. The material used for construction of the loudspeaker membrane should have high Young's modulus and small density. For homogeneous materials, the square root of the ratio of Young's modulus and density is the speed of the longitudinal elastic wave in the material:

$$c_0 = \sqrt{\frac{E}{\rho}} \tag{17.2}$$

where E—Young's modulus, ρ—density of the material.

The c_0 value is a criterion of choice of the material used for loudspeaker diaphragm. However, it is not the only one. The second criterion is the value of internal mechanical losses. It should be as high as possible. Due to the c_0 value, light metals such as titanium or aluminum have the best properties. Unfortunately, they have small internal losses, which cause the loudspeaker's characteristic "metallic" sound, caused by weakly damped resonances. For this reason, materials such as paper or polypropylene are used rather than metals, because they have a high internal loss factor. The industry also uses polypropylene coating for cellulose membranes to reduction of natural vibration. Such materials as Kevlar® or carbon fibers are used in expensive loudspeakers because they have both a high c_0 value and large internal losses [1].

There is a possibility to increase flexural stiffness by increase of the thickness of the membrane. The flexural stiffness increases with third power of the thickness and the stretching one is proportional only to the first power of the thickness. However, for the homogeneous material of the membrane, its mass increases significantly and it causes the decrease of efficiency. Sometimes, the loudspeaker membranes made of a honeycomb structure are applied. It is both thick and light. The aim of this paper is testing of inhomogeneous, layered composite structures for application in loudspeaker membranes. It should be mentioned that for inhomogeneous structures such quantities as Young's modulus, density, loss factor or speed c_0 should be considered only as equivalent to these quantities for homogeneous structures. They are no longer the material parameters, but they depend also on number of layers and their thicknesses.

17.2 Methods of Measurements of Dynamical Properties

The basic dynamical properties of materials are determined by their Young's modulus E, defined as the ratio of stretching stress σ in the rod made of tested material to the relative change of its length, i.e., to strain ε:

$$E = \frac{\sigma}{\varepsilon}[Pa] \tag{17.3}$$

The Young's modulus determined for static or dynamic forces affecting the mechanical system is different, so a static and dynamic modulus should be distinguished. To determine properties of loudspeaker membranes, the dynamic Young's (E) modulus is used. For systems with internal mechanical losses, the dynamic Young's modulus is a complex quantity. Its module is responsible for the elasticity of the material, while the tangent of the phase difference (tan d) between stress and strain determines internal losses. Since the loss angle d is usually very small, it can replace its tangent: $d \approx \tan d$. It is called the loss factor.

Among the many existing methods of measuring Young's modulus, four basic ones can be distinguished [3]:

- A method of measuring dynamic force and dynamic deformation of a small sample, which is treated as a lumped element—a spring. This method gives reliable results only in the very low frequency range [4].
- Torsion-pendulum method that allows measurements in the very low frequency range of materials with low stiffness. The tested sample is a pendulum frame, and the measurement procedure is strictly defined in EN ISO 6721-2 [5].
- Ultrasonic method, in which the propagation speed of mechanical waves in solids in the high frequency range is determined, and solids are treated as unlimited media [3].
- The method of measuring bending vibrations in beams with finite dimensions. This method is used in the low and medium frequency range and is described in EN ISO 6721-3 [6].

For characterizing of vibrational properties of loudspeaker membranes, the last method is most often used, because it gives reliable results in a useful frequency range. The dynamic Young's modulus should be used for calculation of the velocity given by Eq. (17.2).

Measurements of Young's modulus by the method of bending vibrations in beams consist in the determination of vibration resonance curves for a thin cantilever beam with rectangular cross section made of the tested material. The beam is rigidly clamped at one end and free at the second one. A typical measuring system for a beam clamped at one end is illustrated in Fig. 17.2. This method has been proposed by Oberst [7], and it is called the Oberst method.

The free end of the beam is excited for vibrations, e.g., by means of an electromagnetic transducer. Its natural frequencies are explicitly determined by the dynamic Young's modulus. In Fig. 17.3, the first four forms of beam vibration are presented [8].

The natural frequencies of a beam rigidly clamped at one end are given by the formula [8]:

$$f_n = \frac{k_n}{l^2} R_o c_0 = \frac{k_n}{l^2} \frac{h}{\sqrt{12}} \sqrt{\frac{E}{\rho}} \tag{17.4}$$

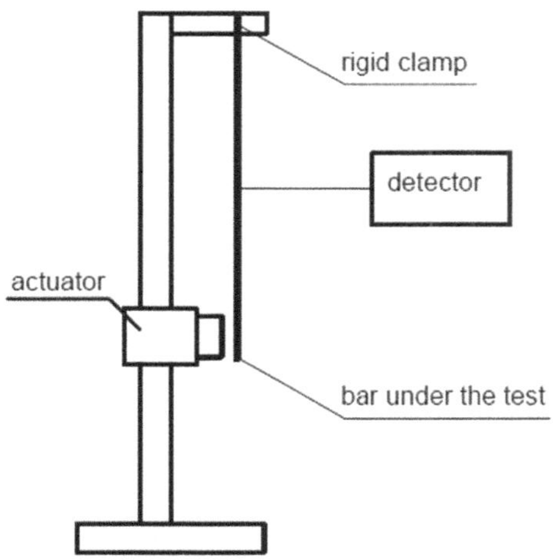

Fig. 17.2 System for measuring the dynamic Young's modulus by the Oberst method

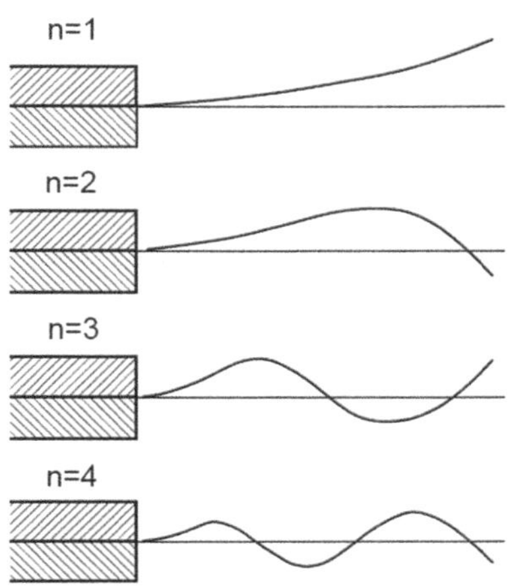

Fig. 17.3 First four modes of a beam clamped at one end and free at the second [8]

where $R_0 = \frac{h}{\sqrt{12}}$—radius of inertia of the cross section of the beam [m]; h and l thickness and length of the beam [m]; speed c_0 is given by the Eq. (17.2); k_n—coefficients for n-th resonance: $k_1 = 0.55966$, $k_2 = 3.507$, $k_3 = 9.8199$, $k_4 = 19.243$, etc. [8]

For a beam with given dimensions made of a material of known density, the dynamic Young's modulus is determined from Eq. (17.4) by measuring the resonant frequencies of the beam. The density can be determined based on the beam's dimensions (volume) and its mass.

The loss factor can be determined by two methods. First one consists in the analysis of the resonance curve of the n-th mode of free vibration from the equation:

$$d = \frac{\Delta f}{f_n} \tag{17.5}$$

where Δf is the width of the resonance curve determined by the decrease (−3 dB) from the maximum value.

In the second method, the loss factor d is determined from the decay of vibration in the beam (Fig. 17.4) by measuring the slope D of such a curve:

$$d = 0.0366\frac{D}{f_n} \tag{17.6}$$

where D—slope of decay of vibration amplitude [dB/s].

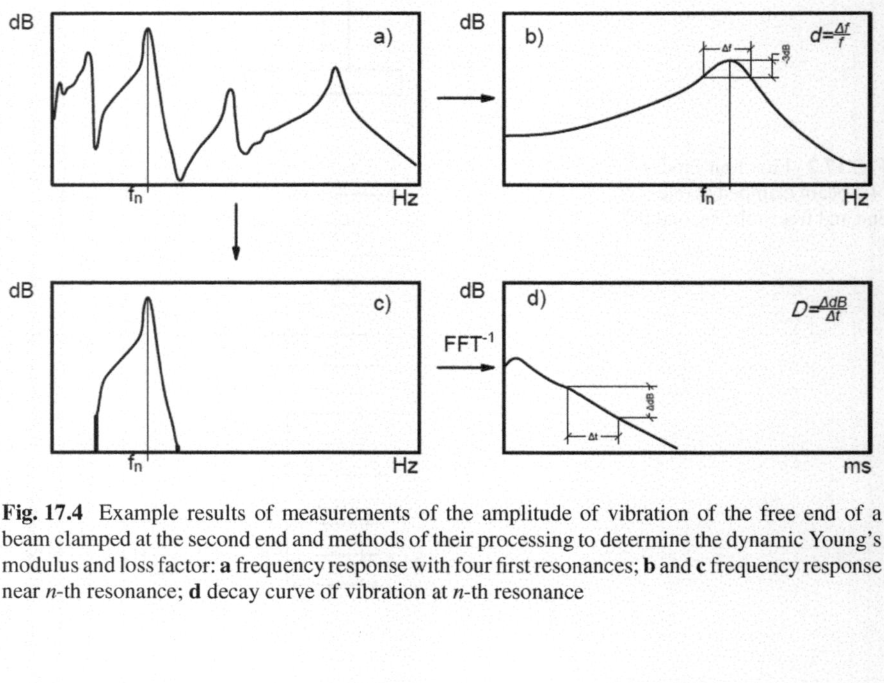

Fig. 17.4 Example results of measurements of the amplitude of vibration of the free end of a beam clamped at the second end and methods of their processing to determine the dynamic Young's modulus and loss factor: **a** frequency response with four first resonances; **b** and **c** frequency response near n-th resonance; **d** decay curve of vibration at n-th resonance

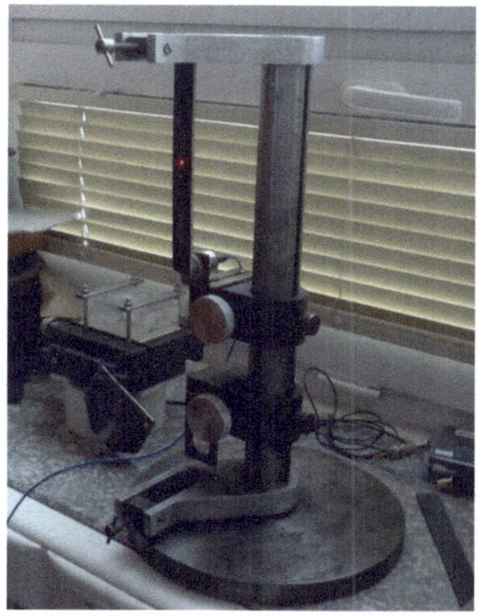

Fig. 17.5 Sample while acoustics tests

The decay curve is obtained by recording vibrations after switching off the exciter that generates vibrations with a frequency equal to the frequency of the n-th resonance or by filtering the beam vibration transmittance function and its inverse Fourier transform (Fig. 17.4).

In this paper, the first method of measurement of loss factor, based on width of the resonance curve, has been chosen. The test arrangement is presented in Fig. 17.5.

The stimulation of the sample is realized by an electromagnetic transducer interacting with a small piece of steel sheet with negligible mass attached to the sample. The vibration of the free end was monitored with a laser vibrometer. The signal of vibration is processed and displayed in a form of a frequency resonance curves.

17.3 Materials

Loudspeaker membranes made of polymer sandwich composites are considered. In these composites, various materials are used for the core. The outer layers were made of thin carbon paper.

The tested composites were based on EPOLAM 2015 epoxy resin cured with EPOLAM 2016 hardener. The outer layers of all samples were made of DONACARBO S-259P carbon paper with a thickness of 0.5 mm. Composite cores are made of four different materials (Table 17.1): balsa wood, Coremat Xi, polyurethane foam and Poraver glass granulate with different ranges of diameters. In addition, three samples were made of only materials used for the core: balsa wood, carbon paper DONACARBO and Coremat Xi. These samples were produced by laminating the materials in the resin. All of 10 prepared samples were used in the study. The structures of the tested composites are presented in Fig. 17.6.

Table 17.1 Types of materials used for cores of the composite samples

Ordinal number	Core material
1	Balsa wood
2	Poraver 0.04–0.125 mm glass granulate
3	Poraver 0.5–1 mm and 2–4 mm glass granulate
4	Polyurethane foam (sample prepared with RTM method)
5	Polyurethane foam (sample prepared by lamination)
6	Coremat Xi (sample prepared with RTM method)
7	Coremat Xi (sample prepared by lamination)

Fig. 17.6 Samples of membrane with different core: **a** balsa wood, **b** Poraver glass granulate 0.04–0.125 mm, **c** Poraver glass granulate 0.5–1 mm and 2–4 mm, **d** polyurethane foam (RTM method), **e** polyurethane foam (hand lamination), **f** COREMAT Xi (RTM method), **g** COREMAT Xi (hand lamination)

Resin Transfer Molding (RTM) is a relatively simple and cheap technology to use. It does not require advanced instrumentation, while allowing for high performance; thus, it is a very good alternative to autoclave molding [9]. The RTM system assumes the use of a rigid form, consisting of two elements, between which non-filtered composite material is laid. Both parts of the mold are clamped together, and then, the resin system responsible for filtering the prepared material is pressed inside [2]. RTM allows the formation of composites of any shape and gives the possibility of mass production [9–11]. To maintain good material flow through the mold, low density resins, such as epoxy resins, are used. The addition of fillers, rubber or thermoplastics, can cause a sudden increase in density, which significantly hinders the movement of the resin in the mold [12]. However, this is not a major problem in the case of sandwich composites, where both mat or fiber covers and the composite core are placed in mold before resin injection. RTM has many varieties, the study used VA-RTM (Vacuum Assisted RTM). With the connector in the mold, a pressure difference is created by a vacuum pump connected to the system, and the resin is drawn into the mold and fills the empty spaces and impregnates the unfiltered fibers. When the process ends, the pressure is disconnected [12]. Currently, research is underway to improve resin flow in the mold and to avoid air voids and fiber deformation [11, 13, 14]. The RTM method with underpressure was used to prepare five types of the samples used in the test.

The next four were made by hand lamination. The laminate was placed between two mold parts and compressed to obtain the thinnest possible samples [2]. All samples were allowed to cure and then placed in an oven at 80 °C for 8 h.

17.4 Results and Discussion

The usefulness of the material used on the speaker membrane is determined by the ratio of Young's modulus (E) to the density of the sample (ρ). The square root of this quotient defines the velocity of the longitudinal wave that propagates in the material. The higher is the speed of the wave propagating in the material, the greater the chance a speaker membrane could be produced from a tested material with satisfactory parameters. The second important parameter is the internal loss factor of the material (d), which should reach values as high as possible. The parameters described above were calculated based on the results obtained in the study (Table 17.2).

Table 17.2 Results of the measurement of longitudinal wave velocity and Young's modulus for the first resonance of bars

Number of specimens	f_r (Hz)	f_d (Hz)	f_g (Hz)	d	c_0 (m/s)	P (kg/m^3)	E (GPa)
1 Balsa wood	54.33	54.02	54.66	0.012	2826.36	957.56	7.65
2 Poraver 0.04–0.125	56.25	55.91	56.63	0.013	2154.29	1187.15	5.51
3 Poraver 0.5–4	54.30	54.00	54.58	0.011	2178.52	818.04	3.88
4 Polyurethane foam (RTM)	50.72	50.41	51.00	0.012	2318.20	753.74	4.05
5 Polyurethane foam (lamination)	61.64	61.33	61.92	0.01	2506.81	577.25	3.63
6 Coremat Xi (RTM)	51.16	50.86	51.47	0.012	2077.14	638.99	2.76
7 Coremat Xi (lamination)	37.70	37.50	37.85	0.093	2325.64	713.91	3.86

f_r—resonance frequency, f_d—lower −3 dB limiting frequency, f_g—upper −3 dB limiting frequency, d—loss factor, c_0—longitudinal wave velocity propagating in material, ρ—density, E—Young's modulus

Table 17.3 Example values of d—internal loss factor and c_0—longitudinal wave velocity propagating in material, for materials used for loudspeaker membranes

Material	c_0 (m/s)	d
Aluminum	5120	0.004
Beryllium	12,600	0.002
Kevlar 49	9500	0.02
Paper	1580–1860	0.04
Polypropylene	1319	0.09
Carbon Fiber	11,600–21,200	0.05
Titanium	4500	0.002
Titanium + Boron	7715	0.04

The results obtained from the measurements were compared with the literature values [8] and placed in Table 17.3. It can be seen that most samples have a loss factor comparable to the Kevlar. Only polypropylene and carbon fiber have a higher loss factor d. However, the c_0 sound velocities obtained in the produced samples significantly exceed the value of the c_0 specified for paper membranes (characterized by a low c_0 value, but still being the dominant material in processing). Also, the c_0 velocity value obtained for polypropylene membranes significantly exceeds. Metal membranes (with an unfavorably low loss factor d) and the ones made of synthetic fibers that perform best in the above classification have greater values of the sound speed c_0. However, it should be noted that the samples made and used for testing have parameters similar to the most popular materials for membranes, i.e., paper and polypropylene. This allows further testing to improve the material and attempt to achieve better properties at the lowest possible thickness.

The measured amplitudes of vibration as the function of frequency were plotted for narrow (up to 100 Hz) and wide frequency range—depending on the material: up to 1 kHz or up to 2 kHz. The width of the range differs because for some materials the first three resonances were not in the 1 kHz range—the range had to be increased in order to record all three resonances. Such a change was needed for samples 2 and 3. From the graphs obtained for the narrower range (more accurate measurement), where only the first resonance was visible, the internal loss coefficient (d) was determined; on the basis of broader ranges, the Young's (E) modulus was determined. The most important results of the research were loss factor of material and the velocity of the longitudinal wave, which propagated in the sample.

The graphs for selected three samples are shown below. These are the characteristics obtained for samples with cores from: balsa wood (Fig. 17.7), which was characterized by the highest speed of the propagating wave, Poraver 0.04–0.125 granules (Fig. 17.8) and sample of Coremat Xi, where the core material was laminated without covers, to check the properties of the core material only. The last sample of core without outer layers, as the only one of three samples made of individual materials, gave results that could be analyzed.

For a sample with a core made of balsa wood, the resonances tested were in the range up to 1 kHz (Fig. 17.7), the internal loss coefficient of the sample reached the

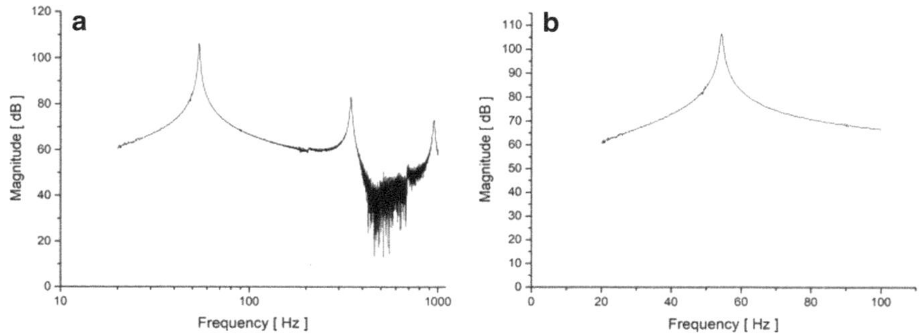

Fig. 17.7 Frequency characteristics of vibration for a balsa wood core sample: **a** wide frequency range; **b** near first resonance

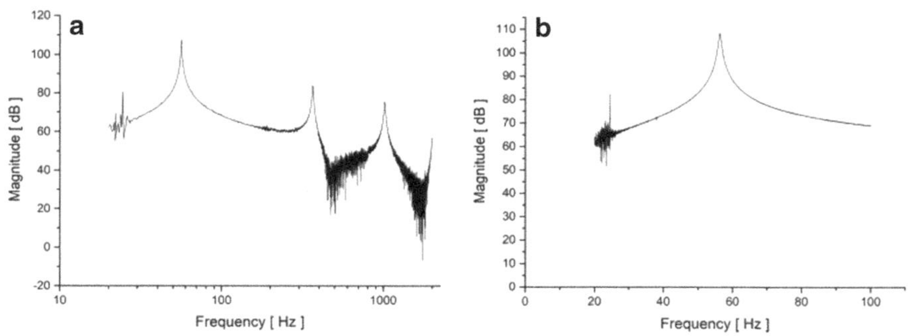

Fig. 17.8 Frequency characteristics of vibration for a sample with Poraver granulate core 0.04–0.125 mm: **a** wide frequency range; **b** near first resonance

level of $d = 0.012$. A similar result was obtained for w sample with a core made of Poraver granules with smaller diameters. As to the longitudinal wave propagating in the sample with balsa wood core, the value was the highest among all tested samples—about $c_0 = 2826.36$ m/s.

In the case of a sample which core was made of a mixture of Poraver granules with different range of diameters, the tested resonance was in the range up to 2 kHz. In the case of the other tested materials, four more samples were characterized by a similar value of internal losses factor—these were samples, whose cores were made of polyurethane foam and Coremat Xi.

Examination of the sample made of material used as the core only was made additionally to compare results with the ones received for sandwich composite. It turned out to be difficult because of the high flexibility of samples. The first resonance occurred at the 8 Hz frequency (Fig. 17.9). For this reason, the values needed for calculations were read only for the first two resonances, and values for the measurement are placed in Table 17.4. The coefficient of internal losses d for Coremat Xi

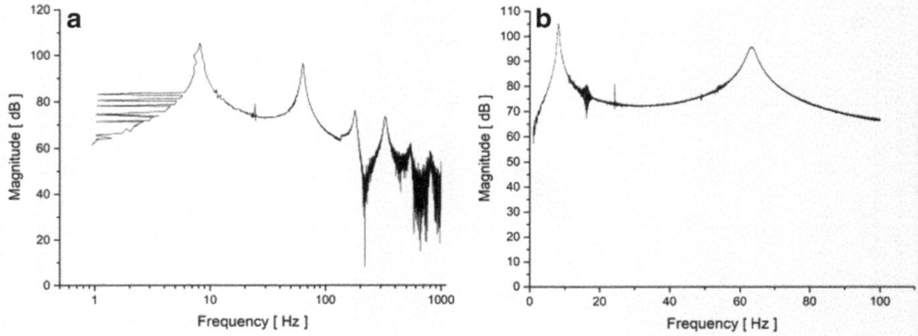

Fig. 17.9 Frequency characteristics of vibration for a Coremat Xi core only, without outer carbon layers: **a** a wide frequency range; **b** first two resonances

Table 17.4 Results of the measurement of longitudinal wave velocity and Young's modulus for the first two resonance for the sample made of Coremat Xi without outer layers—only core material

Number of resonance for Coremat Xi core only	f_r (Hz)	f_d (Hz)	f_g (Hz)	d	c_0 (m/s)	ρ (kg/m^3)	E (GPa)
1	8.17	7.89	8.36	0.057	728.41	398.56	0.21
2	67.21	66.52	67.77	0.038	1262.12	398.56	0.75

f_r—resonance frequency, f_d—lower −3 dB limiting frequency, f_g—upper −3 dB limiting frequency, d—loss factor, c_0—wave velocity propagating in material, ρ—density, E—Young's modulus

without outer layers was the highest among all tested samples and amounted to 0.057. However, the velocity of the longitudinal wave propagation was the smallest.

According to Table 17.2, samples made of polyurethane foam by hand lamination had the lowest density of all prepared materials. The coefficient of internal losses d for all materials turned out to be quite similar—for all seven samples the values oscillate in the range of $d = 0.01$–0.013. With the increase of the Young's modulus the density also increased. The exception to this rule is the sample No. 2 (Poraver 0.04–0.125), which reached the highest density ($\rho = 1187$ kg/m^3) while giving low longitudinal wave velocity obtained in the tests ($c_0 = 2154$ m/s).

17.5 Conclusions

The following conclusions can be drawn based on the research undertaken in the paper.

Higher density is reported for samples, comparably for both types of characterization, where the core was made of the smallest granulated product Poraver 0.04–0.125 mm. Lower density characterized samples in which Poraver glass granules with larger diameters (a mixture of two types) were used. Due to the open pores of grains, it can be concluded that some of them were filled with resin, which caused an increase in mass.

The composite with lower density is a composite with a core made of polyurethane foam, which was made using hand lamination. The density $\rho = 577.25$ kg/m^3.

The Young's modulus study by the Oberst method showed that the largest Young's modulus characterized a composite whose core is made of balsa wood. In addition, the laminate obtained the highest velocity of the longitudinal wave and thus, fulfilled the key relationship that classifies materials for composite membranes—the ratio of Young's modulus to material density. The longitudinal wave speed for samples with balsa wood core exceeded 2800 m/s. For comparison, popular cellulose membrane is characterized by $c_0 =$ 1580–1860 m/s, whereas for high quality carbon fiber membrane $c_0 =$ 11,600–21,200 m/s.

The loss factor d, which was measured for composite membranes, oscillated around $d = 0.01$. This value is comparable with the loss ratio for Kevlar ($d = 0.02$) [1].

From the obtained results, it can also be concluded that the significant influence on some parameters (the velocity of the longitudinal wave propagating in the sample or Young's modulus) of the tested materials has the method of making the samples.

References

1. Colloms M, Darlington P (2018) High performance loudspeakers: optimising high fidelity loudspeaker systems, 7th edn. Wiley. https://onlinelibrary.wiley.com/doi/book/10.1002/9781118706275
2. Beranek LL, Mellow TJ (2012) Acoustics: sound fields and transducers. Elsevier, Oxford
3. EN ISO 6721–1. Plastics. Determination of dynamic mechanical properties. Part 1: General principles
4. Cremer L, Heckl M, Petersson B (2005) Structure-borne sound. Springer, Berlin
5. EN ISO 6721–2. Plastics. Determination of dynamic mechanical properties. Part 2: Tor-sion-pendulum method
6. EN ISO 6721–3. Plastics. Determination of dynamic mechanical properties. Part 3: Flexural vibration—Resonance-curve method
7. Oberst H (1952) About the damping of the bending vibrations of thin sheets by firmly adhering coating. J Acustica 2(4):181–194
8. Dobrucki A, Roszkowski C (1977) Measurement of the viscoelastic constants of the cellulose used for a loudspeaker membrane and their effect upon the electro-acoustic parameters of the loudspeaker. Arch Acoust 2(3):177–188
9. Mazumdar S (2001) Composites manufacturing: materials, product, and process engineering. CRC Press, Florida
10. Okabe T, Oya Y, Yamamoto G, Sato J, Matsumiya T, Matsuzaki R, Yashiro S, Obayashi S (2017) Multi-objective optimization for resin transfer molding process. Compos Part A Appl Sci Manuf 92:1–9
11. Geng Y, Jiang J, Chen N (2019) Local impregnation behavior and simulation of non-crimp fabric on curved plates in vacuum assisted resin transfer molding. Compos Struct 208:517–524
12. Swan S, Yuksel T, Kim D, Gurocak H (2017) Automation of the vacuum assisted resin transfer molding process for recreational composite yachts. Polym Compos 38(11):2411–2424

13. Shi H, Sun B, Liu Q, Yang Z, Yi K, Zhang Y, Fu S (2015) A high ductility RTM epoxy resin with relatively high modulus and Tg. J Polym Res 22(7):134
14. Magagnato D, Seuffert J, Bernath A, Kärger L, Henning F (2018) Experimental and numerical study of the influence of integrated load transmission elements on filling behavior in resin transfer molding. Compos Struct 198:135–143

Chapter 18
Experimental–Numerical Analysis of Radial Compression Test of CFRP/GFRP Composite Rebars

Szymon Duda, Dominika Szymczyk, Paweł Zielonka, Michał Smolnicki, and Grzegorz Lesiuk

Abstract This paper presents the experimental results of composite rebars based on GFRP and CFRP manufactured by a pultrusion system. The off-plane compression strength of rods was determined. This compression test reflects the weakest nature of the composite materials based on the interlaminar compressive strength. The proposed methodology allows us to invariantly describe the experimental transversal strength of the composite materials. As a result, the force–displacement graphs were plotted. The experimental results are compared with computational results using the Finite Element Method. In off-plane compression the matrix carries the load, all the specimens regardless of the length were cracked in the same way, in the middle of the cross-section. It is worth noting that the failure and fracture mechanism plays a crucial role as a material quality indicator in the manufacturing process. The comparison showed differences between the experiment and the FEM results that may come from the imperfection in the structure.

Keywords Composite rebars · Radial compression · FEM analysis

18.1 Introduction

The demand for more advanced materials that exhibit high strength and low density leads to rapid growth in the industry [1]. Composite materials are widely used in many areas of the industry such as civil engineering, biomedical engineering, automotive, aerospace engineering, and others [2, 3].

This paper focuses on the current composites expansion in civil engineering. One of the fundamental material in this area is concrete. This composite material is made from a mixture of two main components: aggregate (sand, gravel, and crushed stones) and paste (cement and water) [4]. Thus the manufacturing process is not complicated, the components are not expensive and quite widely available. Though concrete is highly resistant to compression, it does not perform well under tension

S. Duda · D. Szymczyk · P. Zielonka · M. Smolnicki · G. Lesiuk (✉)
Department of Mechanics, Materials Science and Biomedical Engineering, Wroclaw University of Science and Technology, Wroclaw, Poland
e-mail: Grzegorz.Lesiuk@pwr.edu.pl

G. Lesiuk et al. (eds.), *Structural Integrity and Fatigue Failure Analysis*,
Structural Integrity 25, https://doi.org/10.1007/978-3-030-91847-7_18

[5]. Reinforcing bars are used to obtain higher tensile strength and resistance for cracking and breaking. Conventionally rebars are made from steel as both materials have similar thermal expansion coefficients [6]. However, currently, steel is replaced by FRP (Fibre Reinforced Polymer), which gain more and more popularity nowadays. They exhibit better mechanical properties than typical steel reinforcement rods [7]. Nevertheless, they are extremely dangerous in case of cracking because as brittle material they crack rapidly [8].

Current research on the integrity of the structure of FRP makes this process more predictable. An author such as Urbanski [9] provided information about compression strength for hybrid reinforced rebars but loaded in-plane. Their compression strength totals approximately 46% of the ultimatum strength according to mentioned research. Also, the author described the failure mechanisms of these rebars during tensile tests such as fiber kinking and micro-buckling. Paper submitted by Chisholm [10] evidenced that radial compression test might be applied as a quality control test to evaluate the relative performance of pultruded rod material. Furthermore, the mechanical properties of glass fiber reinforced polymer (GFRP) subjected to tensile, torsion, and bending loading were studied by authors Ahmadi [11] in terms of over braiding. This research showed that these rods have considerably higher shear modulus, but lower tensile modulus and flexural rigidity than those of UD pultruded rods. Finally, authors Jarek [12] and Abbood [13] benchmarked FRP rods available on the market. They showed that the quality of provided rebars varies depending on the manufacturer. Summarizing the axial compression causes failures mostly in fibers, if the radial compression is considered the load is carried by matrix, in that transverse compression strength provide us information about the quality of the structure in terms of cohesion.

Also to have a better understanding of the FPR structure is important to have a solid knowledge of the manufacturing process. In the case of this research, the rebars are manufactured using pultrusion. Pultrusion is a manufacturing process converting fibers and resin into a finished shape provided by a heated die [14]. This process is continuous otherwise to other well-known methods for composite parts manufacturing providing this process less-labor and more automated. In that, it offers the best productivity/cost ratio. Improvement in the control system of the process and understanding the mechanical behaviors of the FPR allow to design and manufacture more advanced and more safe material that has a better quality.

The main goal of this work is to analyze the mechanical behavior of the reinforcing bars, as well as experimentally as through the FEM simulation. After experimented it was possible to compare numerical simulation results with the real behavior of the material.

18.2 Materials and Methods

18.2.1 Material

The reinforcement rods made of carbon and glass fibers in resin were used. They were cut into three types of specimens varied in length between 10 mm, 30 mm, and 50 mm. All of them have the same diameter approximately 10 mm. To increase the accuracy of the measured quantities were used three specimens in each mentioned length. So in the experimental campaign18 specimens were analyzed. For all the specimens were carried out the measurements of the geometry, the results are presented in Tables 18.1 and 18.2.

In further calculation, the real values of the length and the diameter were taken into account (Fig. 18.1).

Table 18.1 Dimensions of the glass fiber reinforcement rod

Length (mm)	Diameter (mm)
10.006	10.009
30.030	10.010
49.957	10.007

Table 18.2 Dimensions of the carbon fiber reinforcement rod

Length (mm)	Diameter (mm)
10.468	9.999
29.904	10.002
49.780	9.994

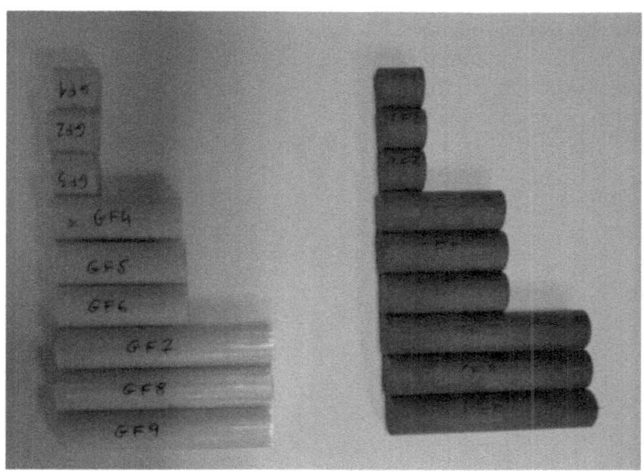

Fig. 18.1 The reinforcing bar used in the experiment

18.2.2 Out-of-Plane Compression

During an experimental part, radial compression test was carried out, that allowed determining the strain–stress curves. The experiment was carried out using the MTS 810 a universal material testing machine. The software system MTS Flex Test Console dedicated to the MTS 810 allows collecting data from the experiment.

Figure 18.2 presents the photography from the experiment. The specimen is put between two compression platen that is getting closer to each other and cause the compression of the specimen. The platen is made from strong and high-quality steel. To be able to distinguish the specimen all are indicated.

All the specimens regardless of the length were cracked in the same way, in the middle of the cross-section. The structure also breaks down on the lateral surface as well as on the top and the bottom of the cylinder. The cracks were not straight, quite the opposite the new surfaces were rough and a little ragged. It is possible to notice the flattening of the cylindrical surface (Fig. 18.3).

Fig. 18.2 The specimen from the glass fiber during the radial compression test

Fig. 18.3 The specimen of the length of 50 mm made from carbon fibers after the radial compression test

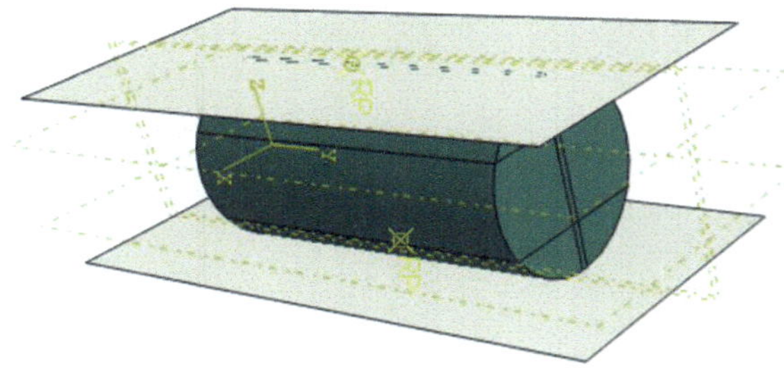

Fig. 18.4 The assembly of specimen and grippers

Table 18.3 Elastic parameters of the GFRP used in the simulations [15]

E_1 (GPa)	E_2 (GPa)	E_3 (GPa)	G_1(GPa)	G_2(GPa)	G_3 (GPa)	ν_{12}	ν_{13}	ν_{12}
36.3	10.8	10.8	4	3	4	0.28	0.09	0.28

18.2.3 Finite Element Method Analysis

Comparing the Finite Element Method (FEM) results with the physical experiment it is required to have the same course of the experiment. However, in FEM some aspects can be simplified to reduce computational cost. In the simulation of the compression test, the specimen made from GFRP was squeezed between two rigid surfaces.

The length of the specimen varies as the experiment used three different lengths (10 mm, 30 mm, and 50 mm) of specimens. Simulations were computed for all mentioned lengths. The stress in the grippers was not analyzed and occurred deflations can be neglected. Thus, they were modeled as a discrete rigid shell so as a perfectly stiffness body. The specimen was represented by a deformable solid body.

The simulation can be successfully computed with general contact; however, this is not the most optimal solution. To decrease computing time the body was partitioned to provide a surface for surface-to-surface. The next partitioning provides an independent cell where the crack occurred and propagated in the real experiment.

The dimension of the part was so selected to cover the specimen and was changed as the specimen length was changed. These parts were modeled in Abaqus software (Fig. 18.4).

The GFRP is an anisotropic material. Thus, parameters need to be specified and assigned in three directions. Material data used in simulations are presented in Table 18.3.

18.3 Results and Discussion

Data established from the experiments are presented in the following charts where is presented the relationship between the force acting on the material and the occurring

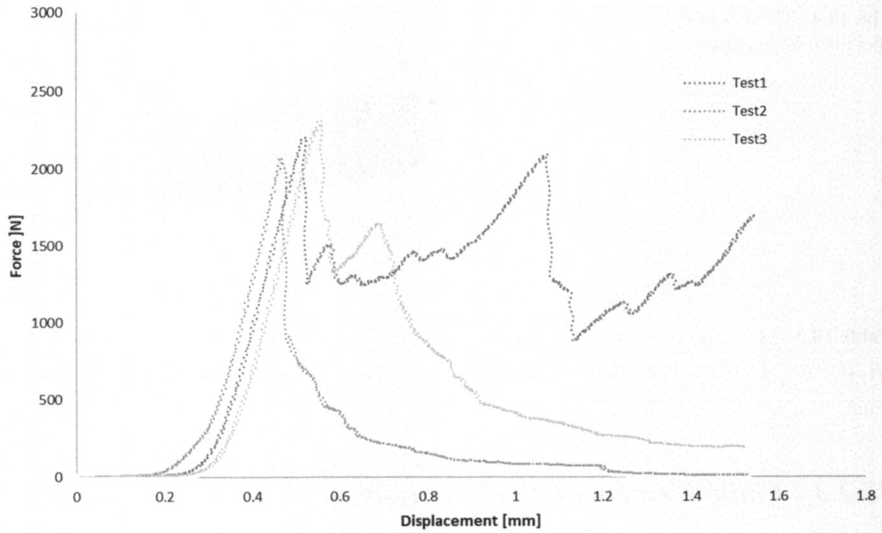

Fig. 18.5 The force–displacement plot for the reinforcing bar made from the carbon fiber which length is equal to 10 mm

displacement. Since there were three experiments for various lengths all three data are on the plot to allow comparison.

a. Carbon fiber reinforced bar

- Length 10 mm (Fig. 18.5)
- Length 30 mm (Fig. 18.6)
- Length 50 mm (Fig. 18.7)

b. Glass fiber reinforced bar with FEM results

- Length 10 mm (Fig. 18.8)
- Length 30 mm (Fig. 18.9)
- Length 50 mm (Fig. 18.10)

According to the established data maximum force for the specimen of the GFRP composite, it is similar for the length of 10 and 30 mm, the value is around 2100 N. However, the difference is spottable for the specimen with a length 50 mm. In this case, the maximum force reached a value above 11,000 N. Nevertheless data from test 1 are different from these for test 2 and test 3, where the maximum force has a value around 8000 N.

The specimen made of the carbon fiber composite reaches the higher value of the maximum forces. Since carbon fibers have higher strength than carbon fiber this behavior is appropriate. In the case of the carbon fiber, reinforcement rods appealed the difference in the maximum force depending on the specimen length. For the length of 10 mm the force reached a value around 2500 N, for the length of 30 mm the force

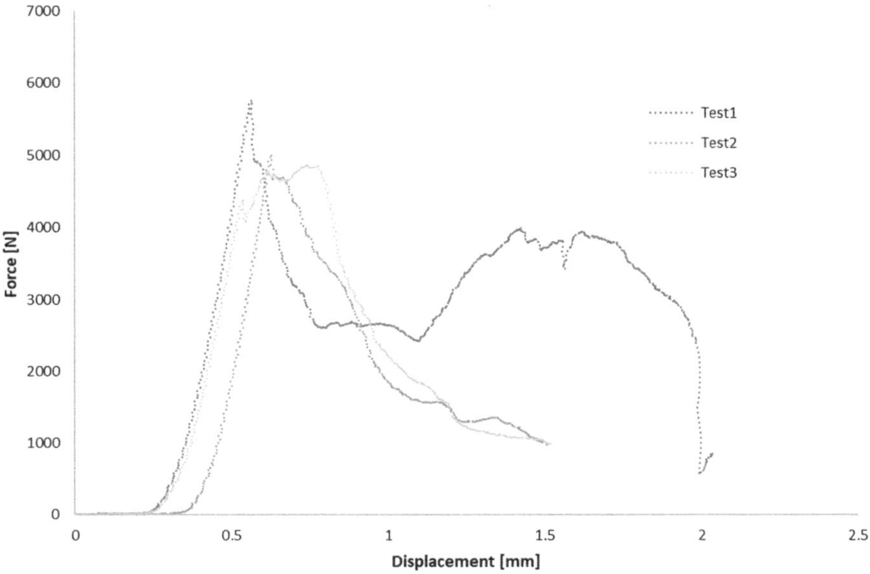

Fig. 18.6 The force–displacement plot for the reinforcing bar made from the carbon fiber which length is equal to 30 mm

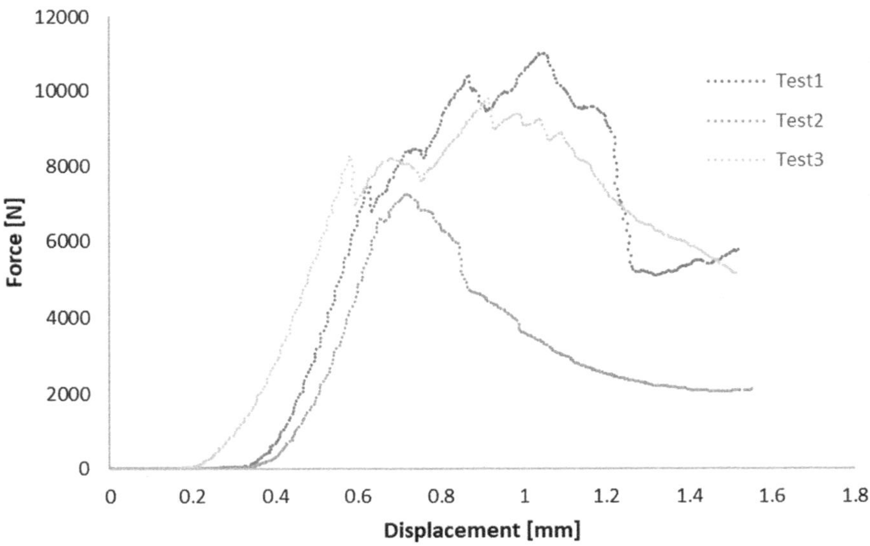

Fig. 18.7 The force–displacement plot for the reinforcing bar made from the carbon fiber which length is equal to 50 mm

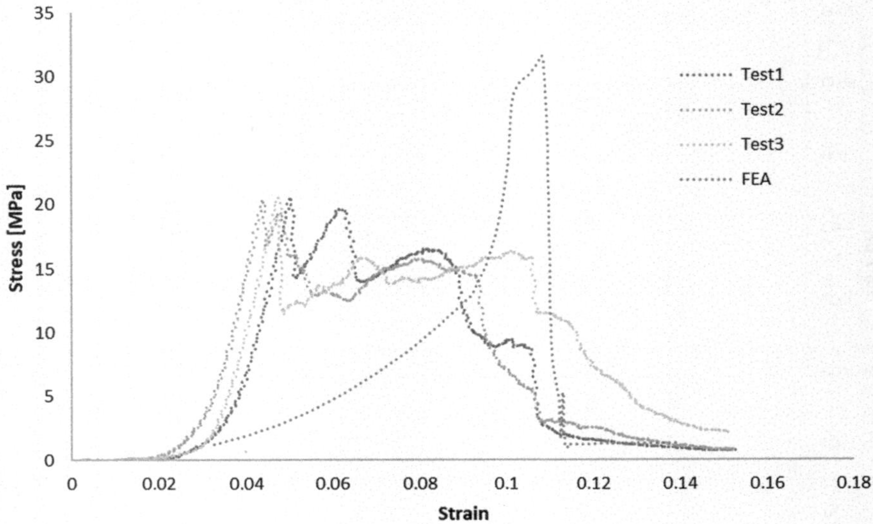

Fig. 18.8 The strain–stress curve for the glass fiber reinforcement rod of the length of the 10 mm

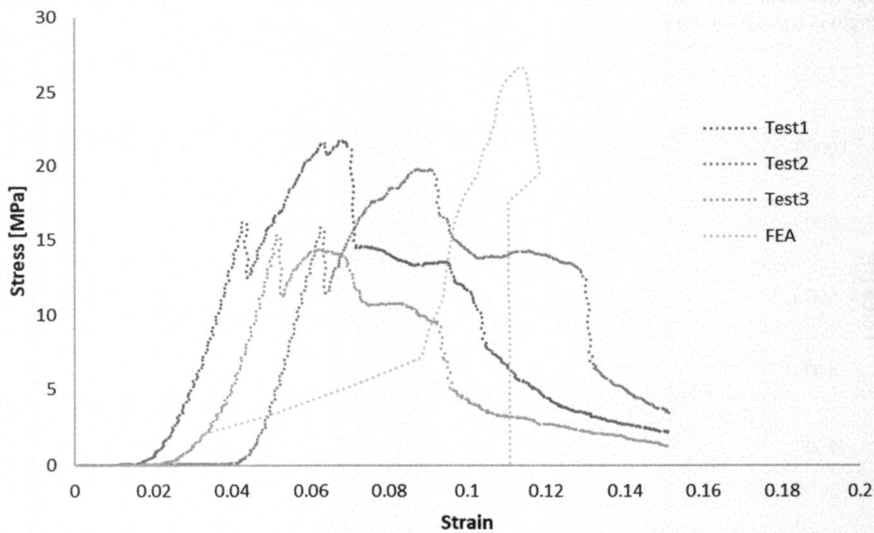

Fig. 18.9 The strain–stress curve for the glass fiber reinforcement rod of the length of the 30 mm

is doubled and is around 5000 N. Nonetheless the experiment data for carbon fiber reinforced bars of the length of 50 mm have the biggest variance. In this case, the maximum force varies from almost 12,000 N to a little above 6000 N. However, the reason for these differences in the results can come from the more imperfection in

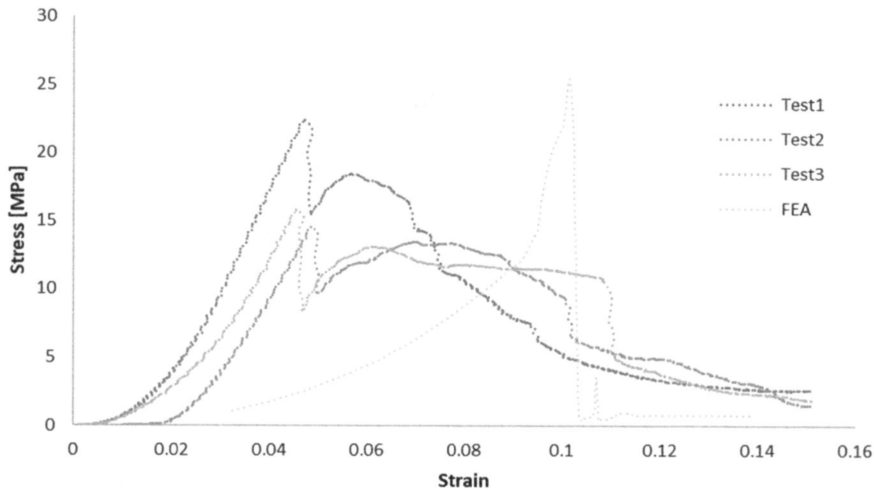

Fig. 18.10 The strain–stress curve for the glass fiber reinforcement rod of the length of the 50 mm

the structure. Specimens made from the glass fiber exhibited differences in the data but mainly for the 50 mm length. Oppositely for the smallest length of 10 mm, the data are overlapped, especially for the linear part when the force rises.

What all the plots have in common is the jump in the data after the maximum force which corresponds to the crack in the specimen. The rest of the breaks and curving are caused by the imperfection in the structure.

18.4 Conclusions

The main goal of this work was to research the mechanical properties of glass fiber and carbon fiber reinforcement rods. In the experimental part, the radial compression tests were carried with different length of the specimen. The results show that:

- carbon fiber rods and glass fiber rods perform similarly. However, carbon fibers have slightly better mechanical properties. Thus they will crack with the same probability.
- according to collected experiment data the tensile stress reaches the value of around 10–14 MPa regardless of the length of the specimen. The maximum force is much greater for longer rods, nevertheless, the formula for the maximum stress includes the length
- among the difference between the FEA result and the physical test is the slope of the curve. Mainly caused by imperfections in the real structures. Taking this into account in analysis may better correspond to the real behavior of the reinforcing bars. Also for better coverage of the plot, used parameters can be established for

the direction for the used reinforcing bar and not the general GFRP reinforcing bar.

- in the case of composite material, an important aspect is fibers volume ratio and quality of used materials. Furthermore, the reinforcement rods can compound their properties because of the microdamage that occurred in the transport or exposure to the temperature.

Acknowledgements The research was financed by the National Centre for Research and Development of Poland (NCBiR) grant number LIDER/40/0219/L-10/18/NCBR/2019.

References

1. (2) (PDF) High Performance Composite Market Size, Trends & Current Growth. Industry Research Report 2018–2025. https://www.researchgate.net/publication/331248202_High_Performance_Composite_Market_Size_Trends_Current_Growth_Industry_Research_Report_2018-2025. Accessed 2 Feb 2021
2. Jinsheng M (2018) Application of composite materials in engine. Mater Sci Adv Compos Mater 1. https://doi.org/10.18063/msacm.v1i1.499
3. Pastuszak PD, Muc A (2013) Application of composite materials in modern constructions. Key Eng Mater 542:119–129. https://doi.org/10.4028/www.scientific.net/KEM.542.119
4. Assistant Professor PH SUBJECT: CONCRETE TECHNOLOGY SUB CODE: 10CV42
5. Li Z (2011) Advanced concrete technology
6. Aydin F (2018) Experimental investigation of thermal expansion and concrete strength effects on FRP bars behavior embedded in concrete. Constr Build Mater 163:1–8. https://doi.org/10.1016/j.conbuildmat.2017.12.101
7. Barcikowski M, Lesiuk G, Czechowski K, Duda S (2021) Static and flexural fatigue behavior of GFRP pultruded rebars. Materials (Basel) 14. https://doi.org/10.3390/ma14020297
8. Boudet JF, Ciliberto S, Steinberg V (1996) Dynamics of crack propagation in brittle materials. J Phys II 6:1493–1516. https://doi.org/10.1051/jp2:1996144
9. Urbański M (2020) Compressive strength of modified FRP hybrid bars. Materials (Basel) 13:1898. https://doi.org/10.3390/ma13081898
10. Chisholm JM, Hahn HT, Williams JG (1989) Diametral compression of pultruded composite rods as a quality control test. Composites 20:553–558. https://doi.org/10.1016/0010-4361(89)90914-2
11. Ahmadi MS, Johari MS, Sadighi M, Esfandeh M (2009) An experimental study on mechanical properties of GFRP braid-pultruded composite rods. Express Polym Lett 3:560–568. https://doi.org/10.3144/expresspolymlett.2009.70
12. Jarek B, Kubik A (2015) The examination of the Glass Fiber Reinforced Polymer composite rods in terms of the application for concrete reinforcement. W: Procedia Eng 394–401, Elsevier Ltd
13. Abbood IS, Aldeen OS, Hasan KF, Jasim MA (2020) Properties evaluation of fiber reinforced polymers and their constituent materials used in structures—a review. Mater Today Proc. https://doi.org/10.1016/j.matpr.2020.07.636
14. Ramôa Correia J (2013) Pultrusion of advanced fibre-reinforced polymer (FRP) composites. In: W: Advanced fibre-reinforced polymer (FRP) composites for structural applications. Elsevier Ltd, pp 207–251
15. (2019) Numerical modelling of glass fiber reinforced polymer (GFRP) Cross Arm. Int J Recent Technol Eng 8:6484–6489. https://doi.org/10.35940/ijrte.D5162.118419

Chapter 19
Numerical Analysis of the New Design of CFRP/GFRP Composite Rebars Subjected to Bending Loading

Michał Smolnicki, Grzegorz Lesiuk, Paweł Stabla, Szymon Duda, and Paweł Zielonka

Abstract This paper presents the numerical simulation of the GFRP rebars subjected to the bending loading. Additionally, the comparison of the mechanical properties of CFRP/GFRP laminates with different fiber volumes is given according to the widely used hypotheses available in the literature. The most suitable hypotheses are the Abolish regarding Kirchhoff's (Shear) modulus and Chamisa's regarding Young's transverse modulus. These parameters were chosen in the further Finite Element Analysis (FEA). The pure bending is used to compare the individual conceptions of the rebar designs. The pure bending is obtained by rotation of 0.262 radians (approximately 15°), since one end of the rod is fixed in all three directions. Profiles with a greater moment of inertia of the cross-section are preferred, which means profiles with larger overall dimensions for instance a ring-type cross-section. Additional elements enriching the cross-section, such as a coil, only slightly transfer the load.

Keywords Hybrid rebars · FEM analysis · Composite materials

19.1 Introduction

Composite materials used as concrete reinforcement are one of the developing paths in civil engineering. The beginning of this technology kicked off in the 1960s. Composite rebars are used for concrete reinforcing in new constructions as well as strengthening in already existed constructions. They may be single or only supplying concrete reinforcement.

Author Portnov et al. [1] summarize present-day used composite rebars for reinforcement. They can be classified as follows:

- round bars with a roughened surface to increase adhesion and friction between rod and concrete,
- round bars with ribs or grooves that increase the mechanical interlocking,

M. Smolnicki · G. Lesiuk (✉) · P. Stabla · S. Duda · P. Zielonka
Faculty of Mechanical Engineering, Wroclaw University of Science and Technology, Smoluchowskiego 25 St., 50-370 Wroclaw, PL, Poland
e-mail: Grzegorz.Lesiuk@pwr.edu.pl

G. Lesiuk et al. (eds.), *Structural Integrity and Fatigue Failure Analysis*, Structural Integrity 25, https://doi.org/10.1007/978-3-030-91847-7_19

- noncircular bars—increasing friction, adhesion, and mechanical interlock.

The analysis of already patent solutions arrives at the following conclusions. The manufacturing capabilities are a limiting factor and the fact that reinforcing with composite materials should be in the geometry of rods, i.e., part with one dominating dimension.

In the paper submitted by Kadioglu et al. [2] the rods with various geometry subjected to axial, bending, and torsional loadings were analyzed. Considering rebars the crucial aspect plays bending and pull-out resistance. The authors of mentioned work were carried out rods with a round cross-section with additional wraps in the shape of ribs and a square cross-section.

In this work numerical analysis using the Finite Element Method is conducted to find the tailored geometry of rods for concrete reinforcing application. The bending test is simulating as a reference to choose the most suitable design for this application. Moreover, to obtain the most suitable mechanical properties of investigated laminate, several hypotheses available in the literature are compared.

19.2 Materials and Methods

The following research assumption was made. The constant cross-sectional area between the analyzed proposals. Research scope—two variations of material:

- Glass Fiber Reinforced Polymer (GFRP)
- Carbon Fiber Reinforced Polymer (CFRP)

Four variations of cross-section were analyzed (Figs. 19.1, 19.2, 19.3 and 19.4).

Table 19.1 shows material properties of matrix—epoxy resin CR144 with proper hardener and accelerator [3].

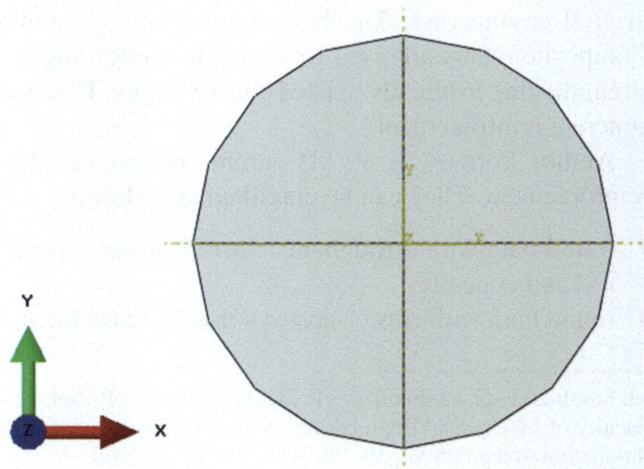

Fig. 19.1 Bar with a round cross-section

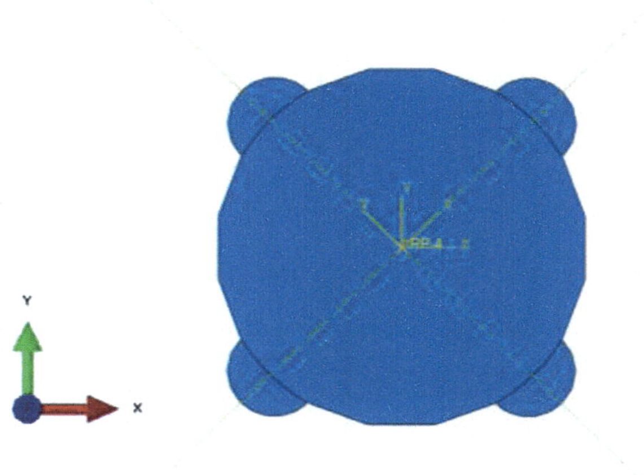

Fig. 19.2 Bar with round cross-section and ribs

Fig. 19.3 Circular rebar with helical wrap

Fig. 19.4 Bar with ring cross-section

Table 19.1 Epoxy resin properties

E (MPa)	2750
$\nu\,(-)$	0.35
G (MPa)	1018.5 (calculated)
$\rho(\frac{g}{cm^3})$	1.21

Table 19.2 Mechanical properties of investigated fibers

	E_1	E_2	E_3	ν_{12}	ν_{13}	ν_{23}	G_1	G_2	G_3
GF		73 GPa			0.2			30.4 GPa	
CF	24.2 GPa		22 GPa	0.2		0.25	91.7 GPa		8.8 GPa

GF—glass fibers, *CF*—carbon fibers

In this paper, two variations of fibers such as glass fiber (roving 4800 tex John Manville) and carbon fiber (roving 3700 tex 50 k) were taken into account. The material properties were implemented according to the material datasheets provided by manufacturers, in case of lacking parameters the data from literature were imported. Carbon fibers are an anisotropy material thus to define their properties three material parameters such as Young Modulus, Poisson ratio, and Shear modulus are necessary for all three perpendiculars to each other's direction. Unidirectional carbon fibers can be considered as transverse isotropic material thus it can be assumed that $E_2 = E_3$; $\upsilon_{12} = \upsilon_{13}$; $G_1 = G_2$. Value of G_3 is known as $\frac{E_2}{2}(1 + \upsilon_{23})$. In the case of glass fibers, they are considered as isotropy material due to their amorphic structure [4]—in that only Young modulus, Poisson ratio, and Shear modulus are needed [5–7]. Material properties of mentioned fibers are shown in Table 19.2.

For composite material, except for material properties of sub materials (i.e., matrix and reinforcement), crucial influence has a volume fraction of fibers in the composite. Material properties of the composite were estimated according to the Representative Elementary Volume (REV). The results from the numerical simulation are presented below as a table of major parameters (Table 19.9.3).

Table 19.3 Material properties for various materials and volume fractions of fibers according to the FEA

	E_1 (MPa)	E_2 (MPa)	ν_{12} (−)	ν_{23} (−)	G_{12} (MPa)	G_{23} (MPa)
Carbon 60%	146,310	7876.8	0.2853	0.4104	2909.4	2792.8
Carbon 70%	170,230	9768.1	0.2759	0.3761	3646.2	3549.8
Carbon 80%	194,150	12,415	0.2665	0.3413	4774.7	4629
Glass 60%	44,918	9946.4	0.2514	0.4058	3645.6	3538.3
Glass 70%	51,939	13,729	0.2373	0.3654	4979.8	5028.9
Glass 80%	58,960	20,672	0.2232	0.3209	7533	7827

There are various hypotheses available in literature allow estimating composite properties according to the characteristic of matrix and fibers. The most popular in used is the rule of mixture, Puck, Halpin–Tsai, Abolish, and Chamisa. Calculations for these rules are presented in the Tables 19.4, 19.5, 19.6, 19.7, 19.8 and 19.9.

To compare individual conceptions of the rod's geometry the bending test was simulated. The forcing in the form of rotation of the plane of the cross-section of the rod was assumed at the level of 0.262 radians, which corresponds to about 15°. This forcing is hard to accomplish during an experiment, but it provides pure bending (only a small influence of different stress around supports). This solution better reflects bending conditions than for instance boundary conditions for a real experiment.

As it is shown in Fig. 19.5 in the case of simulation with boundary conditions reflect four-point bending it is reflected in a proper way in its central part. The

Table 19.4 Material properties made from epoxy resin CR144 and 60% volume fraction of glass fibers according to the various hypothesis

Glass 60%	E_1 (MPa)	E_2 (MPa)	ν (−)	G (MPa)
RoM	44,900	6507.3	0.26	2424.5
Puck		11,902.5		4682.1
Halpin–Tsai		8144.5		3622.9
Abolish		7297.5		
Chamisa		10,802		4052.3

Table 19.5 Material properties made from epoxy resin CR144 and 70% volume fraction of glass fibers according to the various hypothesis

Glass 70%	E_1 (MPa)	E_2 (MPa)	ν (−)	G (MPa)
RoM	51,925	8426	0.245	3149
Puck		17,610		6867.2
Halpin–Tsai		10,799.5		4879.8
Abolish		9426.1		
Chamisa		14,112.8		5322.6

Table 19.6 Material properties made from epoxy resin CR144 and 80% volume fraction of glass fibers according to the various hypothesis

Glass 80%	E_1 (MPa)	E_2 (MPa)	ν (−)	G (MPa)
RoM	58,950	11,949	0.23	4491
Puck		28,786.3		11,177.2
Halpin–Tsai		15,505.2		7070
Abolish		13,291.5		
Chamisa		19,746.2		7515.4

Table 19.7 Material properties made from epoxy resin CR144 and 60% volume fraction of carbon fibers according to the various hypothesis

Carbon 60%	E_1 (MPa)	E_2 (MPa)	ν (−)	G (MPa)
RoM	146,300	5789.5	0.26	2182.5
Puck		10,141.5		4024.3
Halpin–Tsai		6779.1		2898.8
Abolish		6307.3		
Chamisa		8534.3		3270

Table 19.8 Material properties made from epoxy resin CR144 and 70% volume fraction of carbon fibers according to the various hypothesis

Carbon 70%	E_1 (MPa)	E_2 (MPa)	ν (−)	G (MPa)
RoM	170,225	8534	0.245	3270
Puck		13,797.4		5388
Halpin–Tsai		8364.6		3611
Abolish		7693.9		
Chamisa		8534.3		3270

Table 19.9 Material properties made from epoxy resin CR144 and 80% volume fraction of carbon fibers according to the various hypothesis

Carbon 80%	E_1 (MPa)	E_2 (MPa)	ν (−)	G (MPa)
RoM	194,150	9166.7	0.23	3525.6
Puck		19,534.1		7441.7
Halpin–Tsai		10,715.5		4639.8
Abolish		9838.5		
Chamisa		12,650.9		4969.3

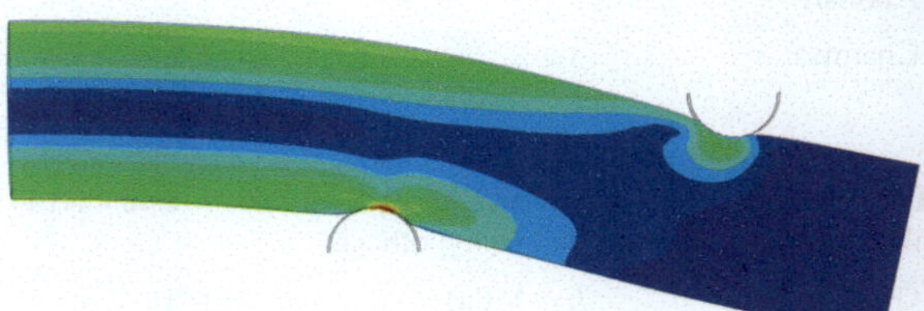

Fig. 19.5 Exemplary field of reduced stresses according to the Huber-Misses for round bar subjected to bending loading reflects four-point bending (only half of the assembly)

influence of the local effects in the area near to the support is difficult to interpretation especially with the lack of composite material failure parameters.

Applied boundary conditions in further analyses are presented in Fig. 19.6.

One end of the bar is fixed in three directions. The second one is connected by a reference point. Then this point is rotated 0.262 radians (approximately 15°).

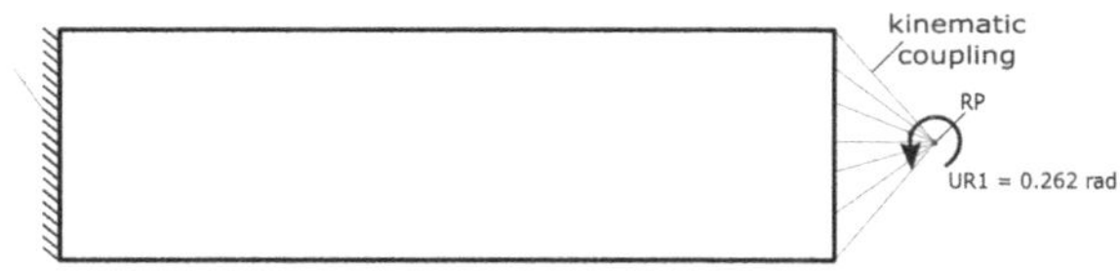

Fig. 19.6 Boundary conditions applied for the simulations (left edge is encastred)

The geometric model was prepared as a 3D solid model. All simulations were conducted in static terms.

For most models, 8-node higher-order elements with hourglass control with reduced integration (Hexa type C3D8R) were used. Only in the case of a bar with a coil, due to the complicated geometry, it was decided to use 10-node tetra elements (C3D10).

19.3 Results and Discussion

The results concerning the determination of material data based on hypotheses and simulation by the finite element method were presented in the previous chapter, as they formed the basis for further work in the field of modeling.

In the case of simple bending analyses, the reaction force needed to bend a 200 mm long section of a bar to an angle of 0.262 radians, i.e., approximately 15°, was determined. Due to the arbitrary selection of these parameters and the assumption as to the linear-elastic material model, the results were presented concerning the reference value for which a rod with a circular cross-section made of CR144 resin reinforced with carbon fiber with a volume of 70% of the total was adopted (Tables 19.10, 19.11, 19.12 and 19.13).

Table 19.10 The momentum needed to deflect round bar in terms of applied fibers and their volume fraction regarding the reference value (round bar, CFRP, 70% fiber content)

Fiber content (%)	CFRP (%)	GFRP (%)
60	86	26
70	100	31
80	114	35

Table 19.11 The momentum needed to deflect round bar with helical wrap in terms of applied fibers and their volume fraction regarding the reference value (round bar, CFRP, 70% fiber content)

Fiber content (%)	CFRP (%)	GFRP (%)
60	87	26
70	101	31
80	115	35

Table 19.12 The momentum needed to deflect ring cross-sectional bar in terms of applied fibers and their volume fraction regarding the reference value (round bar, CFRP, 70% fiber content)

Fiber content (%)	CFRP (%)	GFRP (%)
60	195	60
70	227	70
80	259	79

Table 19.13 The momentum needed to deflect round bar with ribs in terms of applied fibers and their volume fraction regarding the reference value (round bar, CFRP, 70% fiber content)

Fiber content (%)	CFRP (%)	GFRP (%)
60	87	32
70	101	32
80	117	36

As pure bending was simulated—as predicted, the stress maps show the compression part, the neutral axis, and the tension part. The following are exemplary stress maps either in the principal direction or Huber-Misses for simulations carried out with a glass fiber reinforced CR144 resin material (80% by volume) for a bent of about 1.5°. The drawings show a bar with a circular cross-section Fig. 19.7, with

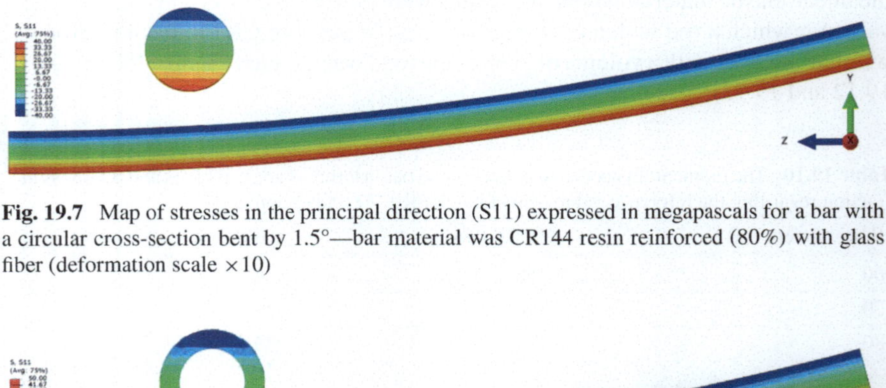

Fig. 19.7 Map of stresses in the principal direction (S11) expressed in megapascals for a bar with a circular cross-section bent by 1.5°—bar material was CR144 resin reinforced (80%) with glass fiber (deformation scale ×10)

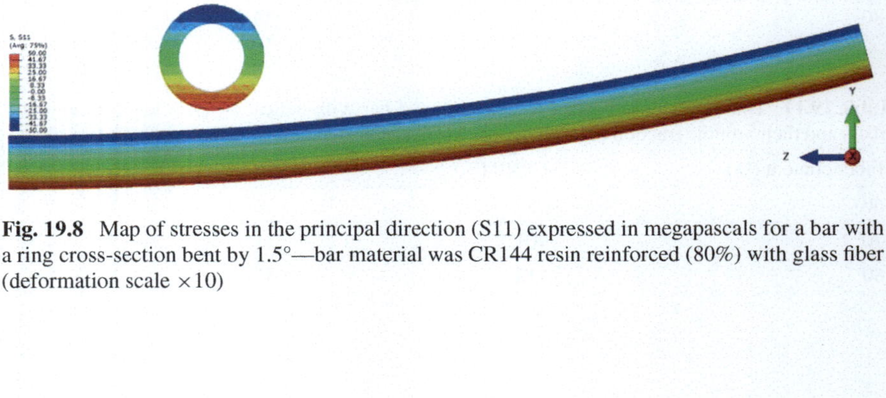

Fig. 19.8 Map of stresses in the principal direction (S11) expressed in megapascals for a bar with a ring cross-section bent by 1.5°—bar material was CR144 resin reinforced (80%) with glass fiber (deformation scale ×10)

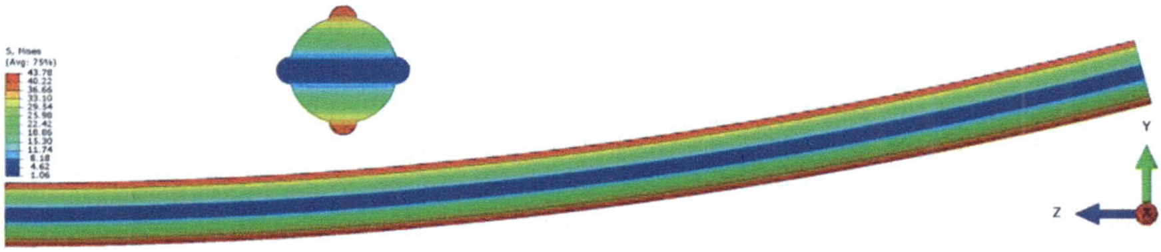

Fig. 19.9 Map of the Huber-Misses stresses expressed in megapascals for a bar with a circular cross-section with protrusions bent by 1.5°—bar material was CR144 resin reinforced (80%) with glass fiber (deformation scale ×10)

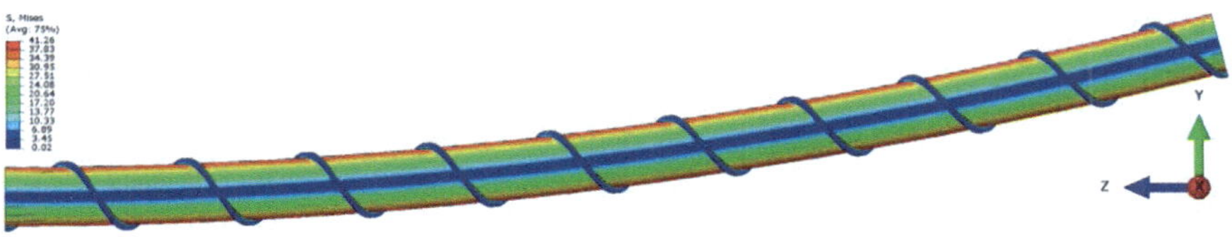

Fig. 19.10 Map of Huber-Misses stresses expressed in megapascals for a bar with a circular cross-section with a coil bent by 1.5°—bar material was CR144 resin reinforced (80%) with glass fiber (deformation scale ×10)

a cross-section of a ring Fig. 19.8, with a circular cross-section with protrusions Fig. 19.9, with a circular cross-section with a coil Fig. 19.10.

According to the theory—taking into account only the bending of the bar—profiles with a greater moment of inertia of the cross-section are preferred, which means profiles with larger overall dimensions—such as, for example, a ring-type cross-section. Additional elements enriching the cross-section, such as a coil, only slightly transfer the load.

19.4 Conclusions

- Using the finite element method, the material data of a composite with a matrix of three-component resin CR144 with carbon and glass fibers for three different proportions of fiber—resin—60%, 70%, and 80% were determined based on the unit cell.
- The data determined in this way were compared with theoretical hypotheses: the rules of mixes, Puck, Halpin–Tsai, Abolish, and Chamisa. In terms of the proportions in which the micromechanical materials were analyzed, the most reliable are the Abolish hypothesis in the field of Kirchhoff's modulus and Chamisa's hypothesis in the field of Young's transverse modulus. They are a good approximation in the range of 60 to 80% fiber volume fraction of the composite if desired.

- For the proposed bar shapes, pure bending and four-point bending were simulated. Four-point bending introduced additional problems related to the destruction of the place of application of boundary conditions and introduced instability of the simulation; therefore, it was decided to use the forced rotation—such boundary conditions ensure pure bending conditions. The influence of shape, material, and fiber content was compared. Simulations showed that profiles with greater moment of inertia are desirable. Furthermore, additional wrapping such as helical only slightly transfers the load.

Acknowledgements The research was financed by the National Centre for Research and Development of Poland (NCBiR) grant number LIDER/40/0219/L-10/18/NCBR/2019.

References

1. Portnov G, Bakis CE, Lackey E, Kulakov V (2013) FRP Reinforcing bars-designs and methods of manufacture (review of patents)
2. Kadioglu F, Pidaparti RM (2004) Composite rebars shape effect in reinforced structures. https://doi.org/10.1016/j.compstruct.2004.01.006
3. Karta materiałowa Biresin CR144/CA141
4. Mounier D, Poilâne C, Bûcher C, Picart P (2012) Evaluation of transverse elastic properties of fibers used in composite materials by laser resonant ultrasound spectroscopy
5. Kv P (2018) KV42 High performance assembled roving series 1–3
6. Kaw AK (2006) Mechanics of composite materials second edition
7. Sudheer M, Somayaji SKRP (2015) Analytical and numerical validation of epoxy/glass structural composites for elastic models. Am J Mater Sci 5:162–168. https://doi.org/10.5923/c.materials.201502.32

Chapter 20
Numerical Analysis of Pultrusion Process of GFRP Composite Rebar

Michał Smolnicki, Paweł Stabla, Szymon Duda, and Paweł Zielonka

Abstract The manufacturing process of composite pultruded rebars requires consideration of various aspects. In this work, the heat transfer at the pultrusion die was investigated. The aim was to analyze the heat transfer from the die to the GFRP composite rebar. The 3D numerical model was prepared to simulate the pulling process of the rebar through the heated die. As a result, detailed heat distribution in the composite was obtained to state, whether the complete curing process may be obtained or not. The numerical models provide the possibility to vary the pulling speed and heating temperature according to the used material composition. The presented approach can be used in the optimization process of heaters placement and their power or heating cycles.

Keywords Pultrusion process · Finite-element method · Composite rebars

20.1 Introduction

Composite materials are widely used structures not only in performing in sometimes extreme conditions structures such as spacecraft, high-pressure composite vessels, sports industry, but also in civil engineering applications. One of the main developing branch in civil engineering in respect to composite materials is composite rebars. Composite rebars exhibit various advantages including high specific strength and the possibility to suit the performance to the specific application by applying any fiber-matrix set. Thanks to the rapid growth of the technology optimization processes, the price of the composite rebars became competitive.

Composite pultruded elements are subjected to different mechanical loadings. In [1] authors investigated the mechanical response of composite rebars on radial compression and bending, both in static and fatigue conditions. On a microscopic scale, the damage and failure of the rebars were indicated. In [2] also a radial compression test was performed. Thanks to the use of the digital image correlation technique

M. Smolnicki (✉) · P. Stabla · S. Duda · P. Zielonka
Wrocław University of Science and Technology, Wrocław, Poland
e-mail: michal.smolnicki@pwr.edu.pl

G. Lesiuk et al. (eds.), *Structural Integrity and Fatigue Failure Analysis*,
Structural Integrity 25, https://doi.org/10.1007/978-3-030-91847-7_20

it was possible to compare the material behavior for the experiment with the finite-element method analysis. Additionally, acoustic emission was involved to better understand the damage mechanisms occurring during the test.

An important issue in the manufacturing process is to properly set the temperature of the die. Many kinds of research were conducted to find the optimal strategy for the specific composite rebar. In [3] the researchers performed a 2D simulation of the temperature and curing profiles of the pultruded rods. Thanks to the combination of the finite-element method with the numerical approximations they found satisfactory results of the profiles of centerline temperature and degree of cure or carbon/epoxy composite rods. In [4] a different approach was taken to obtain the optimal manufacturing parameters. The application of genetic algorithms and simplex method let the author optimize the accuracy of the manufacturing process. Different heating systems of the pultrusion die were investigated in [5, 6]. To determine the amount of energy consumption of the heating system a finite-element analysis was performed. A simulation of "U" composite profile made of glass/polyester resin was conducted and the results were compared with the thermographic profile. Different heating cycles were compared and the optimization led to the decrease of the energy consumption up to 57%. Another study [7] concerned the design process of the new "U" profile pultrusion die. A comprehensive study regarding heat transfer, energy consumption optimization, the material design was conducted to properly design and produce a new product. Moreover, a mechanical study was performed on the produced profiles to confront the application requirements.

In this study, a numerical analysis of heat transfer from the pultrusion die to the composite rebar is conducted.

20.2 Materials and Methods

20.2.1 Pultrusion Die

In the current study, finite-element method analysis is performed to investigate the heat transfer through the existing pultrusion die. The die is shown in Fig. 20.1.

Fig. 20.1 Pultrusion die with the heating sources mounted

Fig. 20.2 3D model of pultrusion die with surface heaters mounted

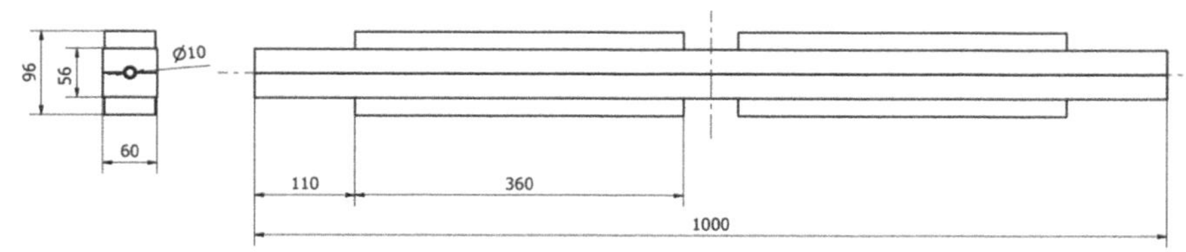

Fig. 20.3 The general dimension of pultrusion die with surface heaters mounted

After accurate measurements, a 3D model was created and is shown in Fig. 20.2. The dimensions of the die with heaters are shown in Fig. 20.3.

20.2.2 Materials

The simulation of the pultrusion process demands a certain number of material data coefficients. The material used in the simulation is glass fiber reinforced plastic and the material data are presented in Table 20.1 [8].

Some of the parameters were calculated according to the following equations:

$$\rho_c = \theta_f \rho_f + \theta_r \rho_r. \tag{20.1}$$

$$c_c \rho_c = \theta_f \rho_f c_f + \theta_r \rho_r c_r. \tag{20.2}$$

$$\frac{1}{k_c} = \theta_f \frac{1}{k_f} + \theta_r \frac{1}{k_r} \tag{20.3}$$

Table 20.1 Properties of used materials [8]

	Material data	Value	Unit
Steel			
k_s	Thermal conductivity	34.5	W/(m*K)
c_s	Specific heat	490	J/(kg*K)
ρ_s	Density	7890	kg/m^3
GFRP			
θ_f	Fiber volume fraction	0.65	–
θ_r	Resin volume fraction	0.35	–
ρ_r	Resin density	1150	kg/m^3
ρ_f	Fiber density	2560	kg/m^3
ρ_c	Composite density	2066	kg/m^3
k_f	Resin thermal conductivity	0.169	W/(m*K)
k_f	Fiber thermal conductivity	1.04	W/(m*K)
k_c	Composite thermal conductivity	0.371	W/(m*K)
c_r	Resin specific heat capacity	1640	J/(kg*K)
c_f	Fiber specific heat capacity	640	J/(kg*K)
c_c	Composite specific heat	835	J/(kg*K)

20.2.3 Finite-Element Method Simulations

The temperature is one of the most important parameters in the pultrusion process. Measurement of the temperature on the surface of the pultrusion die or even inside the die using thermocouple or pyrometers is fairly easy. However, t most important is the temperature inside the rod, which cannot be easily measured. This is the reason to use numerical methods. In this research numerical simulations of heat flow are conducted using the finite-element method. Simulia Abaqus environment was used in version 6.14. The geometric model was based on a prepared 3D model and simplified according to art, i.e., without screw holes. The basic geometry of form modeled as a solid is presented below in Fig. 20.4. The pultruded rod is modeled as a solid cylinder—diameter 10 mm and the length of 4 m.

Material model is based on discussed above material data of steel and glass–epoxy composite. In the case of both materials conductivity, specific heat is specified. Values were chosen to be self-consistent in terms of units (Abaqus has not predefined unit system). Thus mm-*t*-*N*-*s*-*K* system was used in the analysis.

The main purpose of the simulation is to analyze the heat flow inside the die and the rod. However, the movement of the pultruded rod must be somehow realized. Because of that, it is not possible to use simple heat analysis. Instead, coupled thermal-displacement module was used. The most important part of such simulation is to define boundary conditions. The die is fixed and the rod has been taken off 5° of freedom and enforced movement in the x-axis direction—4000 mm. The initial temperature is set for the rod as 20 °C and the die as 250 °C. It represents the initial

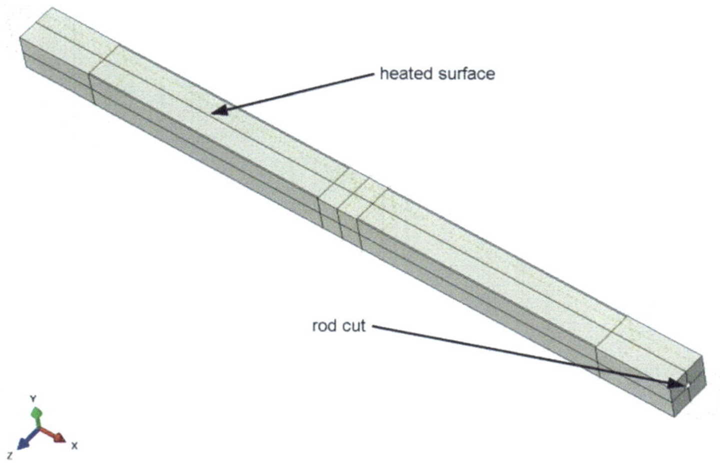

Fig. 20.4 Geometry model of pultrusion die

preheating of the die to the steady-state for the die and the room temperature for the rod material (fibers and epoxy). The initial situation is shown in Fig. 20.5. Thermal conductivity between the die and pultruded rod is set to 3000 $\frac{\mathrm{W}}{\mathrm{m}^2\mathrm{K}}$. Additionally, the heat exchange between the die and the air is represented through convection—defined with a coefficient 0.05. The heating of the die is done by using surface heat flux, which reflects the real heating process.

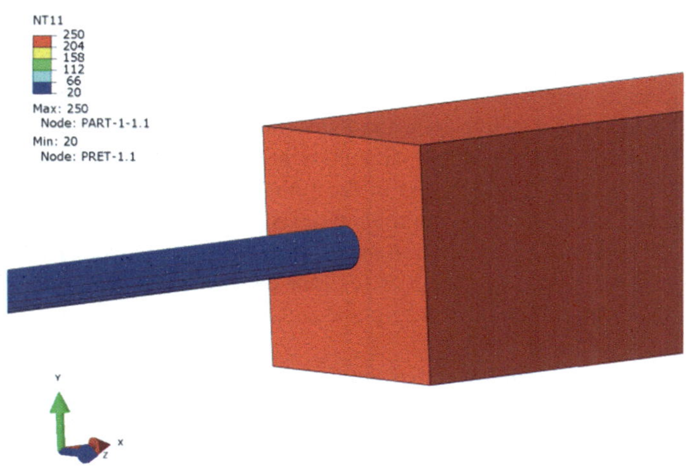

Fig. 20.5 Initial temperature of the rod and die

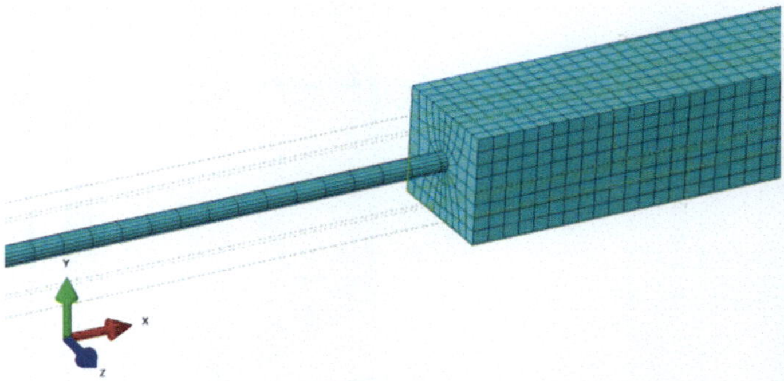

Fig. 20.6 Finite model of the assembly of the die and pultruded rod

In the coupled thermal-displacement simulation special elements are used—an 8-node thermally coupled brick, trilinear displacement and temperature, reduced integration, hourglass control (along with description of Abaqus). Total of over 27,000 elements including 19,128 elements in the die and 8544 elements in the rod were used in the finite model. Finite-element model is presented in Fig. 20.6 below.

The simulation time is set to 240 s, so the velocity of pulling rod through pultrusion die is equal to 1 m/min which is a technological parameter established in the laboratory. The output is set to be taken every 2 s of the process—so at least 120 results set is gathered through the transient process.

20.3 Results and Discussion

In the conducted numerical simulations varied temperature parameters were obtained. The most informative ones are distributions of the temperature inside the die and in the rod. Other gathered parameters are for example heat flux, reaction flux, etc. These values may be used in the future to improve analysis. Below in Fig. 20.7 stress distribution in the pultrusion die is presented. Three fourth of the die is presented to better show distribution inside. Distribution alongside paths (parallel to the x-axis) is presented in Fig. 20.8 for four cases: corner of the die, the outer surface of the die (intersection with *yz*-plane and independently with *xy*-plane), and the inner hole surface.

In the case of the rod, this is important to achieve a high enough temperature in the interior part. This is necessary to the obtained fully cured epoxy matrix of the composite. In Fig. 20.9 the distribution of temperature in the initial phase of the pultrusion process is presented. In Fig. 20.10 whole distribution alongside the x-axis is presented for both center and outer part of the rod.

Fig. 20.7 Stress distribution in pultrusion die

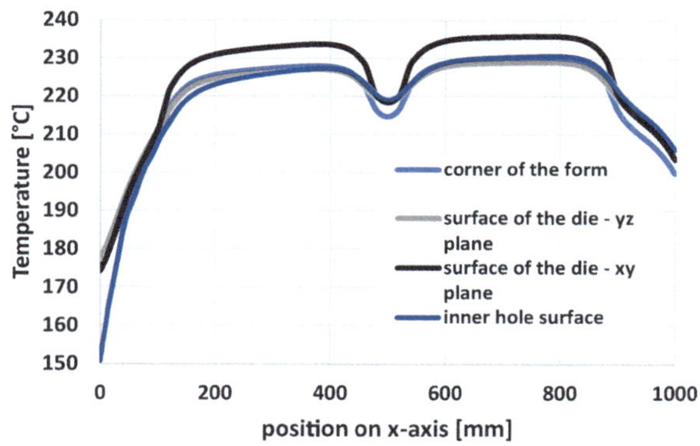

Fig. 20.8 Temperature distribution alongside pultrusion die for different parts (corners, outside surface, inside surface)

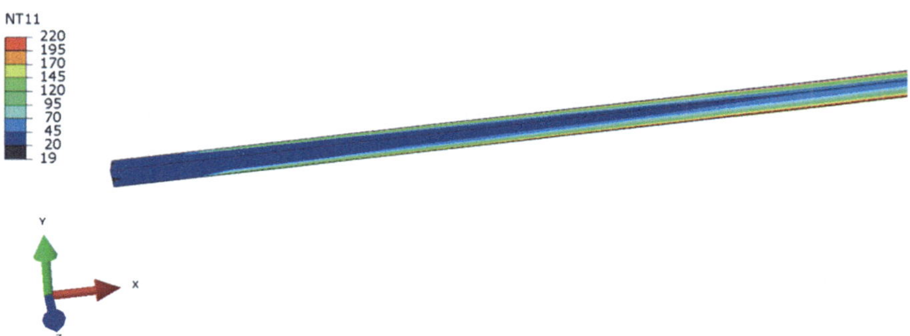

Fig. 20.9 Distribution of temperature in the initial stage of rod inside the form (a quarter of rod is shown)

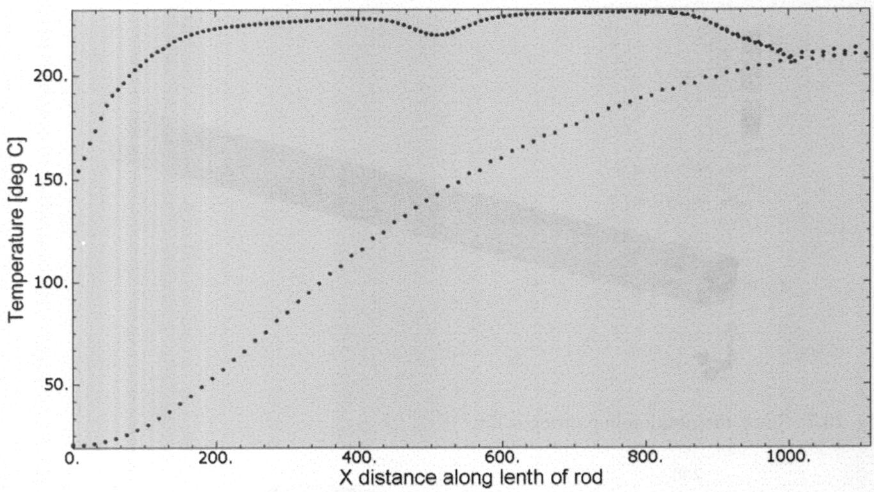

Fig. 20.10 Temperature distribution in pultruded rod—outer surface (purple dots) and inside (blue dots)

20.4 Conclusions

A three-dimensional model of the die (from the pultrusion line) was created during research and then a numerical simulation of the pultrusion process was conducted using a simplified model. The analysis was complex and includes both the die and the rod during the pultrusion process. Full heat distribution of temperature was obtained for both die and the rod. The presented in this paper approach can be used further to optimize the heaters placement (and their number) as well as other parameters like pultrusion process speed, initial die temperature, initial epoxy, and fiber temperature or form material.

Acknowledgements The research was financed by the National Centre for Research and Development of Poland (NCBiR) grant number LIDER/40/0219/L-10/18/NCBR/2019. Calculations have been carried out in Wroclaw Centre for Networking and Supercomputing (http://www.wcss.pl), grant No. 27220656.

References

1. Barcikowski M, Lesiuk G, Czechowski K, Duda S (2021) Static and flexural fatigue behavior of gfrp pultruded rebars. Materials (Basel) 14:1–18. https://doi.org/10.3390/ma14020297
2. Kar NK, Hu Y, Ahn B, Nutt SR (2012) Diametral compression of pultruded composite rods. Compos Sci Technol 72:1283–1290. https://doi.org/10.1016/j.compscitech.2012.05.001

3. Suratno BR, Ye L, Mai YW (1998) Simulation of temperature and curing profiles in pultruded composite rods. Compos Sci Technol 58:191–197. https://doi.org/10.1016/S0266-3538(97)00132-2
4. Carlone P, Palazzo GS, Pasquino R (2007) Pultrusion manufacturing process development: cure optimization by hybrid computational methods. Comput Math Appl 53:1464–1471. https://doi.org/10.1016/j.camwa.2006.02.031
5. Silva FJG, Ferreira F, Costa C, Ribeiro MCS, Meira Castro AC (2012) Comparative study about heating systems for pultrusion process. Compos Part B Eng 43:1823–1829. https://doi.org/10.1016/j.compositesb.2012.01.057
6. Silva FJG, Ferreira F, Ribeiro MCS, Castro ACM, Castro MRA, Dinis ML, Fiúza A (2014) Optimising the energy consumption on pultrusion process. Compos Part B Eng 57:13–20. https://doi.org/10.1016/j.compositesb.2013.09.035
7. Melo D, Nunes JP, Vasconcelos M (2015) Development of a new U 200 pultrusion die to produce GRP composite profiles using different thermosetting resins. Ciência Tecnol dos Mater 27:1–80
8. de Cassia Costa Dias R, Costa ML, de Sousa Santos L, Schledjewski R (2020) Kinetic parameter estimation and simulation of pultrusion process of an epoxy-glass fiber system. Thermochim Acta 690:178636. https://doi.org/10.1016/j.tca.2020.178636

Chapter 21
Fracture Characterization of Hybrid Bonded Joints (CFRP/Steel) for Pure Mode I

Rita Dantas, Anis Mohabeddine, Marcelo de Moura, Raul Moreira, Grzegorz Lesiuk, José Correia, and Abílio de Jesus

Abstract In Europe, several bridges, as Luiz I bridge in Porto, are at the end of their service life, which is pretended to extend due to cultural value of these structures. The application of carbon reinforced polymers (CFRP) in certain regions of the structures under maintenance is a widely used methodology of strengthening. Thus, this work aims to characterize the fracture mechanics properties for pure mode I of hybrid joints, which combine CFRP with steel by means of a structural adhesive. Therefore, asymmetric double cantilever beam (ADCB) type of experimental test was performed to assess pure mode I of fracture. Besides, a numerical study based on a cohesive zone model was also performed, in order to design and evaluate the experimental results.

Keywords Fracture · S355 steel · CFRP · Cohesive models · Mode I

21.1 State of the Art and Methodology

21.1.1 Introduction and Background

There is a large number of steel bridges around Europe which were built in the beginning of the twentieth century and are reaching their life expectation. In Portugal, it is easy to find a couple of examples such as the Luiz I Bridge in Porto or the Eiffel Bridge in Viana do Castelo. Since it is desired to maintain these structures and even extend their service life, the interest around different methodologies of maintenance and strengthening of steel bridges increased and led to several scientific works about this topic [1, 2]. One of the most popular methodologies of strengthening is the application of Carbon Fibre Reinforced Polymers (CFRP) locally in strategic regions of the structures [3–5].

R. Dantas (✉) · A. Mohabeddine · M. de Moura · R. Moreira · J. Correia · A. de Jesus
University of Porto, 4200-465 Porto, Portugal
e-mail: rdantas@inegi.up.pt

G. Lesiuk
Wroclaw University of Science and Technology, 50370 Wrocław, Poland

G. Lesiuk et al. (eds.), *Structural Integrity and Fatigue Failure Analysis*, Structural Integrity 25, https://doi.org/10.1007/978-3-030-91847-7_21

The CFRP are materials of high ultimate strength, low density, low weight, and good resistance to fatigue, which are important characteristics for the rehabilitation of these structures [5]. Thus, these materials increase the stiffness and the strength of steel structural details under repair. Furthermore, since the CFRP patches are connected through a structural adhesive, this strengthening method does not introduce a defect or a critical zone of stress concentration, which can negatively affect the mechanical and fatigue properties, as happens in the case of welding or drilling [3, 4].

Moreover, many scientific studies show that the CFRP application can increase fatigue life, mainly in cases of CFRP with high modulus, and delay fatigue crack propagation. This extension of fatigue life is observed due to the additional stiffness of CFRP that reduces the stress range near the crack tip and the crack opening displacement, which promotes the crack closure [4].

However, it is important to highlight that the several advantages mentioned above and the CFRP strengthening performance will depend on several variables, such as surface preparation and mechanical properties of the CFRP [3]. Besides, usually the weakest point of this method is the interface between the structural detail and the applied CFRP, so the adhesive properties are also important as well as the bonding process [6].

Therefore, this work intends to characterize the static fracture behavior of hybrid joints of CFRP and S355 structural steel. The main focus is pure mode I of fracture mechanics, so Asymmetric Double Cantilever Beam (ADCB) tests were carried out. The experimental campaign, in particular the specimens' geometries, was defined based on numerical studies. Until now, only one structural adhesive was tested, but in the future is intended to evaluate two more adhesives, with the aim of studying the influence of this element on the reinforcement performance.

21.1.2 Materials: Steel, Adhesive, and CFRP

As stated in the previous section, this work aims to evaluate, experimentally, the fracture mechanics properties of hybrid joints, which combine steel and CFRP bonded with a structural adhesive.

Thus, S355 structural steel was selected for the hybrid joint under analysis. This material is a mild steel commonly present on structural elements of bridges and the main mechanical properties of this material, according to Eurocode 3, are listed in Table 21.1 [7].

Regarding the composite substrate of the hybrid joint, it was chosen the S&P C-Laminate HM [8], which is a laminate prefabricated of carbon fibers and resign epoxy, obtained from pultrusion. This material can be applied to reinforce structures made of wood, steel, or concrete. Some of the main mechanical properties of the S&P C-Laminate HM are listed in the Table 21.2.

The CFRP will be bonded to the S355 steel substrate with a structural adhesive fabricated by HUNTSMAN company: Araldite AW4858/Hardener HW4858 [9].

Table 21.1 Main properties of S355 steel [7]

Property	S355 steel
Yield strength (MPa)	355
Ultimate tensile strength (MPa)	510
Poisson's coefficient	0.3
Young's modulus (GPa)	210
Shear modulus (GPa)	81

Table 21.2 Main properties of S&P C-laminate HM [8]

Property	S&P C-laminate HM
Density (g/cm^3)	1.6
Volume of fibers (%)	68
Tensile strength (MPa)	≥2800
Young's modulus (GPa)	≥205
Elongation (%)	≥13.5
Cross-section (mm)	50 × 1.4

Table 21.3 Main properties of araldite AW4858/hardener HW4858 [9]

Property	Araldite AW4858/hardener HW4858
Specific gravity	1.1
Viscosity	Thixotropic (at 25 °C)
Lap shear strength at 23 °C (MPa)	≥28
Tensile strength at 23 °C (MPa)	31
Tensile modulus (MPa)	1600
Elongation at break at 23 °C (%)	7
Flexure strength at 23 °C (MPa)	31
Flexure modulus at 23 °C (MPa)	1650
Shear modulus at 23 °C (MPa)	800

This adhesive is a two-component room temperature curing epoxy paste of high strength and toughness. Furthermore, it is an adhesive extremely resilient, has a long pot life, bonds a wide variety of materials, presents a good moisture resistance, and shows a very high lap shear and peel strength. The main properties of this adhesive can be checked in Table 21.3.

21.1.3 Fracture Mechanics on Hybrid Joints

Pure mode I characterization: asymmetric double cantilever beam test (ADCB)

The double cantilever beam is a type of test frequently used to measure mode I fracture energy (G_{IC}). For hybrid joints, this test can be performed on a specimen formed by two arms glued and with a pre-crack induced of a defined length, as it is portrayed in Fig. 21.1. During this test, a tensile load is applied to the specimen upper arm at the side of the pre-crack, leading to the crack propagation in mode I along the adhesive layer plane.

Since in the case under study the upper and lower arms are of different materials (steel and CFRP), and, consequently, the thickness of each arm probably will be different to ensure mode I, this test is called asymmetric double cantilever beam (ADCB) [10]. The thickness of each arm was defined taking into account the flexure stiffness and strain-based criteria, which will be discussed in the following sections [11, 12].

In order to obtain the G_{IC} from the results of this type of test, it is necessary to define an analytical formulation for the strain energy release rate observed during the experimental test as well as a resistance curve, which plots this variable as function of the equivalent crack length. This formulation is based on the Irwin-Kies relation [Eq. (21.1)], concept of compliance [Eq. (21.2)], and Castigliano theorem [Eq. (21.3)], which are summarized by the equations bellow, respectively [10]:

$$G_T = \frac{P^2}{2B}\frac{\mathrm{d}C}{\mathrm{d}a} \tag{21.1}$$

$$C = \frac{\delta}{P} \tag{21.2}$$

$$\delta = \frac{\mathrm{d}U}{\mathrm{d}P} \tag{21.3}$$

where U is the strain energy, P is an applied load, δ is the displacement, C is the compliance, B is the thickness, and a is the crack length. Therefore, the strain energy of the specimen due to bending and shear stress, assuming a plane strain state, was

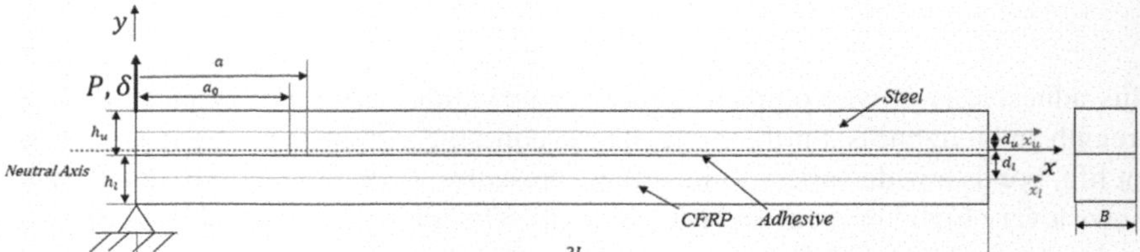

Fig. 21.1 Representation of ADCB test type

defined by the following equation:

$$U_s = U_{\text{Bending}} + U_{\text{Shear}} = \int_0^a \frac{(Px)^2}{2D_u}\mathrm{d}x + \int_0^a \frac{(Px)^2}{2D_l}\mathrm{d}x + \int_0^a \int_{-\frac{h_l}{2}}^{\frac{h_l}{2}} \frac{\tau_l^2 B}{2G_{13(l)}}\mathrm{d}y\mathrm{d}x + \int_0^a \int_{-\frac{h_l}{2}}^{\frac{h_l}{2}} \frac{\tau_u^2 B}{2G_{(u)}}\mathrm{d}y\mathrm{d}x \quad (21.4)$$

where $D_u = \frac{E_u I_u}{(1-\nu_{(u)}^2)}$ is the bending stiffness of the upper arm (steel), $D_l = \frac{E_l I_l}{(1-\nu_{13(u)}^2)}$ is the bending stiffness of the lower arm (CFRP), $G_{(u)}$ is the shear modulus of the upper arm, $G_{13(l)}$ is the shear modulus 13 of the lower arm, τ_u is the shear stress throughout the upper arm and τ_l is the shear stress throughout the lower arm.

The first two integrals represent the strain energy bending and by solving them it was achieved the following bending compliance:

$$C_{\text{Bending}} = \frac{a^3}{3}\left(\frac{1}{D_u} + \frac{1}{D_l}\right) \quad (21.5)$$

Regarding the shear's compliance, first of all, the shear stresses in the upper and lower arm were defined as follows:

$$\tau_u = \frac{3P}{2h_u}\left(1 - \frac{4y^2}{h_u^2}\right) \quad (21.6)$$

$$\tau_l = \frac{3P}{2h_l}\left(1 - \frac{4y^2}{h_l^2}\right) \quad (21.7)$$

Consequently, by integrating the last two terms of Eq. (21.4), the shear compliance was calculated:

$$C_{\text{Shear}} = \frac{6a}{5B}\left(\frac{1}{h_u G_{(u)}} + \frac{1}{h_l G_{13(l)}}\right) \quad (21.8)$$

Then, the total compliance of the ADCB specimen can be written as follows:

$$C = C_{\text{Bending}} + C_{\text{Shear}} = \frac{a_{\text{eq}}^3}{3}\left(\frac{1}{D_u} + \frac{1}{D_l}\right) + \frac{6a_{\text{eq}}}{5B}\left(\frac{1}{h_u G_{(u)}} + \frac{1}{h_l G_{13(l)}}\right) \quad (21.9)$$

Thus, the total compliance was combined with (Eqs. 21.1, 21.2, and 21.3), leading to the following total strain energy release rate definition:

$$G_T = \frac{P^2}{2B}\left(a_{\text{eq}}^2\left(\frac{1}{D_u} + \frac{1}{D_l}\right) + \frac{6}{5B}\left(\frac{1}{h_u G_{(u)}} + \frac{1}{h_l G_{13(l)}}\right)\right) \tag{21.10}$$

As can be observed in equations above, a new variable is introduced: the equivalent crack length (a_{eq}), which is obtained from the correction of the real value of crack length (a), to include the effect of Fracture Process Zone (FPZ). FPZ is the region behind the crack tip in which the material suffers a softening damage. Furthermore, this region influences the compliance and the strain energy release rate. Therefore, the formulation presented accounts for the effect of FPZ, by solving Eq. (21.9) for a_{eq} as function of the compliance, and then applying in Eq. (21.10) the values obtained. Moreover, this formulation has a great advantage in comparison to other approaches such as the Compliance Calibration Method, Direct Beam Theory, or Corrected Beam Theory; it does not require the measurement of crack length throughout the experimental test, since the equations rely only on the compliance, which is calculated from direct measurement of the load and displacements applied [13].

Cohesive Zone Models

During the last decades, the strength of adhesive joints as well as the crack and damage growth in their interfaces have been evaluated and simulated by combining cohesive zone models with finite element modeling. These approaches pretend to overcome the inability of linear elastic fracture mechanics (LEFM) to simulate the large process and plastic zone ahead of the crack tip, which is present in adhesives and can influence the crack initiation and propagation [14–16].

The Cohesive Zone Models are based on traction–separation laws which can be of different shapes, depending on the material or interface behavior that is pretended to simulate. The triangular and trapezoidal CZM shapes are the most widely used for adhesives, but there are also linear-parabolic, polynomial, and exponential laws which can be found in the literature (Figs. 21.2 and 21.3) [14, 15, 17].

Therefore, it is defined a cohesive zone composed by two cohesive surfaces, which are connected by a traction–separation law, that defines the relation between

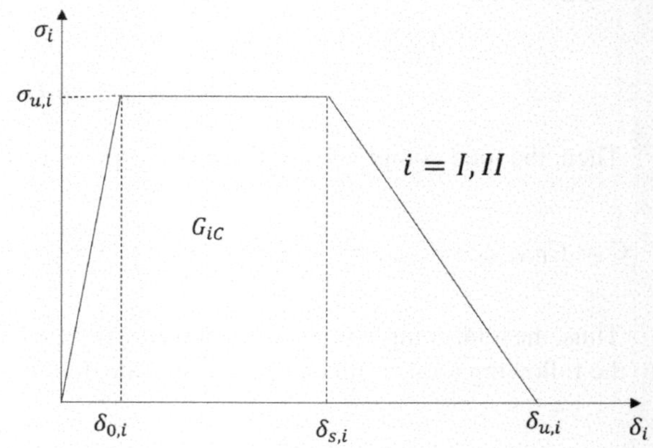

Fig. 21.2 Trapezoidal shape law

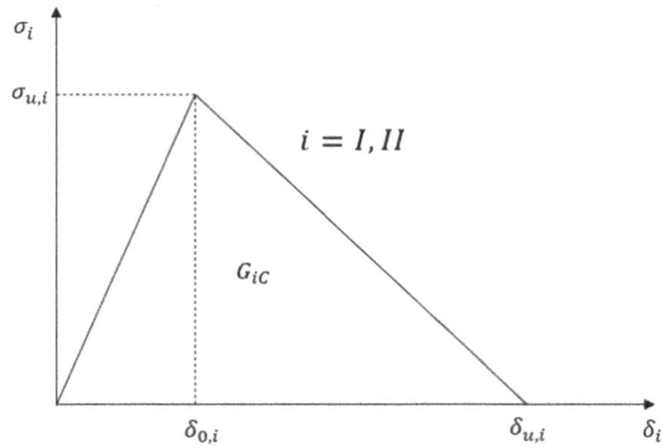

Fig. 21.3 Triangular shape law

the cohesive traction and the relative displacement of the two cohesive surfaces. Thus, the material fails when the cohesive surfaces, accordingly to the traction–separation law, are completely separated. This cohesive zone is modeled ahead of the crack tip in order to simulate the damage process zone and leads to the progressive crack opening when certain values of traction and relative displacement are achieved [18]. This technique implies the identification of a pre-defined crack path, since it is required the definition of cohesive surfaces or elements, in the case of finite elements modeling, in which will happen the crack growth and propagation [15].

These approaches can simulate not only the strength evolution of the adhesive, but also the softening and degradation of the material until the final failure. This progression of material properties is defined in the different traction–separation laws and is dependent of the fracture energy associated to material and type of failure. For each adhesive/interface can be defined a traction–separation law for fracture under pure mode I or pure mode II, as well as under mixed-mode I + II loading.

As it can be seen in Figs. 21.2 and 21.3, both triangular and trapezoidal laws show an initial linear relation which represents the material elastic behavior until the moment when is achieved the cohesive strength/ultimate strength of a certain mode ($\sigma_{u,I}$ and $\sigma_{u,II}$), to which corresponds the damage onset displacement ($\delta_{0,I}$ and $\delta_{0,II}$). Then, for the triangular shape begins the stress softening until the final failure when the crack propagates, at the failure displacement ($\delta_{u,I}$ and $\delta_{u,II}$). The final relative displacement is defined by G_{IC} or G_{IIC}, since these variables represent the area under cohesive zone models laws.

Regarding the trapezoidal shape law, the behavior described is very similar to the triangular law unless in the additional plateau section present after the elastic region. Thus, the damage process is divided into two phases [15]:

- a plateau region, which initiates at the cohesive strength ($\sigma_{u,I}$ and $\sigma_{u,II}$), when the $\delta_{0,I}$ and $\delta_{0,II}$ displacements are reached. During this softening phase, the relative displacement increases at constant stress until the displacements $\delta_{s,I}$ or $\delta_{s,II}$ where the stress softening initiates;

- a second linear region, which starts at $\delta_{s,I}$ or $\delta_{s,II}$. At this phase of softening the stress decreases and the relative displacement increases until the final displacement ($\delta_{u,I}$ and $\delta_{u,II}$) where is observed the complete failure, leading to the crack opening.

The trapezoidal law is more suitable to describe the fracture behavior of ductile adhesives [15].

21.2 Numerical Simulation

21.2.1 Introduction and Methodology

The first step of this work was to specify the geometry of ADCB specimen. The geometry was defined with the aim of minimizing the level of mode-mixity during crack propagation, in order to obtain almost pure mode I of fracture during the experiment test. Since CFRP and S355 steel, according to the information present in the technical sheets, show a very similar Young's modulus, the thickness of upper and lower arms of the specimens should be similar. In this case, the material that imposes the thickness is the CFRP, since it is a prefabricated laminate of defined thickness.

Currently, there are two criteria of design which are commonly applied to minimize the mode-mixity: the flexure stiffness-based criterion and the strain-based criterion [11, 12]. The first criterion establishes as design condition the matching of flexure stiffness of upper and lower arms, which is summarized by the following equation:

$$E_l h_l^3 = E_u h_u^3 \tag{21.11}$$

On the other hand, Wang et al. in [12] proposed the strain-based criterion which defines the equation bellow as the condition to minimize mode-mixity:

$$E_l h_l^2 = E_u h_u^2 \tag{21.12}$$

Thus, according to the strain-based criterion, for a CFRP's thickness of 1. 4 mm, the steel arm should be designed with a thickness of 1.38 mm, while by applying the flexure stiffness criterion is suggested a thickness of 1.39 mm. Therefore, the steel arm was defined with a thickness of 1.4 mm, which is equal to the thickness of the CFRP arm. However, in order to verify specimens' geometry defined, a finite element analysis was carried out.

Therefore, a numerical model which simulates the ADCB experimental test was developed and analyzed in ABAQUS. The methodology followed is described in the scheme of Fig. 21.4 and can be summarized into the following steps:

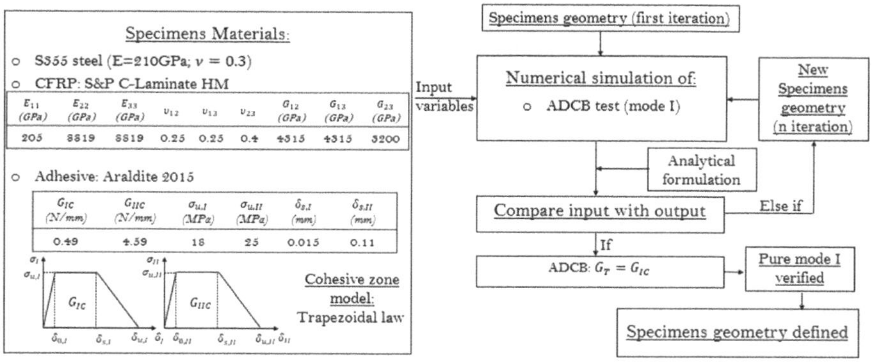

Fig. 21.4 Scheme of the methodology

- The geometric and mechanical properties of the substrates and adhesive are introduced in the numerical model as input values;
- The *P*-*δ* curves are obtained from the numerical models as output load-displacement values;
- The analytical formulation developed is applied to the output values, in order to obtain the resistance curves. These curves plot the total strain energy released as function of the equivalent crack length, throughout the simulated experimental test;
- The output values of G_T are compared to the fracture energy values of pure mode I introduced as input in the numerical model. If the final value of G_T is similar to G_{IC} in ADCB simulation, the specimens are correctly designed to obtain, experimentally, the adhesive pure properties of fracture. On the other hand, if this condition is not verified, the geometry should be redefined, and the previous steps repeated.

21.2.2 Description of Numerical Model

As stated before, the numerical analysis was based on a numerical model which aims to simulate ADCB experimental test. Since fracture mechanics properties of the Araldite AW4858/Hardener HW 4858 are not known, the numerical simulation was characterized assuming the properties of Araldite 2015, which is another epoxy adhesive expected to be very similar to the adhesive under study. Regarding the substrates, they were defined as completely elastic and with the properties present on the scheme of Fig. 21.4. However, for the CFRP, once again, only the properties in the direction of fibers are known, so other properties were defined based on another CFRP characterized in [10].

The numerical model consists of two arms of different materials: The upper arm was defined as steel and the lower as CFRP. Besides, the substrates were simulated

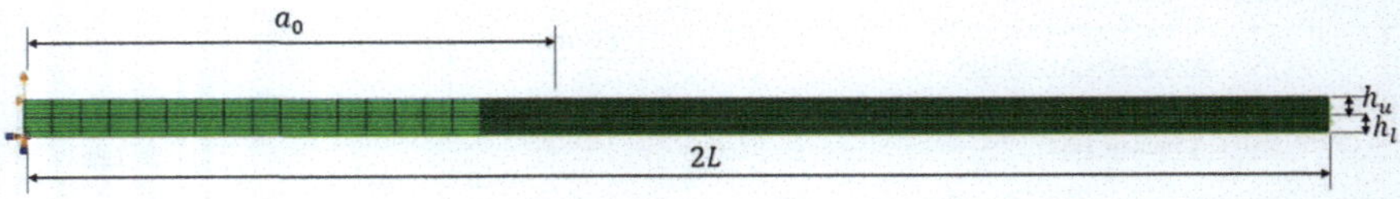

Fig. 21.5 Numerical model of ADCB specimen

with CPE8 elements of ABAQUS, i.e., eight-node plane strain elements. Moreover, each arm was modeled with four elements along the thickness, while along the length, per each row of elements, there are 420 elements. The size of elements is not constant, but around the expected crack path the mesh was refined to a minimum size is 0.19 mm.

Besides, the adhesive layer was simulated with cohesive elements of 6 nodes and null thickness, which are compatible with the elements of the arms. The cohesive elements were defined according to the trapezoidal cohesive zone laws for pure mode I and mode II defined in Fig. 21.4. This element was implemented in the numerical model by a subroutine developed in FORTRAN and denominated USER ELEMENT.

The boundary conditions defined can be seen in Fig. 21.5. At the first node, located at the left button, the vertical and horizontal displacement were constrained. Furthermore, in the same row of nodes, but at the top of the upper arm, it was imposed a vertical displacement ($\delta_{\mathrm{DCB}} = 10$) and constrained the horizontal displacement.

The damage initiation was defined by the quadratic criterion presented below [19, 20]:

$$\begin{cases} \left(\frac{\sigma_I}{\sigma_{u,I}}\right)^2 + \left(\frac{\sigma_{II}}{\sigma_{u,II}}\right)^2 = 1, & \text{if} \sigma_I > 0 \\ \sigma_{II} = \sigma_{uII}, & \text{if} \sigma_I \leq 0 \end{cases} \quad (21.13)$$

On the other hand, to predict the fatigue crack growth was considered the linear energetic criterion:

$$\frac{G_I}{G_{IC}} + \frac{G_{II}}{G_{IIC}} = 1 \quad (21.14)$$

In order to obtain the best geometry to perform the DCB experimental tests, a couple of different configurations for the numerical model were studied. As can be checked in Table 21.4, the initial crack length (a_0) was the only parameter modified between models, so three different lengths were studied: $a_0 = 40$ mm, $a_0 = 50$ mm, and $a_0 = 60$ mm. Moreover, also in Table 21.4, it is mentioned the values of increment and penalty stiffness that were considered and allowed the best results.

Table 21.4 Main parameters of the ADCB numerical models (see Fig. 21.5)

Parameter	Model 1	Model 2	Model 3
a_0 [mm]	40	50	60
$2L$ [mm]	100	100	100
B [mm]	20	20	20
h_u [mm]	1.4	1.4	1.4
h_l [mm]	1.4	1.4	1.4
Penalty [N/mm^3]	1E6	1E6	1E6
Increment	0.0001	0.0001	0.0001

21.2.3 Results

The load-displacement curves of the ADCB test are the main output of the performed simulations. Then, the equivalent crack length, for each time step, was calculated based on the compliance (obtained from the load and displacement) and using Eq. (21.9). Finally, G_T was calculated as function of a_{eq} by applying Eq. (21.10), and it was plotted as resistance curves. Since these curves converge to a plateau close to the input value of G_{IC} defined in the trapezoidal law, it was concluded that the specimen geometry was well defined and can be introduced a crack length between 40 and 60 mm (Figs. 21.6 and 21.7).

As conclusion of the numerical study conducted, the geometry of ADCB specimens was defined as it is portrayed in Fig. 21.8. Moreover, it is important to highlight that the real length of the specimens is a bit larger to account for the setup configuration. The specimens are attached to the gripping system through a pair of blocks and mechanical pin. Therefore, the effective length can only be measured from the point where the displacement is applied to the specimen.

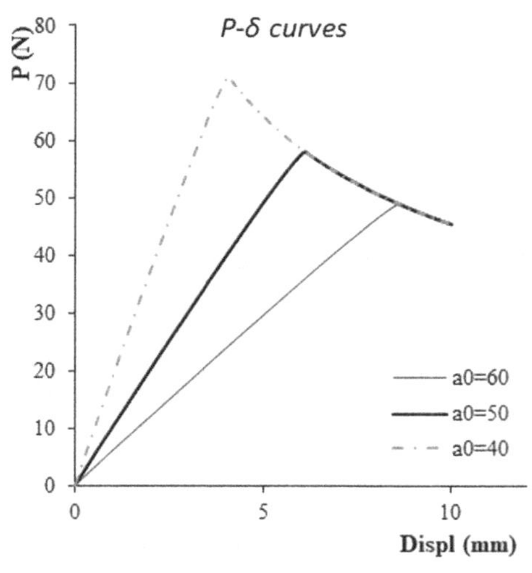

Fig. 21.6 Numerical *P*-*δ* curves

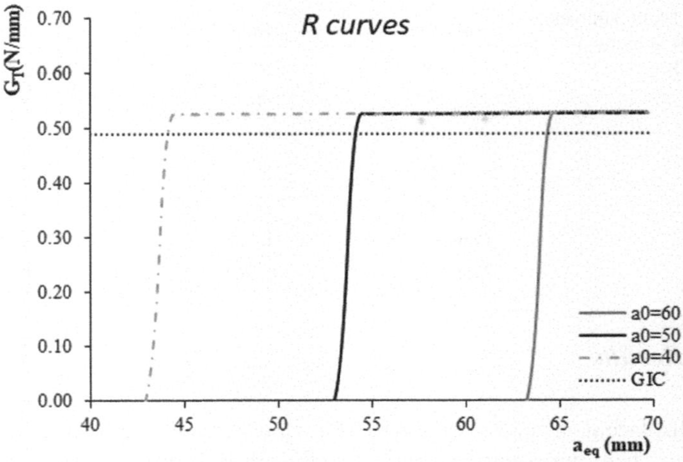

Fig. 21.7 Numerical resistance curves

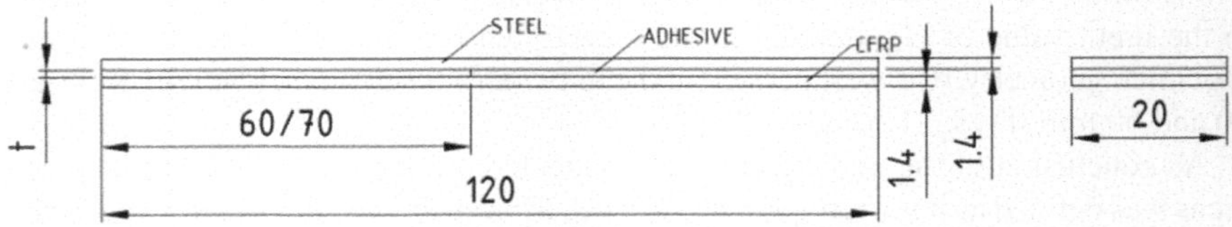

Fig. 21.8 Geometry defined for ADCB specimen

21.3 Experimental Work

21.3.1 Specimens Preparation

Within the scope of this experimental campaign, four ADCB specimens were tested. The preparation of specimens is a laborious and very manual work. First, the specimens were polished and cleaned with an alcohol solution. The adhesive thickness was defined with stainless steel calibrated tapes and the pre-crack shape with razor blades. Thus, it was ensured a uniform thickness of 0.2 mm of adhesive. Both parts of the joint—steel and CFRP substrates—were covered with adhesive. In this process was used a spatula to ensure a uniform and plane distribution of the adhesive applied. Moreover, all the specimens were fabricated and let to cure for a couple of days inside a steel mold, which ensures a uniform thickness of the adhesive and avoids porosities and voids inside it.

Fig. 21.9 ADCB test setup

21.3.2 Experimental Tests and Results

The experimental tests were performed on a MTS810 machine with a loading cell of 10kN coupled. The ADCB specimen was attached to the gripping system of the testing machine by means of blocks and mechanic pins (Fig. 21.9). During the test, the lower grip fixes the specimen, while the upper grip starts to move, vertically, apart at 1 mm/min speed until the two arms of the specimen are completely separated (final failure). The displacement and load measured by the loading cell were recorded along the experimental tests.

The four specimens were tested following the procedure described above. The results obtained were consistent and similar in all the four specimens tested: Instead of a cohesive failure, it was observed the failure of the CFRP (Figs. 21.10). The crack propagated through the laminate and it was observed fiber-bridging.

In order to evaluate the results obtained, the numerical model was adapted for the real geometrical dimension of each specimen. The experimental and numerical P-δ curves were plotted and compared as can be seen in Figs. 21.11, 21.12, 21.13, 21.14, 21.15, 21.16, 21.17 and 21.18. This comparison shows the specimens fail before the numerical model prediction, which is in accordance with the substrate failure observed in the specimens. Thus, it is possible to conclude that, probably, the mode I fracture energy of the CFRP is lower than the adhesive fracture energy. This conclusion is emphasized by the Resistance curves plotted for each specimen, where can be observed a maximum value of G_T around 0.3 N/mm.

However, the experimental P-δ curves show a lower slope than the numerical curve. This fact raises some suspicious about the Young's modulus of the CFRP indicated by supplier in technical sheets. Therefore, as future work it is expected to perform a couple of experimental tests, to verify the properties present on the technical sheet and characterize in more detail this material.

Fig. 21.10 ADCB specimens tested

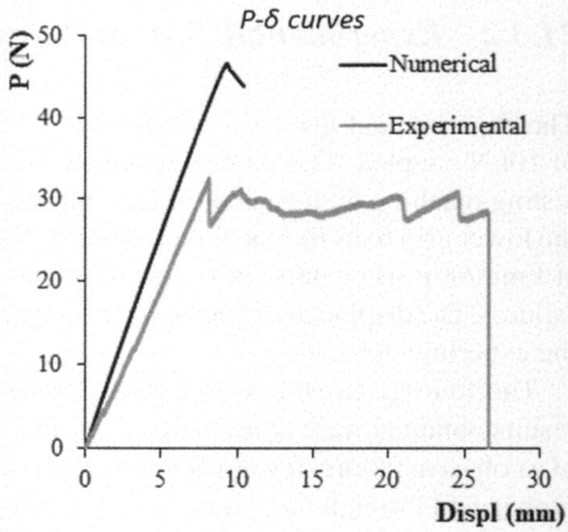

Fig. 21.11 *P*-*δ* curve of DCB_1 specimen

21.4 Conclusions and Future Work

This work aimed to characterize and determine the fracture properties of hybrid joints, of CFRP and S355 steel, under pure mode I loading. Besides, it was intended to obtain the fracture energy of the structural adhesive Araldite AW 4858/Hardener HW 4858 for the pure mode.

Therefore, an experimental campaign including asymmetric double cantilever beam was conducted. This campaign and the design of fracture specimens were supported by a finite element analysis based on a cohesive zone model approach combined with numerical simulation.

In the fracture tests, it was not observed a cohesive failure, which turned impossible to determine G_{IC} of the adhesive under analysis. Thus, it was possible to conclude

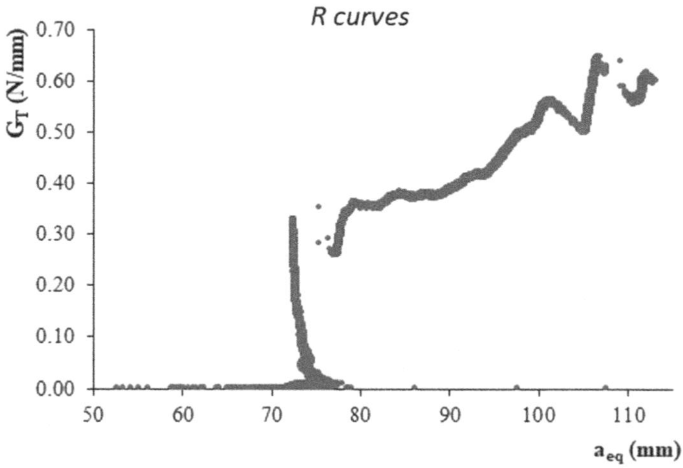

Fig. 21.12 Resistance curve of DCB_1 specimen

Fig. 21.13 *P-δ* curve of DCB_2 specimen

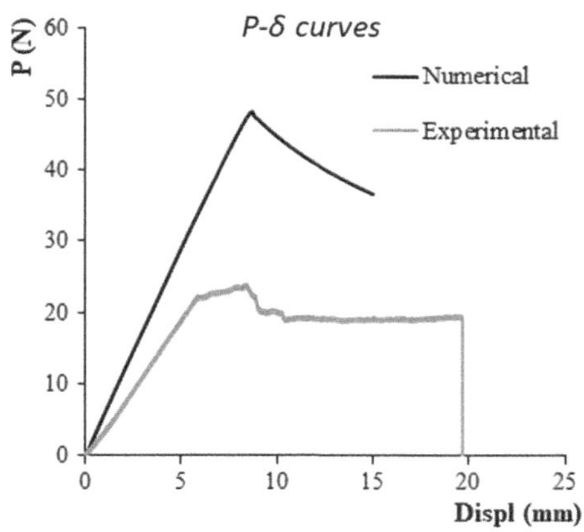

that the CFRP is weaker than the adhesive and also to determine the fracture energy of the joint. Consequently, it can be settled that the fracture energy of these hybrid joints is around 0.3 N/mm and dictated by the interlaminar fracture of the carbon-epoxy laminate.

In order to determine the G_{IC} of the Araldite AW 4858/Hardener HW 4858, as future work, it is expected to repeat the experimental campaign, but this time for different joints to overcome the problems found. Moreover, instead of hybrid joints, it will be fabricated and tested joints only of steel.

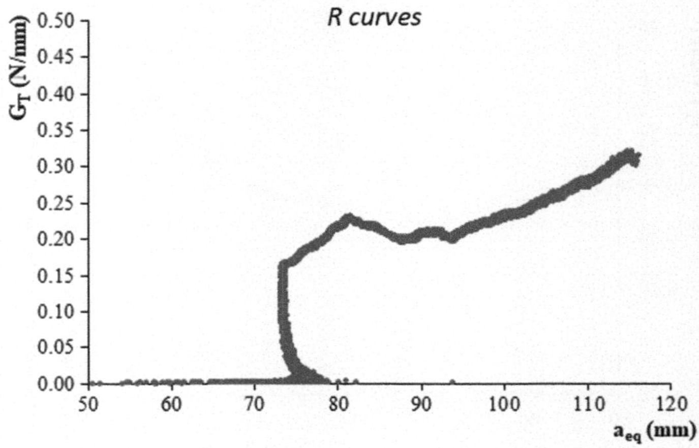

Fig. 21.14 Resistance curve of DCB_2 specimen

Fig. 21.15 P-δ curve of DCB_3 specimen

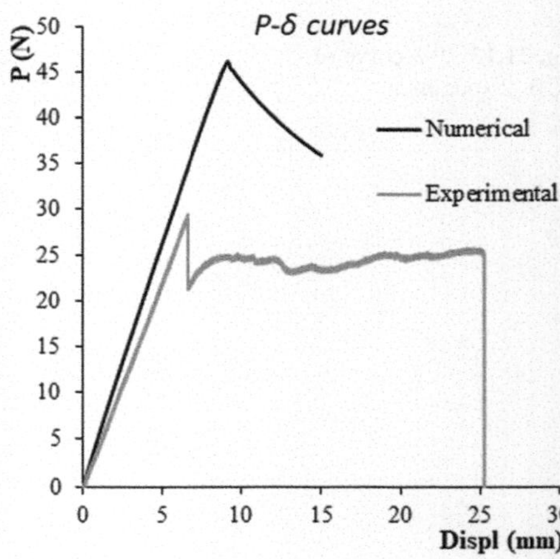

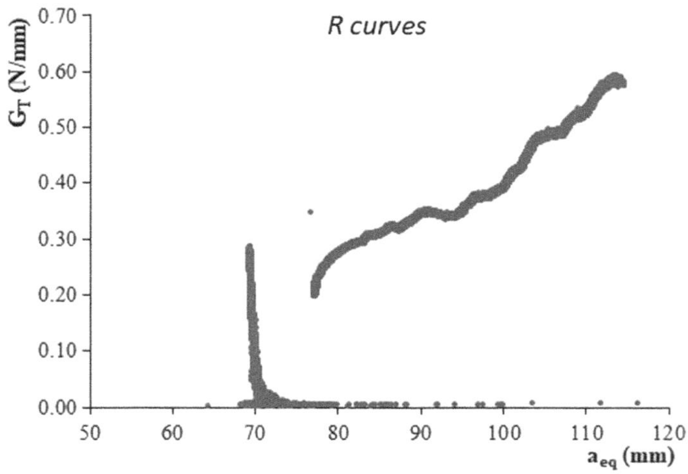

Fig. 21.16 Resistance curve of DCB_3 specimen

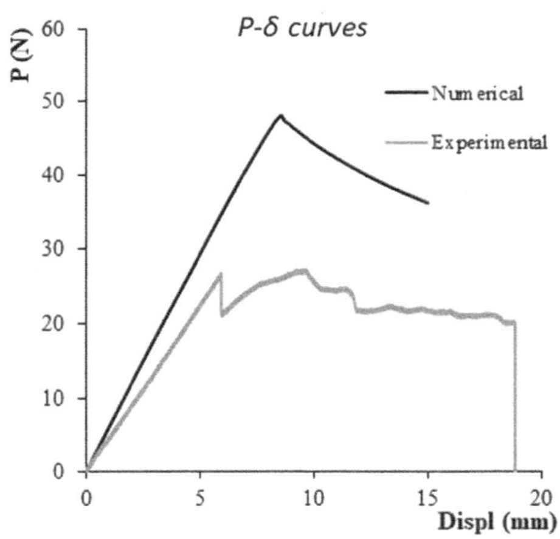

Fig. 21.17 *P*-*δ* curve of DCB_4 specimen

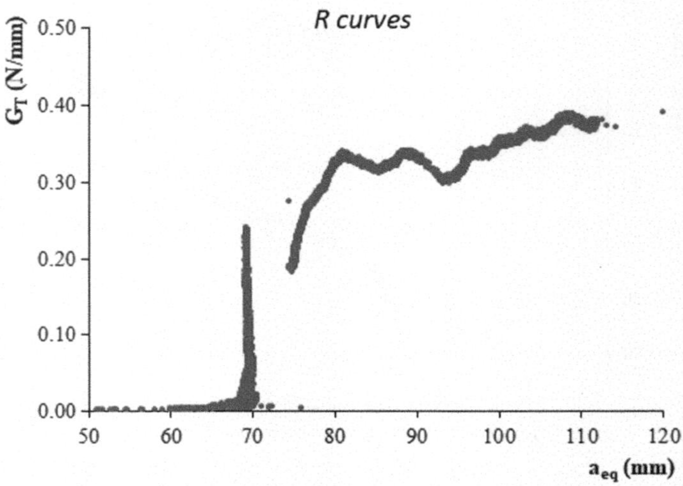

Fig. 21.18 Resistance curve of DCB_4 specimen

References

1. Moy SSJ, Bloodworth AG (2007) Strengthening a steel bridge with CFRP composites. Proc Inst Civ Eng Struct Build 160:81–93
2. Pang B, Yang P, Wang Y, Kendall A, Xie H, Yurong Z (2015) Life cycle environmental impact assessment of a bridge with different strengthening schemes. Int J Life Cycle Assess 1300–1311
3. Jones SC, Civjan SA, Asce M (2003) Application of fiber reinforced polymer overlays to extend steel fatigue life. J Compos Constr ASCE 7:331–338
4. Mohabeddine A, Correia JAFO, Castro JM, Montenegro P, Jesus AMP De, Calçada RAB (2020) Numerical investigation on the fatigue life of non- cracked metallic plates repaired with bonded CFRP. EUROSTEEL2021. Sheffield
5. Heshmati M, Haghani R, Al-emrani M (2017) Durability of bonded FRP-to-steel joints : effects of moisture, de-icing salt solution, temperature and FRP type. Compos Part B 153–167. Elsevier Ltd.
6. Bocciarelli M, Colombi P, Fava G, Poggi C (2009) Fatigue performance of tensile steel members strengthened with CFRP plates. Compos Struct [Internet] 87:334–343. Elsevier Ltd. Available from: https://doi.org/10.1016/j.compstruct.2008.02.004
7. Comité Europeu de Normalização. Eurocódigo 3—Projecto de estruturas de aço Parte 1–1: Regras gerais e regras para edifícios
8. S&P. S & P C-Laminate (2018)
9. HUNTSMAN. Araldite ® AW4858/ Hardener HW4858 (2016)
10. Moreira RDF, Moura MFSF De, Silva FGA (2020) A novel strategy to obtain the fracture envelope under mixed-mode I + II loading of composite bonded joints. Eng Fract Mech [Internet]. 232:107032. Elsevier. Available from: https://doi.org/10.1016/j.engfracmech.2020.107032
11. Khoshravan M, Asgari Mehrabadi F (2012) Fracture analysis in adhesive composite material/aluminum joints under mode-I loading; experimental and numerical approaches. Int J Adhes Adhes [Internet]. 39:8–14. Elsevier. Available from: https://doi.org/10.1016/j.ijadhadh.2012.06.005
12. Wang W, Lopes Fernandes R, Teixeira De Freitas S, Zarouchas D, Benedictus R (2018) How pure mode I can be obtained in bi-material bonded DCB joints: a longitudinal strain-based criterion. Compos Part B Eng 153:137–148

13. De MMFSF (2006) Numerical simulation of the ENF test for the mode-II fracture characterization of bonded joints fracture characterization of bonded joints. J Adhes Sci Technol 20:37–52
14. Campilho RDSG, Banea MD, Neto JABP, Silva LFM (2013) Modelling adhesive joints with cohesive zone models: effect of the cohesive law shape of the adhesive layer. Int J Adhes Adhes 44:48–56
15. da Silva LFM, Campilho RDSG (2012) Advances in numerical modelling of adhesive joints. Springer, Berlin
16. Elices M, Guinea GV, Gómez J, Planas J (2002) The cohesive zone model: advantages, limitations and challenges. Eng Fract Mech 69:137–163
17. Sun CT, Jin Z-H (2012) Fracture mechanics. Elsevier, Oxford
18. Zhang J, Shang D, Sun Y, Wang X (2018) Multiaxial high-cycle fatigue life prediction model based on the critical plane approach considering mean stress effects. Int J Damage Mech 27:32–46
19. Campilho RDSG (2009) Repair of composite and wood structures. University of Porto
20. Pereira FAM (2015) Identification of cohesive laws of cortical bone under mode I, mode II and mixed-mode I+II. University of Porto

Part V
Experimental Identification, Reliability and Structural Analysis

Chapter 22
Durability Assessment of Multiaxial Strain Loads for Rural Road Condition in Time–Frequency Domain

Nazirul Muhaimin Hamzi, Salvinder Singh Karam Singh, Shahrum Abdullah, Lennie Abdullah, Mohammad Rasidi Mohammad Rasani, Abdul Hadi Azman, and Noorsuhada Md. Nor

Abstract This paper aims to characterise the multiaxial strain loading signal in order to assess the durability of the coil spring in the time–frequency domain. Coil spring experiences vertical axial loading, direct shear force, and torsion during movement when subjected to cyclic loading. The experiment was set to capture random strain signal in the different angle at x-axis (0°), 45°, and y-axis (90°) from the coil spring at operating conditions. The time history of random strain signal was assessed based on kurtosis and the power spectrum density in order to analyse strain signals in time and frequency domain. The life is predicted through the strain-life models namely Coffin-Manson (CM), Smith–Watson–Topper (SWT), and Morrow model and is correlated using linear regression to obtain the relationship of the strain signal at 0°, 45°, and 90° axis. The results indicated that the strain at 45° axis shows the highest vibrational energy content followed by the strain at 90° axis and 0° axis. The fatigue life prediction for the CM, Morrow, and SWT models showed the same trend for the strain signals at 0°, 45°, and 90° axis. The correlation of strain life at 0° and 90° axis showed a good agreement with R^2 value exceeding 0.9 except for 45° axis. This shows that the strain life at 45° axis is affected by the torsion. This study showed a successful establishment of the relationship between the strain signal at 0°, 45°, and 90° axis using linear regression.

Keywords Power spectral density · Multiaxial strain · Strain life · Linear regression · Fatigue life

---

N. M. Hamzi · S. S. K. Singh (✉) · S. Abdullah · L. Abdullah · M. R. M. Rasani · A. H. Azman
Department of Mechanical and Manufacturing Engineering, Universiti Kebangsaan Malaysia, UKM, 43600 Bangi, Selangor, Malaysia
e-mail: salvinder@ukm.edu.my

N. Md. Nor
Civil Engineering Studies, Universiti Teknologi MARA, Cawangan Pulau Pinang, Kampus Permatang Pauh, 13500 Permatang Pauh, Pulau Pinang, Malaysia

G. Lesiuk et al. (eds.), *Structural Integrity and Fatigue Failure Analysis*, Structural Integrity 25, https://doi.org/10.1007/978-3-030-91847-7_22

22.1 Introduction

Multiaxial fatigue life is an important criterion for consideration when designing automotive components structures. Fatigue failure always leads to serious loss of material properties and the failure can occur without warning [1]. However, the prediction of multiaxial fatigue damage is still a complicated problem compared to uniaxial fatigue damage due to many factors especially in analysing the coil spring fatigue life. The uniaxial strain loads on a coil spring analysis only consider the strain in a single direction, which is the tension or compression. In contrast, multiaxial strain loads consider the tension–compression force, shear force, and torsion effect. The uniaxial strain considers the loadings as proportional, however, multiaxial strain considers the strain loadings as non-proportional [2].

From the previous study [3], the finite element analysis of a coil spring showed that the stress on the inner surface of the helical shape of the spring is critical compared to the outer surface due to load applied on the spring. The fatigue crack is expected to initiate and propagate from the critical surface to the non-critical surface in which the direction is parallel to the cross section area of the coil spring. However, research revealed that coil spring in a real situation experience crack propagation not parallel to the plane of a cross section area of the coil spring when it fails [4]. Therefore, the difference in direction of crack propagation showed that the fatigue failure of a coil spring is not only caused by compression tension load but also caused by shear force and torsion of the spring. Therefore, it is interesting to investigate the behaviour of the multiaxial load in coil spring during its working life to prevent failure of coil spring.

Multiaxial fatigue criteria have been proposed from the previous studies in analysing the multiaxial fatigue failure of materials. Stress Scale Factor (SSF) method is used to predict multiaxial fatigue lives compared to the fatigue lives to the critical plane approach [5]. Stress Scale Factor (SSF) criterion also is used to estimate fatigue damage under a multiaxial constant loading block and random loading block spectra [6]. The critical plane-based criterion is used to determine the fatigue life of metallic materials under cyclic tension and torsion loading and the new expression is proposed in order to determine the orientation of the critical plane [7]. The critical plane approach is based on the Fatemi–Socie model and is used to predict the fatigue life on the tubular welded structures for torsional fatigue assessment [8]. A study by Xu et al. [9] used a new critical plane multiaxial fatigue model parameter to characterise the influence of shear stress and non-proportional hardening in order to assess the fatigue life of material alloys.

Multiaxial strain random signals have crucial effects on fatigue life of coil spring. Valles et al. [10] investigate the cause of the fracture of one of the spring flanges under torsion loads. In a study by Manouchehrynia et al. [11], the group proposed a mathematical model to estimate the strain-life probabilistic modelling of the coil spring under random strain loads. Kong et al. [12] studied the relationship between spring stiffness and fatigue lives by using power law regression for durability analysis.

Fatigue strain loading data of leaf spring by using power spectral density and root mean square value was used in fatigue life prediction [13].

This study aims to characterise the behaviour of multiaxial strain loading signal in order to assess the durability of the coil spring in the time–frequency domain. The strain signal at x-axis, 45°, and y-axis is analysed to show that the shear force, torsion, and tension–compression are correlated. Strain signal at x-axis refers to 0° while strain signal at y-axis refers to 90°. All strain signals at 0°, 45°, and 90° axis are in the same plane. The correlation between the strain signal at 0°, 45°, and 90° axis showed a good agreement and the result can be used for future studies.

22.2 Methodology

Figure 22.1 shows the durability assessment process in analysing the fatigue life of multiaxial strain loads for rural road condition. The strain gauge is connected to a data acquisition instrument and the random strain signal was captured. The behaviour of the random strain signal at 0°, 45°, and 90° axis was assessed using the kurtosis and power spectrum density. The fatigue life of the strain signal at 0°, 45°, and 90° axis was predicted, and the fatigue life obtain was correlated using a simple linear regression method.

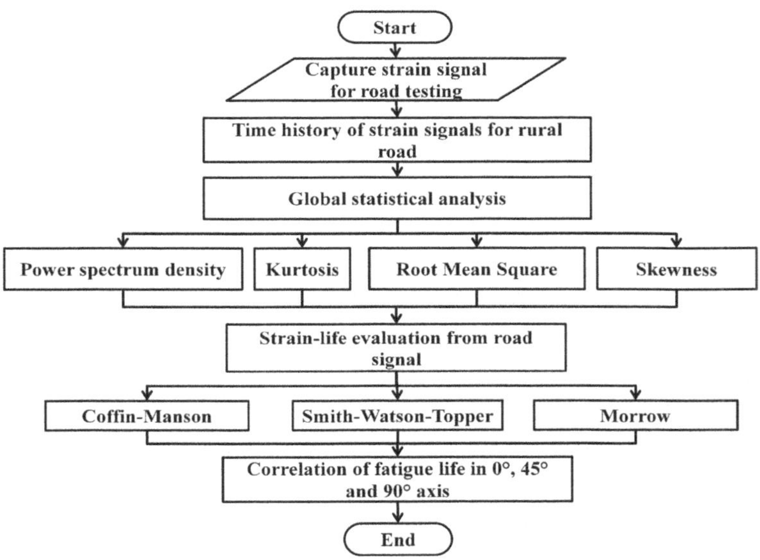

Fig. 22.1 The durability assessment process of analysing fatigue life of multiaxial strain loads for rural road condition

22.2.1 A Signal Acquisition Process

Figure 22.2 illustrates the techniques for the signal acquisition from the experimental setup for fatigue life prediction under random strain loads. The rosette strain gauge that consists of 0°, 45°, and 90° axis was used to measure the strain at 0°, 45°, and 90° axis in order to determine the behaviour of the multiaxial strain signals. The rosette strain gauge was attached on a coil spring. The rural road is curvy and contains pothole. The values of the kurtosis and root mean square (RMS) were extracted from the time-history loading signal. A kurtosis value below three indicates that the data is stationary while above three indicate the data is non-stationary, high damage, and high amplitude of the signal data [13]. The power spectrum density analysis was determined in order to assess the range of the frequencies with high amplitude.

The fatigue life of the strain signal was predicted through strain-based fatigue model, CM, SWT, and Morrow. CM model is superior when the mean stress is zero. The Morrow mean stress correction model is more suitable when the load is predominantly compressive whereas the SWT model is more suitable when the load is predominantly tensile [12].

The fatigue life of the strain signal was predicted through strain-based fatigue model, Coffin-Manson (CM), Smith–Watson–Topper (SWT), and Morrow. For the Coffin-Manson (CM) model, the equation is as shown below:

$$\varepsilon_a = \frac{\sigma'_f}{E}(2N_f)^b + \varepsilon'_f.(2N_f)^c \tag{22.1}$$

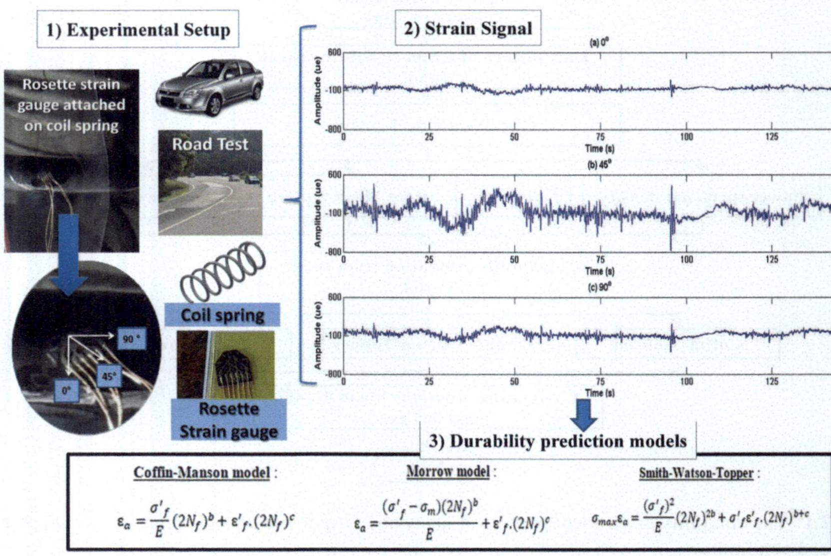

Fig. 22.2 Signal acquisition process of analysing fatigue life of multiaxial strain loads

where σ'_f is fatigue strength coefficient while ε'_f is the fatigue ductility coefficient and ε_a is the total strain amplitude of the signal, while b and c are the fatigue strength exponent and fatigue ductility exponent, respectively. The number of cycles to failure is represented by N_f. The CM model does not consider the mean stress effect while the Morrow strain-life model is predominant when the signal is in compressive behaviour. The equation of the model is as shown below:

$$\varepsilon_a = \frac{(\sigma'_f - \sigma_m)(2N_f)^b}{E} + \varepsilon'_f.(2N_f)^c \tag{22.2}$$

where σ_m is the mean stress and E refers to Young's modulus. The SWT strain-life model is predominant when the signal is in tension behaviour. The equation of the model is represented as follow:

$$\sigma_{\max}\varepsilon_a = \frac{\left(\sigma'_f\right)^2}{E}(2N_f)^{2b} + \sigma'_f\varepsilon'_f.(2N_f)^{b+c} \tag{22.3}$$

where $\sigma_{\max}$ is the maximum stress.

22.3 Results and Discussion

Figure 22.2 shows the time history of the strain signal at 0°, 45°, and 90° axis. The strain signal at 45° axis showed higher amplitude compared to the strain signal at 0° and 90° axis. This is due to the twisting of the coil spring wire is more dominant compared to the tension and compression of the coil spring wire during working condition. However, the strain signal at 0° and 90° axis seem to inverse when the 0° and 90° axis (x and y axis) are perpendiculars. Mean value of the strain at 0°, 45°, and 90° axis were −3343 µε, −3116 µε, and −3228 µε respectively.

The RMS and kurtosis values as shown in Fig. 22.3 showed the total vibrational energy content and the sensitivity to high amplitude events in the data [14]. Figure 22.3 illustrates the kurtosis and RMS of the strain data. The strain signals possess kurtosis values exceeding 3, indicating high damage experienced by the coil spring [15]. All three strain signals data showed kurtosis value exceeding 3 and this showed high damage in the strain signal and non-stationary. Therefore, all of the strain signals data are non-stationary since the data have high amplitude value. The strain at 0° axis gives the highest value of kurtosis at 4.27 µε followed by strain at 45° and 90° axis which is 4.03 µε and 3.83 µε respectively. Meanwhile, strain at 45° axis provides the highest value of RMS at 7395 µε followed by strain at 90° and 0° axis which is 4189 µε and 3763 µε. Thus, strain at 0° axis shows the highest kurtosis with lower vibrational energy content due to the constant amplitude of the strain data. As the strain at a 45° axis provides the highest value of RMS, the high

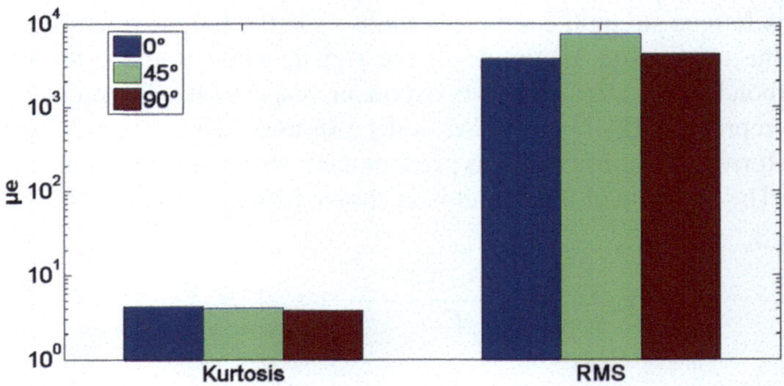

Fig. 22.3 RMS and kurtosis parameters of coil spring for strain signals at 0°, 45°, and 90° axis

vibrational energy content can contribute to the high amplitude data of the strain signal where the damage can occur.

Figure 22.4 shows that the power spectral density (PSD) of the random strain data is in the frequency range of 0–5 Hz indicating that the vibration occurred at low frequency. The strain signal at 45° axis has the highest energy at 8.06 × 10^6 $\mu\varepsilon^2$/Hz followed by the strain at 90° (1.21 × 10^6 $\mu\varepsilon^2$/Hz) and 0° (4.86 × 10^5 $\mu\varepsilon^2$/Hz) axis. Therefore, there is a significant difference in the energy content at 0°, 45°, and 90° axis. Furthermore, the area under the curve in the PSD has the same trend with the RMS value. Therefore, based on the time and frequency domain analysis, the strain signal at 45° axis has the highest vibrational energy contents.

Figure 22.5 shows the fatigue life prediction that was computed using strain-life models of Coffin-Manson (CM), Morrow, and Smith–Watson–Topper (SWT). The compression condition is exhibited on the coil spring whereby the mean values for the signals showed a negative value. This can be related to the Morrow model, which

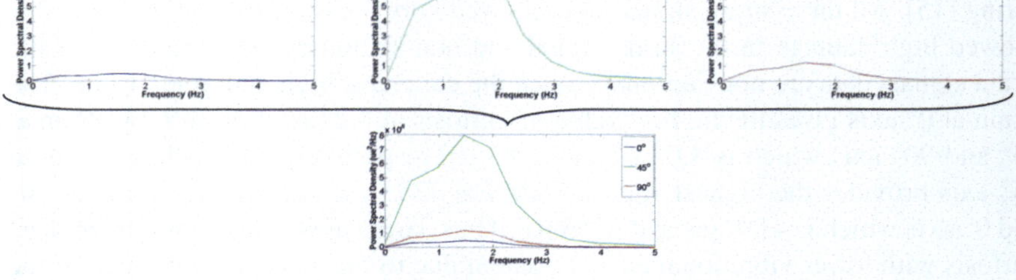

Fig. 22.4 Power Spectrum Density (PSD) of four strain signals at 0°, 45°, and 90° axis

Fig. 22.5 Fatigue life of multiaxial strain gauge of each axis

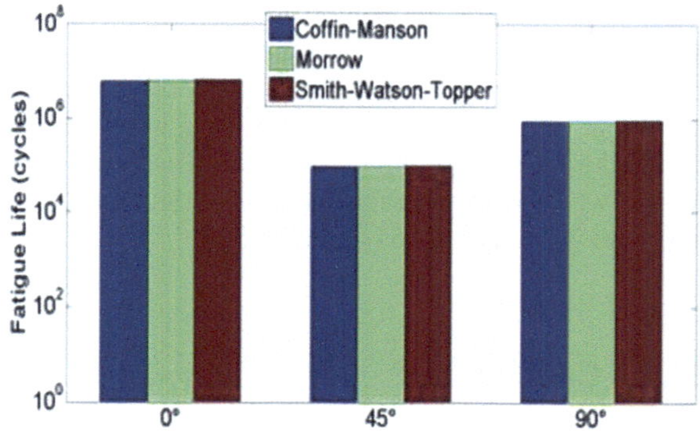

is more predominant to compressive. Based on the Morrow model, the fatigue lives of the strain at 0°, 45°, and 90° axis were 6.07×10^6, 9.55×10^4, and 8.87×10^5 cycles/block, respectively. The fatigue life at 45° axis has the lowest fatigue life in which it contained the highest vibrational energy. For the strain signals at 0°, 45°, and 90° axis, the fatigue life prediction for the CM, Morrow, and SWT model has the same trend with the fatigue life of the low strain signal at 45° axis followed by the signal at 90° and 0° axis. The fatigue life at 45° axis is the lowest due to the varying internal stress from the torsion effect in the coil spring that caused the load to be non-proportional.

The data correlation of fatigue life at 0°, 45°, and 90° axis is displayed in Fig. 22.6. R^2 values exceeded 0.8 showing that the fatigue lives were well correlated to each other. The models are reliable to predict fatigue life if the R^2 value is higher than 0.8 and more than 0.9, which can be classified as 'good' and 'very good', respectively [3]. The fatigue life at 45° axis shows a relationship with the 0° and 90° axis with R^2 values of 0.81 and 0.87, which these values are below 0.9. Meanwhile, the fatigue life at 0° axis showed a good relationship with fatigue life at 90° with R^2 values of 0.99, which is the highest among the three correlations. The correlation of fatigue life at 45° axis shows R^2 value below 0.9 due to the torsional effect in the coil spring. The helical shape of the coil spring makes the wire of the spring are affected due to the torsion.

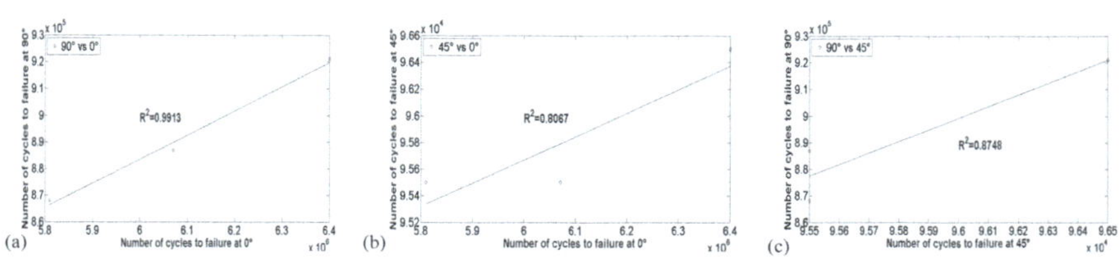

Fig. 22.6 Correlation of fatigue life for different strain axis: **a** 90° versus 0°, **b** 45° versus 0°, **c** 45° versus 90°

Therefore, there are significant differences between the strain signal at 0°, 45°, and 90° axis since due to the differences in R^2 value among the three correlation. The correlation shows that the different angle of strain load resulting in the change in the behaviour of the multiaxial strain signals, which is related to the response of the coil spring during the operating condition. The strain at 45° axis is predominant to the torsion whereby the correlation of the fatigue life at 45° axis towards the fatigue life at 0° and 90° axis only 81 and 87%.

22.4 Conclusion

This paper aims to characterise the behaviour of multiaxial strain loading signal to assess the durability of the coil spring in the time–frequency domain. The following conclusions can be made based on the durability assessment of the multiaxial strain loads. Strain at 0° axis shows the lowest vibrational energy content followed by strain at 90° axis and 45° axis, which is 3763, 4189, and 7395 $\mu\varepsilon$ based on the RMS value which shows significant difference in the vibrational energy content.

The fatigue life prediction for the CM, Morrow, and SWT model has the same trend for the strain signals at 0°, 45°, and 90° axis. Based on the Morrow model, the fatigue lives of the strain at 0°, 45°, and 90° axis were 6×10^6, 9×10^4, and 8×10^5 cycles/block, respectively. The fatigue life at 45° axis shows a relationship with 0° and 90° axis with R^2 values of 0.81 and 0.87, respectively. The relationship between fatigue life at 0° and 90° axis is the best with R^2 value of 0.99 showing that the shear force is not significantly affecting the failure. The torsion affects the failure since the R^2 value is below 0.9. Therefore, this study shows that torsion affecting failure more than the shear force.

Acknowledgements The authors would like to express their gratitude and thanks to Universiti Kebangsaan Malaysia for funding this research under the Research University Grant Scheme GUP-2018–077 and DIP–2019–015.

References

1. Sonsino C, Heim R, Melz T (2016) Lightweight-structural durability design by consideration of variable amplitude loading. Int J Fatigue 91:328–336
2. Zhu S, Yu Z, Correia J, Jesus A, Berto F (2018) Evaluation and comparison of critical plane criteria for multiaxial fatigue analysis of ductile and brittle materials. Int J Fatigue 112:279–288
3. Kong YS, Abdullah S, Schramm D, Omar MZ, Haris SM (2019) Correlation of uniaxial and multiaxial fatigue models for automobile spring life assessment. Exp Tech 44(2):197–215
4. Pastorcic D, Vukelic G, Bozic Z (2019) Coil spring failure and fatigue analysis. Eng Fail Anal 99(1):310–318
5. Freitas MD, Reis L, Meggiolaro MA, Castro J (2017) Stress scale factor and critical plane models under multiaxial proportional loading histories. Eng Fract Mech 174:104–116

6. Anes V, Caxias J, Freitas M, Reis L (2017) Fatigue damage assessment under random and variable amplitude multiaxial loading conditions in structural steels. Int J Fatigue 100:591–601
7. Carpinteri A, Kurek M, Łagoda T, Vantadori S (2017) Estimation of fatigue life under multiaxial loading by varying the critical plane orientation. Int J Fatigue 100:512–520
8. Hemmesi K, Farajian M, Fatemi A (2017) Application of the critical plane approach to the torsional fatigue assessment of welds considering the effect of residual stresses. Int J Fatigue 101:271–281
9. Xu S, Zhu SP, Hao YZ, Liao D, Qian G (2018) A new critical plane-energy model for multiaxial fatigue life prediction of turbine disc alloys. Eng Fail Anal 93:55–63
10. Valles M, García-Martínez M, Pastor A (2019) Study of a torsion spring fracture Valles. Eng Fail Anal 98:150–155
11. Manouchehrynia R, Abdullah S, Singh SSK (2020) Fatigue reliability assessment of an automobile coil spring under random strain loads using probabilistic technique. Metals 10(1):1–24
12. Kong YS, Abdullah S, Omar MZ, Schramm D (2018) Generation of artificial road profile for automobile spring durability analysis. J Kejuruteraan 30(2):123–128
13. Abdullah L, Singh SSK, Abdullah S, Azman AH, Ariffin AK, Kong YK (2018) The needs of power spectral density in fatigue life prediction of heavy vehicle leaf spring. J Mech Sci Technol 34(6):2341–2346
14. Zhao YG, Zhang XY, Lu ZH (2018) A flexible distribution and its application in reliability engineering. Reliab Eng Syst Saf 176:1–12
15. Jianzheng J, Zaigang C, Wanming Z, Tao Z, Yifan Y (2020) Vibration characteristics of railway locomotive induced by gear tooth root crack fault under transient conditions. Eng Fail Anal 108:104285

Chapter 23
Mode Superposition Techniques for a Priori High Stress Detection and Fatigue Hotspot Identification

Carsten Strzalka, Manfred Zehn, and Dragan Marinkovic

Abstract For the evaluation of a structure's fatigue life, the knowledge of detailed stress histories plays a crucial role. In industrial application, the finite element method (FEM) has become the most widely applied tool for numerical stress- and subsequent fatigue analyses. While most computational cost is spent for the evaluation of time-dependent global stress fields, the areas that are relevant for durability analyses are limited to highly stressed local regions (hot spots). For the reduction of processed data, a subsequent fatigue analysis can be carried out using so-called hot spot filters, pointing out highly stressed elements. Available methods can be summarised as posteriori methods, as they root in the calculated stress histories or modal contributions. In this paper, a novel approach is developed, based on the a priori superposition of modal fields by means of appropriate prediction of maximum modal contributions. Main influences on modal contributions are identified and suitable parameters for superposition are summarised. As modal fields are scalable, special attention is paid to mode normalisation. Combining the developed approach with material data, appropriate threshold values for hot spot detection are presented. For validation, numerical fatigue analyses are carried out on a complex FE-model from automotive industry.

Keywords Fatigue · Hot spot filter · High stress detection · Durability analysis

23.1 Introduction

Just as the static strength, also the durability under dynamic loading is an essential property of structural components, as the unforeseen fatigue failure can cause catastrophic accidents or a malfunction [1]. In addition to cost-intensive experimental investigations, numerical methods, as the finite element method (FEM), have gained high importance over the last decades [2]. While the estimation of the fatigue life of simple components under uniaxial loading is well understood, increasing research effort is spent on the investigation of multiaxial fatigue phenomena in real vibrating

C. Strzalka (✉) · M. Zehn · D. Marinkovic
Technical University Berlin, Str. des 17. Juni, 10623 Berlin, Germany
e-mail: carsten.strzalka@tu-berlin.de

G. Lesiuk et al. (eds.), *Structural Integrity and Fatigue Failure Analysis*,
Structural Integrity 25, https://doi.org/10.1007/978-3-030-91847-7_23

structures. The existing fatigue criteria are numerous, including equivalent stress criteria; energy-based parameters as well as critical plane approaches [3, 4]. As these criteria are often based on complex analyses of stress histories, like change of principal stress directions and cycle counting, additional computational cost is high [5]. To overcome this, efficient methods exist to reduce computational cost and processed data. Among others, well-established methods are based modal decoupling of the equations of motion of the discretised FEM system [6]. Independent from time- or frequency-domain methods, the knowledge of a component's stress is crucial for subsequent durability analyses. Therefore, one of the main challenges in reliably predicting fatigue life of components under complex boundary and loading conditions is the accurate determination of detailed stress fields. Looking at the results of industrial stress analyses, one can see that typically high stress concentration is locally limited to critical regions of only a small fraction of the whole system (around 1% in automotive industry) [7]. For reduction of the effort spent on the evaluation of calculated stress histories for fatigue analysis, subsequent analysis steps can be limited to only highly stressed regions of the system. Despite the high potential benefit of the knowledge of highly stressed areas, only few research aims to these investigations and is mostly limited to posteriori methods by evaluation of existing stress histories. Huang et al. [7, 8] apply modal transient methods for the durability analysis of a vehicle body structure. The applied approach is based on a modal stress superposition method. For the identification of critical elements, an element prescan is performed for elements exceeding a threshold value of von Mises stress in a short peak loading event. Veltri [9] describes a method for the a priori prediction of stress concentration based on CMS-methods, namely the Craig-Bampton method, for the durability analysis of an all-terrain vehicle frame subjected to road-load time histories. Information about highly stressed elements is gained from selected fixed interface normal modes, while the influence of external loading is captured by a set of constrained modes. Zhou et al. draw conclusions about fatigue critical regions by investigating modal von Mises stress fields of plate-type structures subjected to random base excitation [10]. Later the contributions of dominant modes have been interpreted from mass modal participation factors of a notched elbow structure and a subsequent random vibration fatigue analysis has been performed on the identified hot spots [11].

In this paper, a novel approach is developed, based on the a priori superposition of modal fields by means of appropriate approximation and prediction of maximum modal contributions for force excited systems. For reliable a priori hot spot identification, mathematically and physically consistent procedures are derived using a weighted superposition of appropriate modal fields. The paper is structured as follows: In Sect. 23.2 the basic equations of modal stress superposition are derived. In Sect. 23.3, appropriate modal fields for superposition and investigations on mode normalisation are summarised. The influence of the structures external loading is investigated in Sect. 23.4, leading to the proposed concept of damage participation for a weighted a priori superposition. For reliable high stress and fatigue hot spot identification threshold values are given in Sect. 23.5. For validation, the proposed approach is applied to an industrial example of a twist beam rear axle in Sect. 23.6, followed by a conclusion in Sect. 23.7.

23.2 Basic Equations

The detailed modelling of structural components using the finite element method results in a large number of second-order differential equations. Solving the system's eigenvalue problem, with the system's mass matrix $\boldsymbol{M}$ and stiffness matrix $\boldsymbol{K}$, the ith eigenvalue ${\omega_i}^2$ and the corresponding eigenvector $\boldsymbol{\varphi}_i$ can be calculated.

$$\left(\boldsymbol{K} - \omega_i^2 \boldsymbol{M}\right)\boldsymbol{\varphi}_i = 0 \tag{23.1}$$

From established approximation methods in structural dynamics, it is well known, that a component's dynamic displacement field can be approximated by linear superposition of the contributions of the system's eigenvectors $\boldsymbol{\varphi}_i$ and modal coordinates q_i [6].

$$\boldsymbol{u} = \sum_{i=1}^{r} q_i \boldsymbol{\varphi}_i \tag{23.2}$$

Just as these displacement modes, also the so-called strain modes $\boldsymbol{\Psi}_i$ are intrinsic dynamic characteristics of the vibrating structure. The strain modes are calculated as the derivative of the displacement modes. For the FE-discretised structure, this is achieved by a differential operator $\boldsymbol{D}$ applied to the displacement modes.

$$\boldsymbol{\Psi}_i = \boldsymbol{D}\boldsymbol{\varphi}_i \tag{23.3}$$

The structure's dynamic strain response $\boldsymbol{\varepsilon}$ can therefore be approximated by linear superposition of the contributions of modal strain fields [12].

$$\boldsymbol{\varepsilon} = \sum_{i=1}^{r} q_i \boldsymbol{\Psi}_i \tag{23.4}$$

Assuming linear elastic and homogeneous material, the corresponding stress response $\boldsymbol{\sigma}$ is calculated using Hooke's matrix $\boldsymbol{H}$.

$$\boldsymbol{\sigma} = \sum_{i=1}^{r} q_i \boldsymbol{H}\boldsymbol{\Psi}_i \tag{23.5}$$

23.3 Modal Fields of Consideration

A component's failure mode is always a function of material, boundary and loading conditions. As in static strength analysis different theories and yield criteria have been developed for specific materials, these findings can be transferred to fatigue failure of dynamically stressed components [13]. Motivated by modern damage parameters as well as classical strength theories, different modal field values can be taken into consideration, whereas for any a priori superposition approach, these fields need to conform some restrictions. The scalability of the mode shape results in a change of the sign of any signed field variable with change of sign of the scaling factor. For applicability in a priori superposition, the sign of any scaling factor must have no influence on the resulting modal field. Therefore, following classical theories, the considered modal fields can be the von Mises equivalent stress field, as well as the modal field of absolute maximum principal stress. In addition to that, energy-based theories can be adapted, like the modal strain energy density field and more sophisticated derivatives [3]. For the presented investigations it is assumed that the component's modal von Mises stress field is an appropriate indicator for possible high stress and fatigue critical regions, as it is well established in strength analysis and widely accepted as multiaxiality criterion in fatigue analysis. For specific materials, loading and boundary conditions, the considered modal fields can easily be adapted. For localisation and indication of high stress regions in the finite element model, the modal fields will be evaluated on element level for each mode.

23.3.1 Mode Normalisation

Eigenvectors calculated from modal analysis using commercial FE-tools are usually normalised to maximum displacement or modal mass. Considering the simple example of a beam structure shown in Fig. 23.1, the effects on the resulting modal fields can be generally exposed. The cantilever beam (length of 1 m, square cross-section of 10×10 mm^2) has a support at its end and is discretised with 15 linear beam elements, type B31.

With mode normalisation to modal mass (or maximum displacement) the resulting modal fields show an increase in maximum local modal stress and strain with increasing mode number, as depicted in Fig. 23.1a for the first 10 von Mises stress modes. The amplitudes of maximum stress within each mode can even differ in the order of magnitude between higher and lower (often dominant) modes. As local maxima of higher modes would overlay or mask the dominant global maxima, a general stress mode superposition using that type of mode normalisation is not applicable for a priori superposition. In the presented approach, mode normalisation to unit modal stiffness is proposed.

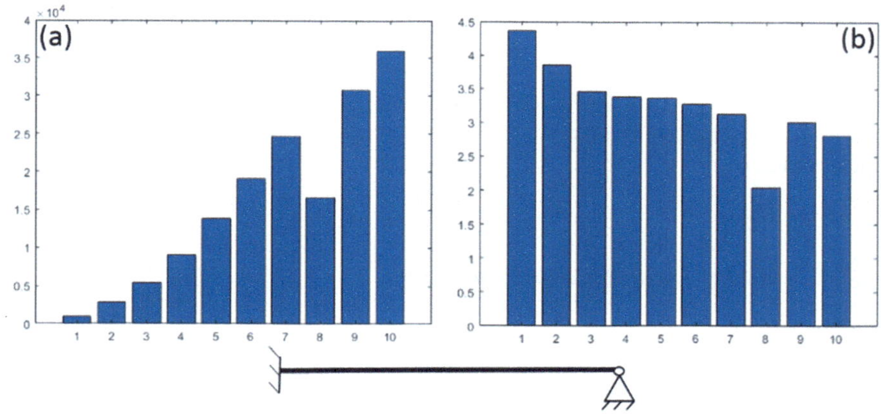

Fig. 23.1 Maximum elemental modal von Mises stress for the first 10 modes: **a** mass normalisation, **b** stiffness normalisation

$$\overline{\varphi}_i = \frac{1}{\sqrt{c_i}}\varphi_i; \quad \overline{\Psi}_i = \frac{1}{\sqrt{c_i}}\boldsymbol{\Psi}_i \tag{23.6}$$

With the modal stiffness c_i calculated from the system's stiffness matrix $\boldsymbol{K}$ and corresponding eigenvectors $\boldsymbol{\varphi}_i$:

$$\boldsymbol{\varphi}_i^T \boldsymbol{K} \boldsymbol{\varphi}_i = c_i \tag{23.7}$$

From Eq. (23.7) a connection from modal stiffness can be drawn to the strain energy content of each mode. Normalisation of the mode shapes to unit modal stiffness in this case can be interpreted as normalisation with respect to the strain energy. Looking at the theories of fatigue and damage accumulation, we find a direct connection from, e.g. the strain energy density amplitude to the induced structural damage [4]. With each mode normalised to the maximum strain energy, we therefore assume equal damage potential for each mode. The resulting maximum element stress shows a more uniform distribution of amplitudes in the same order of magnitude for each mode, as seen in Fig. 23.1b.

23.4 Loading Influence

As a mechanical system's response is always a function of its excitation, the influence of loading position as well as loading direction plays a crucial role in the development of an a priori mode superposition approach. In the following, the modal load is proposed as a weighting coefficient from the system's decoupled equations of motion in Eq. (23.8), with separation of the forcing function into the force-direction vector $\boldsymbol{f}$ and a time-dependent vector $\boldsymbol{p}(t)$.

$$\ddot{q}_i + \omega_i^2 q_i = \frac{1}{m_i}\boldsymbol{\varphi}_i^T \boldsymbol{f}\boldsymbol{p}(t) \tag{23.8}$$

From the steady state response of the SDOF equation of motion it follows, that the maximum amplitude of the modal coordinates is directly proportional to the modal load coefficient.

23.4.1 Modal Damage Participation

Combining the results of mode normalisation and loading influence, the modal damage participation of each mode is developed. The main assumption therefore is, that the maximum structural stress results from the superposition of the maxima of modal coordinates of each corresponding mode. From Eqs. (23.5), (23.6) and (23.8), we find the modal damage participation factor Γ_i, as an approximation of maximum modal coordinates and modal damage potential and a weighted stress mode superposition can be performed for high stress and fatigue hot spot identification.

$$\Gamma_i = \frac{\boldsymbol{\varphi}_i^T \boldsymbol{f}}{m_i c_i}; \quad \boldsymbol{\sigma} = \sum_{i=1}^{r} \Gamma_i \boldsymbol{\Psi}_i \tag{23.9}$$

23.5 Threshold Values

For proper evaluation of the proposed algorithm and for the assessment of critical locations from the superimposed stress fields, reliable threshold values marking an element as hot spot element will be introduced in the following. At this point, connections can be drawn to material data and the underlying fatigue theory applicable for the specific material and type of loading. As the considered modal fields are scalable, this can only be achieved in a relative way. For the presented investigations, the Coffin-Manson equation is taken into consideration, as it is valid for both, high cycle (HCF) and low cycle fatigue (LCF) application.

$$\frac{\Delta\varepsilon}{2} = \frac{1}{E}\sigma'_f\left(2N_f\right)^b + \varepsilon'_f\left(2N_f\right)^c \tag{23.10}$$

With the elastic and plastic terms, threshold values for occurring stress can be determined from the slope of stress-life or strain-life curves for a fraction of maximum occurring damage (e.g. half damage relative to maximum damage) from the fatigue strength exponent b or the fatigue ductility exponent c, respectively. The threshold stress is calculated from each element's stress compared to the maximum occurring stress where A is the factor of damage compared to maximum (e.g. $A = 2$ for half damage).

$$\sigma_{\text{th,HCF}} = \frac{\sigma_{el}}{\sigma_{\max}} = A^{-b}; \quad \sigma_{\text{th,LCF}} = \frac{\sigma_{el}}{\sigma_{\max}} = A^{-c} \tag{23.11}$$

23.6 Industrial Application

For the verification of the proposed method, a complex example from automotive industry is presented. A large majority of engineering structures belong to the group of shells [14, 15] and the chosen example reflects this fact quite well. The considered rear axle is depicted in Fig. 23.2. The full FE-model (courtesy of Volkswagen AG) consists of approximately 44,000 linear shell elements and 5000 solids, with altogether over 300,000 DOF. For the fatigue-life estimation, a multiaxial load case is defined, consisting of two non-simultaneous impact loadings in Z and Y-direction, as depicted in Fig. 23.2. To account for the structure's stiffness in different directions, the amplitudes are adjusted to show reasonable effects in both Y and Z-direction.

The structure's dynamic response is calculated using direct time integration. The undamped displacement response of the loaded node is depicted in Fig. 23.3a, whereas the resulting von Mises stress history at the centroid of one critical element is shown in Fig. 23.3b.

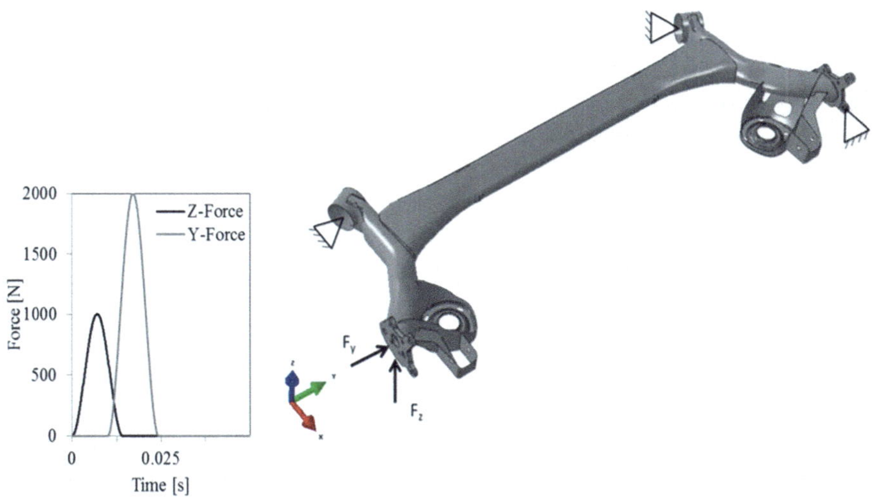

Fig. 23.2 Twist-beam rear axle with boundary conditions and loading

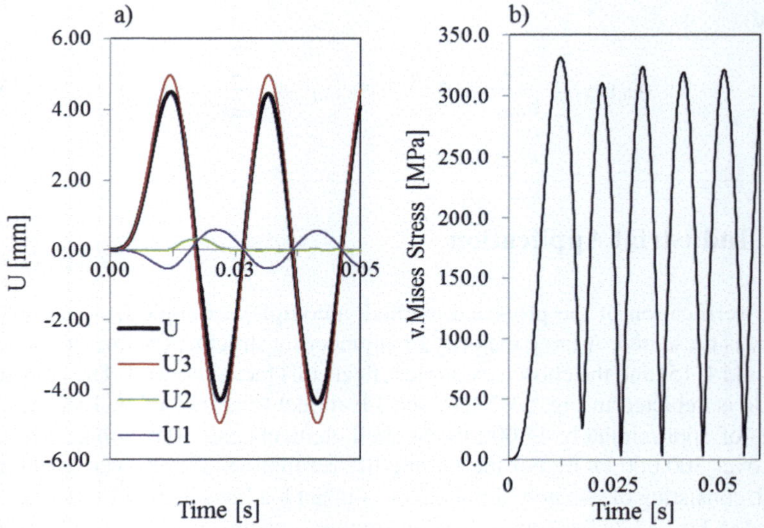

Fig. 23.3 **a** Displacement response and **b** corresponding von Mises stress response

23.6.1 Fatigue Life Estimation

For the validation of the proposed method, a numerical fatigue analysis is carried out using commercial software MSC-Fatigue. According to the loading conditions, an LCF analysis for crack initiation is performed for SAE Steel grade 1045. For the assessment of the fatigue life, one loading event is taken as the structure's vibration response over a period of $T = 0.5$ s. A contour plot of the induced damage per loading event and corresponding fatigue critical elements is depicted in Fig. 23.4.

For the evaluation of true hot spot elements, a threshold value of half damage relative to maximum occurring damage is chosen.

23.6.2 Results

For validation of the proposed algorithm, modal von Mises stress superposition has been performed in several different configurations. The considered frequency range is captured by superposition of 60 modes, representing a maximum eigenfrequency of 2198 Hz. Figure 23.5a shows the results for a plain superposition with no weighting factor for the mass normalised as well as for the stiffness-normalised modal von Mises stress fields, with identical results. For the presented investigations, a threshold value of half damage is chosen, with a fatigue ductility exponent of $c = 0.5$ for SAE grade 1045 steel in the LCF analysis, as proposed in Sect. 23.5.

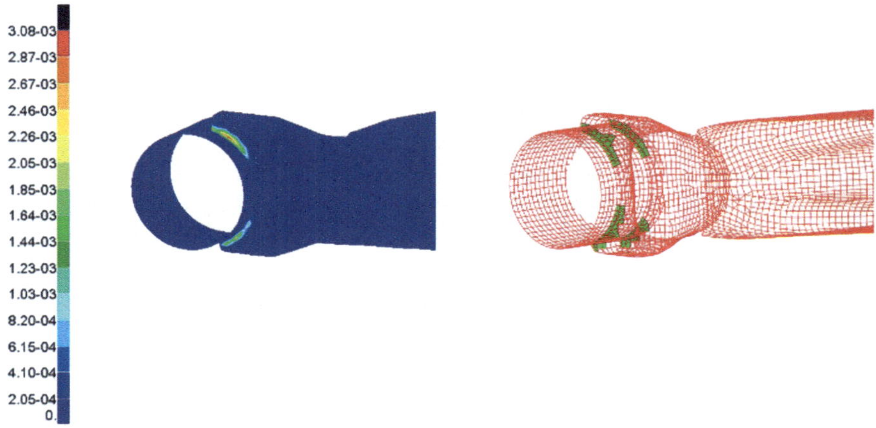

Fig. 23.4 Contour plot of damage and corresponding fatigue hot spot elements

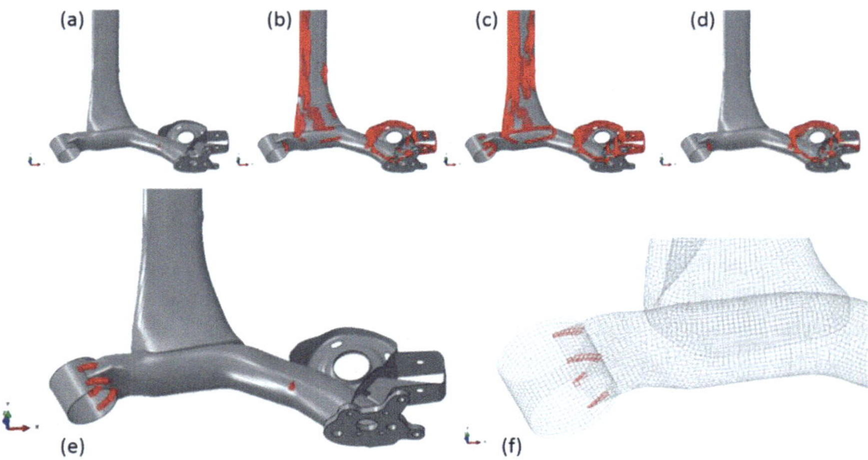

Fig. 23.5 Results of hotspot identification: **a** no weighting, **b** no weighting, threshold 0.15 **c** stiffness normalisation, threshold 0.15 **d** weighting with modal load coefficient threshold 0.15 **e** proposed weighting and threshold **f** reference solution

$$\sigma_{\mathrm{th,LCF}} = \frac{\sigma_{el}}{\sigma_{\max}} = 2^{-0.5} \qquad (23.12)$$

Figure 23.5a, b, c and d show, that no reliable information of high stressed regions or hot spot elements can be gained. To include true critical elements (see Fig. 23.5f for reference), threshold values would need to be decreased to 0.15 of maximum damage, resulting in an increase of false positive identification in Fig. 23.5b, while still not all critical elements are captured. As depicted in Fig. 23.5c, normalisation to modal stiffness does not improve these results, neither does a weighting with the modal load coefficient of mass normalised stress fields in Fig. 23.5d. These investigations show, that no efficient or reliable result can be achieved with conventional superposition techniques.

By application of the proposed modal damage participation, Fig. 23.5e shows that all critical elements could be identified with only a small number of false positive identification, validating the approach in combination with the proposed threshold values as reliable and efficient.

23.7 Summary and Conclusion

In this paper, a strategy for the identification of highly stressed regions, so-called hot spots, in complex components under multiaxial loading conditions has been presented. The goal of the investigation was to identify these fatigues critical regions prior to cost-intensive dynamic calculations by means of modal von Mises stress superposition on element level. Driven by investigations on mode normalisation and loading influence, a modal damage participation factor was developed for a weighted superposition of modal stress contributions. Starting with a transient dynamic FE-analysis of a component from industrial application, a fatigue analysis was carried out to build the bases for the validation of the proposed algorithm. For the identification of the fatigue analysis driven true hot spot elements as well as for the detection of potential hot spots in the proposed algorithm, threshold values were introduced and investigated.

The presented investigations show that by application of the proposed strategy a hot spot identification can be reliably performed on complex structures under multiaxial loading. In contrast to recent approaches, by introduction of the modal damage participation, reliability and performance can be increased to a great extent. As the only necessary input for the identification is the system's eigenvectors and the acting forces, the algorithm is applicable in early stages of product development. As the identification is performed prior to transient dynamic calculations, besides the hot spot filtering in fatigue analysis the result can furthermore act as an input to submodelling approaches, partitioned model order reduction [16] or optimisation algorithms.

References

1. Schijve J (2001) Fatigue of structures and materials. Kluwer Academic Publishers, Dordrecht
2. Nguyen VA, Zehn M, Marinković D (2016) An efficient co-rotational FEM formulation using a projector matrix. Facta Univ Ser Mech Eng 14(2):227–240
3. Fatemi A, Shamsaei N (2011) Multiaxial fatigue: an overview and some approximation models for life estimation. Int J Fatigue 33:948–958

4. Ince A, Glinka G (2014) A generalized fatigue damage parameter for multiaxial fatigue life prediction under proportional and non-proportional loadings. Int J Fatigue 62:34–41
5. Zhu S-P, Yu Z-Y, Correia J, De Jesus A, Berto F (2018) Evaluation and comparison of critical plane criteria for multiaxial fatigue analysis of ductile and brittle materials. Int J Fatigue 112:279–288
6. Craig R, Kurdila A (2006) Fundamentals of structural dynamics, 2nd edn. John Wiley & Sons Inc., Hoboken, New Jersey
7. Huang L, Agrawal H, Borowski V (1997) Durability analysis of a vehicle body structure using modal transient methods. In: Proceedings of the 15th international modal analysis conference, pp 407–414
8. Huang L, Agrawal H (2001) Method of identifying critical elements in fatigue analysis with von Mises stress bounding and filtering modal displacement history using dynamic windowing. U.S. Patent Nr. 6,212,486
9. Veltri M (2016) FEM techniques for high stress detection in accelerated fatigue simulation. J Phys Conf Ser 744
10. Zhou Y, Fei Q, Wu S (2017) Utilization of modal stress approach in random-vibration fatigue evaluation. Proc Inst Mech Eng part G J Aerosp Eng 231(14):2603–2615
11. Zhou Y, Tao J (2019) Theoretical and numerical investigation of stress mode shapes in multiaxial random fatigue. Mech Syst Signal Process 127:499–512
12. Yam LY, Leung TP, Li DB, Xue KZ (1996) Theoretical and experimental study of modal strain analysis. J Sound Vib 191(2):251–260
13. Stephens RI, Fatemi A, Stephens RR, Fuchs HO (2000) Metal fatigue in engineering. John Wiley & Sons Inc., Hoboken, New Jersey
14. Rama G, Marinkovic D, Zehn M (2018) High performance 3-node shell element for linear and geometrically nonlinear analysis of composite laminates. Compos B Eng 151:118–126
15. Marinković D, Rama G, Zehn M (2019) Abaqus implementation of a corotational piezoelectric 3-node shell element with drilling degree of freedom. Facta Univ Ser Mech Eng 17(2):269–283
16. Kerfriden P, Goury O, Rabczuk T, Bordas SPA (2013) A partitioned model order reduction approach to rationalise computational expenses in nonlinear fracture mechanics. Comput Methods Appl Mech Eng 256:169–188

Chapter 24
On Stochastic Method for Scale-Structural Failure Estimating and Structure Durability at Safety Operational Loading

E. B. Zavoychinskaya

Abstract This work is devoted to the problem of creation of reliability criterion taking into account technogenic and anthropogenic factors in relation to the assessment of pipeline operation safety. There are proposed relations for the function of the structure failure probability through the failure probability of its similar structural elements. The structural element durability is found taking into account the results of the analysis of the probability of metal failure at a certain level of accumulated defects. Assuming that the failure probability should not exceed its acceptable value (the criterion of structural reliability), the relation for finding the structure life is written. The classification of damaging factors at destruction of structural elements of product pipelines has been carried out. The ratios for social, industrial and environmental risks used in the calculation practice are given. The acceptable values of risks are considered. A safe operation criterion in the form of an inequality is proposed, in which the structural risk does not exceed an acceptable value, multiplied by a coefficient determined by the calculated and acceptable values of industrial, social and environmental risks. The criterion is a theoretical generalization of the known relationships used in design practice of investment projects on the construction and operation of various structures of product pipelines.

Keywords Structural reliability · Safe operation · Failure probability

24.1 Introduction

The purpose of this study is to develop the experimental and theoretical foundations of the stochastic method for assessing of longevity and diagnostic periods of the technical conditions of various long structures whose elements are at the internal pressure of the pumped product, the action of mass forces, temperature field and natural-climatic and technogenic influences. The methodological basis for the

E. B. Zavoychinskaya (✉)
Theory of Elasticity Department, Mechanical and Mathematical Faculty, Moscow State University Named After M.V. Lomonosov, Leninskie Gori, GSP-1, Moscow 119991, Russia
e-mail: elen@velesgroup.com

G. Lesiuk et al. (eds.), *Structural Integrity and Fatigue Failure Analysis*,
Structural Integrity 25, https://doi.org/10.1007/978-3-030-91847-7_24

proposed method of durability estimation is the works of scientists on the fundamental scientific problem of the technogenic safety of structure operation [1–5]. Here are considered three main technogenic spheres such as people, constructions and the environment, and, accordingly, are introduced the concepts of technogenic and anthropogenesis risks, namely the probability of injury to people (the social risk), the probability of destruction of industrial objects (the industrial risk) and the probability of flora and fauna destruction (the ecological risk) located in a potentially dangerous zone near the structure at destruction. The modern problems of safety estimation are to formulate the criteria of structural reliability, to study of the probabilities of the appearance and spread of injury factors at the structure destruction, to estimate of technogenic and anthropogenesis risks based on the operation data, to establish the acceptable risks, to develop of numerical modeling for durability and residual life finding taking into account the risks, to create of longevity management methods and to calculate of potential economic damage at the construction and operation.

The nature of the change in operating loads, a significant heterogeneity of the mechanical characteristics of materials, a variation in design technological factors, as well as the need to take into account of technological and operational defects, leads to need to use the probabilistic methods for estimation of the durability and crack resistance of structural elements. The probabilistic parameters of material properties (characteristics of crack opening, Woehler curve, Coffin-Manson equation and Paris relation, yield stress, etc.) are entered in the durability calculation, and random stationary loading processes are considered.

One of the main modern directions is the creation of algorithms for predicting the residual life of structural elements based on the established patterns of failure processes at the micro-, meso- and macro-levels.

24.2 The Criterion of Structural Reliability

Here is introduced the function $Q = Q(\tau), 0 \leq Q \leq 1, \ \tau \in [0, t]$ is determined as a probability of structure failure (the structural risk) at a time τ.

Structures such as product pipelines are divided into sections according to the functional and design principle: linear sections with branches and looping, crossings over natural and artificial obstacles (roads, railways, air crossings over water obstacles, ravines, underwater crossings, etc.), connection nodes of other structures, construction of gas and oil metering stations, gas recovery units, units of starting and receiving of cleaning devices, design of head and intermediate pumping stations, etc. So there is considered the probability $Q_k = Q_k(\tau)$ of k-section failure, $k = 1, \dots K$. The sections consist of a large number of similar q-elements $n_{k,q}, q = 1, \dots \mathbb{Q}$, of the same type (base metal, longitudinal and ring welded joints, tee connection, diversion, adapter, bottom) for which the probability $Q_{k,q} = Q_{k,q}(\tau)$ of q-element k-section failure is introduced. It is assumed that the failure probability $Q = Q(\tau)$ is determined through the failure probabilities $Q_k = Q_k(\tau)$ of k-section failure according as follows:

$$\text{optimistic scenario } Q(\tau) = \sum_{k=1}^{K}\left(\frac{Q_k(\tau)}{1 - Q_k(\tau)}\right)\prod_{k=1}^{K}[1 - Q_k(\tau)] \tag{24.1}$$

$$\text{pessimistic scenario } Q(\tau) = 1 - \prod_{k=1}^{K}[1 - Q_k(\tau)] \tag{24.2}$$

Relation (24.1) determines the sum of independent events, namely k-section failure in the absence of the remaining section failure, relation (24.2) is the sum of independent events at least of k-section failure. Analogically we have the following:

$$\text{optimistic scenario } Q_k(\tau) = \sum_{q=1}^{\mathbb{Q}}\left(\frac{Q_{k,q}(\tau)}{1 - Q_{k,q}(\tau)}\right)\prod_{q=1}^{\mathbb{Q}}\left[1 - Q_{k,q}(\tau)\right] \tag{24.3}$$

$$\text{pessimistic scenario } Q_k(\tau) = 1 - \prod_{q=1}^{\mathbb{Q}}\left[1 - Q_{k,q}(\tau)\right] \tag{24.4}$$

Let k-section consists of $n_{k,q}q$-elements of linear dimensions l_q, $q = 1, \ldots \mathbb{Q}$, with life $t_{f,k,q}$. It is proposed to describe the probability $Q_{k,q} = Q_{k,q}(\tau)$ by a function of the Poisson distribution type on the first failure of q-element:

$$Q_{k,q}(\tau) = \varphi_{k,q}(\tau)e^{1-\varphi_{k,q}(\tau)}, \quad \varphi_{k,q}(\tau) = \lambda_q l_q n_{k,q}\frac{\tilde{t}}{t_{f,k,q}}\tau, \; q = 1, \ldots \mathbb{Q}, \; k = 1, \ldots K. \tag{24.5}$$

In the relation (24.5) a parameter $\tilde{t}$ is the economically and socially acceptable structure life, assigned by the design standards. For example, it lies within 35–45 years for main pipelines, 60–65 years for tie pipelines, 15–20 years for field pipelines. Parameter λ_q is the intensity of the q-element failure flow, namely the number of q-element failure per unit of time (year) and per unit of length (km), known from the statistics of destruction at operation of similar structures in similar natural and climatic conditions. For tee connections, the number of failures per unit time attributed to the total number of elements is considered as λ_q, and relation (24.5) does not include the value l_q.

Relation (24.5) includes the durability of q-element k-section $t_{f,k,q}$, which is determined using a stochastic model of scale-structural fatigue [6–8]. The durability at operation loading is described by random processes taking into account the potential stochastic element failure, random mechanical loading, random environmental influences, etc. Therefore, a stochastic approach and methods of the theory of random processes and statistical analysis are chosen. As a toolkit for numerical experiments and solving practical problems, finite element methods of the ANSYS are used. There is proposed a stochastic method of safe operation assessing on the basis of a system of safe operation criteria for long structures, taking into account the determination of the structural element life using a stochastic model of fatigue scale-structural failure [9–12]. The model predicts the probability that a certain function of the structure, depending on the density and size of multilevel defects, reaches the limiting values during loading. The basic experiments for identifying of material functions are based on the standard tests of materials for long-term and fatigue strength.

The criterion of structural reliability is formulated as follows:

$$Q(\tau) \le \tilde{Q}, \tag{24.6}$$

on condition $t_{f,k,q} \ge \tilde{t}$, $k = 1, \dots K$, $q = 1, \dots \mathbb{Q}$, the probability $Q = Q(\tau)$ is determined by (24.1)–(24.5), $\tilde{Q}$ is acceptable structural risk according to design standards. The structure life is determined by the equation:

$$Q(t_f) = \tilde{Q} \tag{24.7}$$

The probability of failure $Q_R = Q_R(\tau)$, $t_R \le \tau$ (t_R is the total time of all R standard diagnostics), after carrying out R diagnostics and replacing structural elements with unacceptable defects is determined through the function of the probability $Q_{k,R} = Q_{k,R}(\tau)$ for k-section is expressed through functions (24.1), (24.2) as follows:

$$\text{optimistic scenario } Q_R(\tau) = \sum_{k=1}^{K} \left(\frac{Q_{k,R}(\tau)}{1 - Q_{k,R}(\tau)} \right) \prod_{k=1}^{K} [1 - Q_{k,R}(\tau)] \tag{24.8}$$

$$\text{pessimistic scenario } Q_R(\tau) = 1 - \prod_{k=1}^{K} [1 - Q_{k,R}(\tau)] \tag{24.9}$$

Correspondingly for $Q_{k,R} = Q_{k,R}(\tau)$ here can be written the following:

$$\text{optimistic scenario } Q_{k,R}(\tau) = \sum_{r=1}^{R} \left(\frac{Q_{k,r}(\tau)}{1 - Q_{k,r}(\tau)} \right) \prod_{r=1}^{R} [1 - Q_{k,r}(\tau)] \tag{24.10}$$

$$\text{pessimistic scenario } Q_{k,R}(\tau) = 1 - \prod_{r=1}^{R} [1 - Q_{k,r}(\tau)] \tag{24.11}$$

where the probability of k-section failure revealed during r-diagnostics $Q_{k,r} = Q_{k,r}(\tau)$ is expressed through the probability of q-element k-section failure revealed during r-diagnostics $Q_{k,r,q} = Q_{k,r,q}(\tau)$ in the form:

$$\text{optimistic scenario } Q_{k,r}(\tau) = \sum_{q=1}^{\mathbb{Q}} \left(\frac{Q_{k,r,q}(\tau)}{1 - Q_{k,r,q}(\tau)} \right) \prod_{q=1}^{\mathbb{Q}} \left[1 - Q_{k,r,q}(\tau)\right] \quad (24.12)$$

$$\text{pessimistic scenario } Q_{\kappa,r}(\tau) = 1 - \prod_{q=1}^{\mathbb{Q}} \left[1 - Q_{k,r,q}(\tau)\right] \quad (24.13)$$

The failure probability $Q_{k,r,q} = Q_{k,r,q}(\tau)$ for q-element of k-section on all defects detected by r-diagnostics, $1 \le r \le R$, is set through $Q_{k,r,q,j} = Q_{k,r,q,j}(\tau)$ as follows:

$$\text{optimistic scenario } Q_{k,r,q}(\tau) = \sum_{j=1}^{J} \left(\frac{Q_{k,r,q,j}(\tau)}{1 - Q_{k,r,q,j}(\tau)} \right) \prod_{j=1}^{J} \left[1 - Q_{k,r,q,j}(\tau)\right] \quad (24.14)$$

$$\text{pessimistic scenario } Q_{\kappa,r,q}(\tau) = 1 - \prod_{j=1}^{J} \left[1 - Q_{k,r,q,j}(\tau)\right] \quad (24.15)$$

The probability $Q_{k,r,q,j} = Q_{k,r,q,j}(\tau)$ of failure of $n_{k,q}q$ element k-section due to defects of j-type, $j = 1, \ldots J$, revealed by r-diagnostics, $r = 1, \ldots R$, is determined by the following relations:

$$Q_{k,r,q,j}(\tau) = \varphi(\tau) e^{1-\varphi(\tau)}, \quad \varphi(\tau) = \lambda_{r,q,j} l_q n_{k,q} \frac{\tilde{t}}{\Delta t_{f,k,q,j}} \tau, \quad q = 1, \ldots \mathbb{Q}, \quad k = 1, \ldots K. \quad (24.16)$$

In (24.8) l_q is the linear size of q-element, $\lambda_{r,q,j}$ are the coefficients of failure flow, i.e., the number of q-element failure on a defect of j-type, revealed by r-diagnostics or known from the failure statistics of failure (in this case $r = 1$) per unit of time (year) per unit of length (km), $\Delta t_{f,k,q,j}$ is the residual life of q-element of k-section on j-failure.

The criterion of structural reliability for structure after regulatory R diagnostics and replacing structural elements with unacceptable defects is also determined by (24.6) as follows:

$$Q_R(\tau) \le \tilde{Q}, \quad (24.17)$$

where $Q_R = Q_R(\tau)$ is the structural risk of k-section in the loading interval $[t_R, t]$ according (24.8)–(24.16), $\tilde{Q}$ is acceptable structural risk according to design standards.

Finally the residual life after regulatory diagnostics and replacing structural elements with unacceptable defects is determined from the Eq. (24.7) on the equation:

$$Q_R(\Delta t_f) = \tilde{Q} \quad (24.18)$$

24.3 The Safe Operation Criterion

Based on the analysis of literature and regulatory documents, the following main negative factors of damage from pipeline destruction are identified: toxic effects from the outflow of pumping toxic liquids and gases ($i = 1$); thermal effect from ignition of a gas jet flowing out from a through crack ($i = 2$); thermal effect when a cloud of gas-air mixture ignites ($i = 3$); shock air waves from gas and gas combustion product expansion ($i = 4$); defeat from the scattering of destroyed structure fragments ($i = 5$). These factors arise with a probabilities J_i, $i = 1, \dots 5$, accordingly, which are determined by the regulatory statistics of the appearance of negative factors at destructions of similar structures.

Here are considered the known notions of the social risk $I_1 = I_1(\tau),\ \ 0 \le I_1 \le 1,\ \tau \in [0, t]$ (the probability of injury to people, located in a potentially dangerous zone near the failure structure), the industrial risk, $I_2 = I_2(\tau),\ 0 \le I_2 \le 1,\ \tau \in [0, t]$ (the probability of destruction of industrial objects) and the ecological risk $I_3 = I_3(\tau),\ \ I_4 = I_4(\tau)\ \ 0 \le I_3 \le 1,\ \ 0 \le I_4 \le 1,\ \ \tau \in [0, t]$ (the probability of flora and fauna destruction accordingly) located in a potentially dangerous zone near the structure at destruction. Standards for structures determine the acceptable values of social, industrial and environmental risks $\tilde{I}_m,\ \ m = 1, \dots 4, m = 1, \dots 4,$ during the acceptable life $\tilde{t}$.

According to the developed approach [6–9] the safe operation criterion into account social, industrial and environmental risks according during the structure operation

$$Q(\tau)\mathrm{I}_m(\tau) \le \tilde{Q}\tilde{\mathrm{I}}_m,\quad m = 1, \dots 4, \tag{24.19}$$

$$\text{optimistic scenario } I_m = \sum_{i=1}^{5} \left(\frac{J_i I_{m,i}}{1 - J_i I_{m,i}} \right) \prod_{i=1}^{5} [1 - J_i I_{m,i}], \tag{24.20}$$

$$\text{pessimistic scenario } I_m = 1 - \prod_{i=1}^{5} [1 - J_i I_{m,i}], \tag{24.21}$$

$$I_{m,i} = \max \left\{ \int_0^R \int_0^{2\pi} \rho_m(r, \theta) I_i(r, \theta, \tau) r dr d\theta : \ \ 0 \le \tau \le t_i \right\},\quad m = 2, 3, \tag{24.22}$$

$$I_{m,i} = \max \left\{ \int_0^R \int_0^{2\pi} \rho_m(r, \theta, \xi) I_i(r, \theta, \tau) r dr d\theta : \ 0 \le \xi \le t_p;\ 0 \le \tau \le t_i \right\},$$
$$m = 1, 4, \tag{24.23}$$

$$I_1(r,\theta,\tau) = a_1 * \ln\left\{\left(\frac{D(r,\theta,\tau)}{D_1}\right)^2 \frac{\tau}{t_1}\right\}, \quad 0 \le \tau \le t_1, a_1, D_1 = \text{const}, \quad (24.24)$$

$$I_2(r,\tau) = a_2 * \ln\left\{\left(\frac{q(r,\tau)}{q_2}\right)^{4/3} \frac{\tau}{t_2}\right\}, \quad 0 \le \tau \le t_2, a_2, q_2 = \text{const}, \quad (24.25)$$

$$I_3(r,\tau) = a_3 * \ln\left\{\left(\frac{q(r,\tau)}{q_3}\right)^{4/3} \frac{\tau}{t_3}\right\}, \quad 0 \le \tau \le t_3, a_3, q_3 = \text{const}, \quad (24.26)$$

$$I_4(r,\tau) = a_4 * \ln\left(\left(\frac{p_0}{p(r,\tau)}\right)^{\alpha} + \left(\frac{I_0}{I(r,\tau)}\right)^{\beta}\right),$$
$$0 \le \tau \le t_4, a_4, p_0, I_0, \alpha, \beta = \text{const}, \quad (24.27)$$

$$I_5(r) = a_5 + b_5\left(\frac{mv(r,\tau)^2}{I_0}\right), \quad 0 \le \tau \le t_5, a_5, b_5, I_0 = \text{const}, \quad (24.28)$$

where (r, θ) is the polar coordinate system centered at the point of $i-$negative factor origin, $\rho_m = \rho_m(r, \theta, t)$ is the distribution function of people, objects, flora and fauna density in the zone $[0, R]$ depending on time, R is the radius of $i-$negative factor action, t_1 is time period of $i-$negative factor action. Function $D = D(r, \theta, \tau)$ in (24.24) is the concentration (referred to a unit of volume) of toxic substance at a point (r, θ) at a time τ, depending on gas density, average wind speed, intensity and duration of emissions and are determined by hydro-aerodynamics methods, constants (a_1, D_1, t_1) lie in the following range: $0.2 \le \alpha_1 \le 2.5, \quad 7 < -\alpha_1 \ln(D_1^2 t_1) < 60$. Function $q = q(r, \tau)$ in (24.25) is the heat flow (per unit surface) at the point r at the time τ, t_2 is the total time of jet burning. In (24.26) $q = q(r, \tau)$ is the heat flow (per unit surface) at the point r at the time τ, t_3 is the life of the fireball, constant a_3 is chosen equal $a_3 = 2.5$. In (24.27) functions $p = p(r, \tau)$ and $I = I(r, \tau)$ are the impulse and the maximum overpressure on the wave front, t_4 is the life of the fireball, constants are chosen as follows: $a_4 = -0.2, \quad p_0 = 40$ MPa, $I_0 = 450$ kg*m/s, $\alpha = 7.5, \quad \beta = 11.5$. In (24.28) m and $v = v(r, \tau)$ are, respectively, the mass and velocity of the fragment, t_5 is the time of fragment flight, and they are found on the solution of the problem about the shock destruction of pressure vessels, parameters a_5 and b_5 are chosen in the view $a_5 = 10.5, b_5 = -21$.

For gas pipelines of the Russian Federation, the federal acceptable risks of operation during an acceptable life are as follows: $I_1^* = (2 \times 10^{-4} - 2 \times 10^{-5})\left[\frac{\text{number of people}}{\text{km}}\right] * L_0$, $I_2^* = (10^{-3} - 10^{-4})\left[\frac{\text{number of objects}}{\text{sq. km}}\right] * S_0$, $I_3^* = (10^{-1} - 10^{-2})\left[\frac{\text{number of flora representatives}}{\text{sq. km}}\right] * S_0$, $I_4^* =$

$\left(10^{-2}-10^{-3}\right)\left[\frac{\text{number of fauna representatives}}{\text{sq. km}}\right] * S_0$, L_0 is the pipeline length, S_2 is the area of a potentially dangerous zone.

The time of dangerous operation of structure is found as a solution of the following equation according to (24.19):

$$Q\left(t_{f,m}\right)=\frac{\tilde{\mathrm{I}}_m}{\mathrm{I}_m}\tilde{Q} \quad m=1,\ldots 4, \tag{24.29}$$

In this case, the following inequality: $t_{f,m} \le t_f$ is fulfilled, where t_f is the structure life, determined by Eq. (24.7) without taking into account the social safety of its operation. The life $t_{f,m}$ $m=1,\ldots 4$, the risk-adjusted longevity considered social, industrial and environmental risks correspondingly.

The proposed safe operation criterion was applied for predicting of the longevity of oil and gas pipelines. A number of Conclusions on the life and residual life of various designs of gas and oil pipelines with a certain level of accumulated defects were prepared [6–9].

References

1. Zavoychinsky BI (1992) Durability of main and technological pipelines (theory, calculation methods, design). Nedra, Moscow
2. Klyuev VV (ed) (2003) Engineering. Encyclopedia. T. IV-3. Machine reliability
3. Makhutov NA (2018) Russia's safety. Law, social, economical, scientific and technical aspects. MGF Knowledge, Moscow, p 1016
4. Makhutov NA, Matvienko YuG, Romanov AN (eds) (2018) Problems of strength, technogenic safety and structural material science, Lenand, Moscow
5. Makhutov NA (2017) Safety and risks: system research and development. Nauka, Novosibirsk
6. Zavoychinsky BI, Zavoychinskaya EB, Volchanin AV (2012) Probabilistic assessment of the residual operational safety of long structures. Directory Mag Eng J 7:41–46; 12:33–36
7. Zavoychinsky BI, Giller GP, Zavoychinskaya EB (2002) Recommendations for assessing of the safety and durability of gas pipelines during design. IRC Gazprom, Moscow
8. Zavoychinsky BI, Tutnov IA, Zavoychinskaya EB (2000) Recommendations for assessing of the safety and durability of main gas pipelines during design. IRC Gasprom, Moscow
9. Zavoychinskaya EB (2018) Fatigue large-scale structural failure and durability of structures under proportional loading processes, Author's abstract. Dr. Dissertation, Moscow, Ltd. "Genesis", p 46. https://istina.msu.ru/publications/book/153957031/
10. Zavoychinskaya EB (2019) About the criterion of scale-structural metal fatigue at complex stress state. In: Proceedings of the XII all-Russian congress on fundamental problems of theoretical and applied mechanics, vol 3, Ufa, pp 607–609. https://doi.org/10.22226/2410-3535-2019-congress-v3
11. Zavoychinskaya EB (2019) Fatigue fracture of metals under complex stress state with consideration of structural changes. Moscow Univ Mech Bull 74(2):36–40. https://doi.org/10.3103/S002713301902002X
12. Zavoychinskaya EB (2020) On the theory of scale structural fatigue of metals at the proportional loading. J Phys 1431:012024–012032. https://doi.org/10.1088/1742-6596/1431/1/012024

Chapter 25
Experimental Study of the Rolling Friction Coefficient in Highly Loaded Supports of Rotary Kilns

Igor Kuzio, Volodymyr Gursky, Pavlo Krot, Radoslaw Zimroz, and Tatyana Sorokina

Abstract The results of rolling friction experimental research of two cylindrical rollers on the special testing stand, which imitates the operation of a rotating kiln support unit, are presented in this article. The rolling friction moment is determined using the special plate clutch on the driving roll shaft. The rolling peripheral velocity, radial load and the rollers contact surfaces condition are the variable parameters. The actual values of the rolling friction coefficient under different rolling conditions are calculated. The regularities of rolling friction coefficient changes from the rolling speed, oil viscosity and contact pressure from the radial load, which is corresponding to the real values in the basic support units of industrial rotary kilns, are established. The wear intensity effect on the rolling friction moment is determined. The range of the rolling friction coefficient, which should be taken into account in the calculations of power losses in support units of large-dimension rotational assemblies, are recommended.

Keywords Rotary kiln · Rolling friction coefficient · Tyre wear · Contact pressure · Rollers support unit

I. Kuzio · V. Gursky
Lviv Polytechnic National University, Lviv 79013, Ukraine
e-mail: kuzo@polynet.lviv.ua

P. Krot (✉) · R. Zimroz
Faculty of Geoengineering, Mining and Geology, Wroclaw University of Science and Technology, 50-370 Wroclaw, Poland
e-mail: pavlo.krot@pwr.edu.pl

R. Zimroz
e-mail: radoslaw.zimroz@pwr.edu.pl

T. Sorokina
Admiral Makarov National University of Shipbuilding, Mykolaiv 54025, Ukraine
e-mail: super_tan-sorokina@ukr.net

G. Lesiuk et al. (eds.), *Structural Integrity and Fatigue Failure Analysis*,
Structural Integrity 25, https://doi.org/10.1007/978-3-030-91847-7_25

25.1 Introduction

The large-dimension rotating technological equipment for firing, grinding, granulating, drying and mixing various raw materials are used in many industrial fields. This equipment includes the followings: rotate kilns, dryers, mills, granulators, grizzlies, which can have from 2 to 9 support units, in which the process of tyres ring rolling over the support rollers takes place during the operation (Fig. 25.1).

The total weight of such plants can be from 100 to 5000 ton and the total load F on the one supporting unit approaches to 10 MN. The rotation speed n depends on assemblies purposes and is 1–20 *rpm*, but due to a big diameter, the linear rolling speed v is 0.12–1.8 m/s in the contact zone. In the general case, the rolling and sliding frictional losses in the support units can be up to 30% of the total drive power (1000–2000 kW).

Start-ups of such heavy rotating plants cause significant transient dynamics in the gears and drivetrain shafts [1] as a result of their excessive wear [2]. In particular, due to the body geometric axis curving and different changes of its position, the stress state in the shell is significantly changed and the support loads rise, which greatly increases the contact pressures between the supports elements [3]. The distribution of the contact pressures along the contact length is irregular [4] due to the implicit gaps [5], which are impossible to control by the methods known for drivelines of heavy machinery [6] and vibration diagnostics techniques [7]. Methods of motor current analysis [8] can only be useful in the assessment of energy consumption. In this case, the support rollers misalignment concerning the tyres (non-parallelism of their axes of rotation) can further increase the contact pressure to $p = 1000$–1600 MPa, which leads to additional losses due to the frictional process. Typical patterns of tyres wear due to different factors of kiln operation and maintenance are shown in Fig. 25.2.

The influence of shell axis deviation on the radial forces and bending moments in supports is represented in Fig. 25.3a. The difference in the support position of even 4–5 mm may cause a significant increase in loads [10]. The optimization problems are solved in [11] to minimize the loads on the supports.

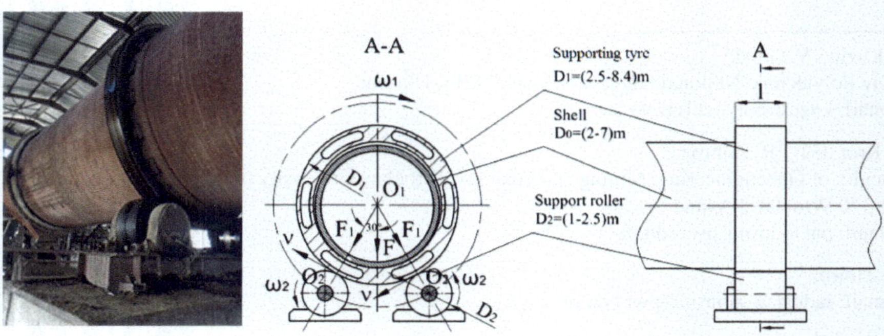

Fig. 25.1 **a** Rotary kiln and **b** scheme of support unit with rollers and tyre

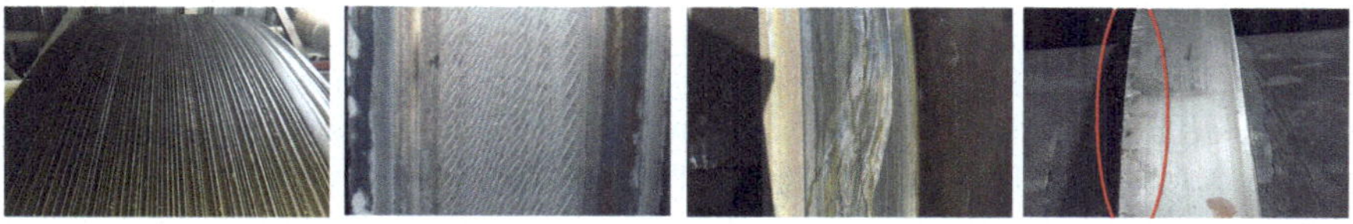

(a) (b) (c) (d)

Fig. 25.2 Types of wear in the rotary kiln tyres [9]: **a** groove-type pattern when the drum is out of alignment; **b** spalling-type pattern due to excessive wheel skewing; **c** gouging-type wear from obstruction on the tyre; **d** wear due to lubrication penetration into metal causing cracks

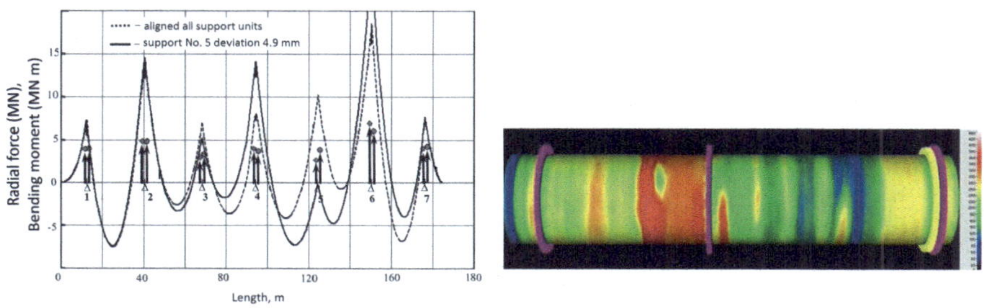

(a) (b)

Fig. 25.3 Radial forces with bending moments in seven supports on the kiln shell (**a**); the heating of the kiln sections (www.hgh-infrared.com) (**b**)

The loads increase due to the kiln shell geometric axis changes and that is why according to Hertz's theory [12] the contact pressures increases. The geometrical characteristics of the contact zone can be determined following the work [13], taking into account the main force and mechanical characteristics of the support unit materials.

It is necessary to pay attention to the changes in the contact pressures during the rolling and corresponding hysteresis losses in the contact zone [14]. Besides, the operation of these assemblies is accompanied by a high-temperature gradient along the kiln length (Fig. 25.3b), which has a negative impact on strength [15], reliability and energy consumption. In particular, the redistribution of contact pressures and increasing the rolling resistance moment in the contact pair takes place [16].

The temperature accelerates an appearance of cracks in the shell, the support rollers are worn out quickly, and corrosion and fatigue cracks take place on their working surfaces [17, 18]. Therefore, the rotational assemblies support units require very thorough and complex analysis, where the condition of the surface and its real roughness class should be taken into consideration [19].

The following parameters, such as the rolling and sliding friction coefficients, which are obtained in specific technical conditions, are generally used in the calculations of energy consumption in the rotational assembly and the required drive power estimation. The noted parameters may not fully correspond to the actual operating conditions of a particular assembly and its support units. For example, it was established that the rolling friction coefficient nonlinearly depends on the normal load and rolling speed, and the last makes the extreme characteristics view for the rolling friction coefficient [20]. The authors of available researches interpret the recommended rolling friction coefficients as not dependant on the geometric, force and kinematic parameters. Therefore, experimental research of rolling friction is necessary for kiln models verification and multiaxial fatigue life prediction [21].

This article represents the main results of experimental research of the rolling tyre interaction with the support rollers at the specially made test stand. During the research, it is established the regularities of the rolling friction coefficient changes from such factors as rolling speed, radial loading, condition of a surface and oil viscosity with an estimation of contact pressures, which are corresponding to the real conditions on the industrial equipment.

25.2 The Methodology of Research

A typical scheme of two cylindrical rollers motion, which can take place in different technical systems, is shown in Fig. 25.4.

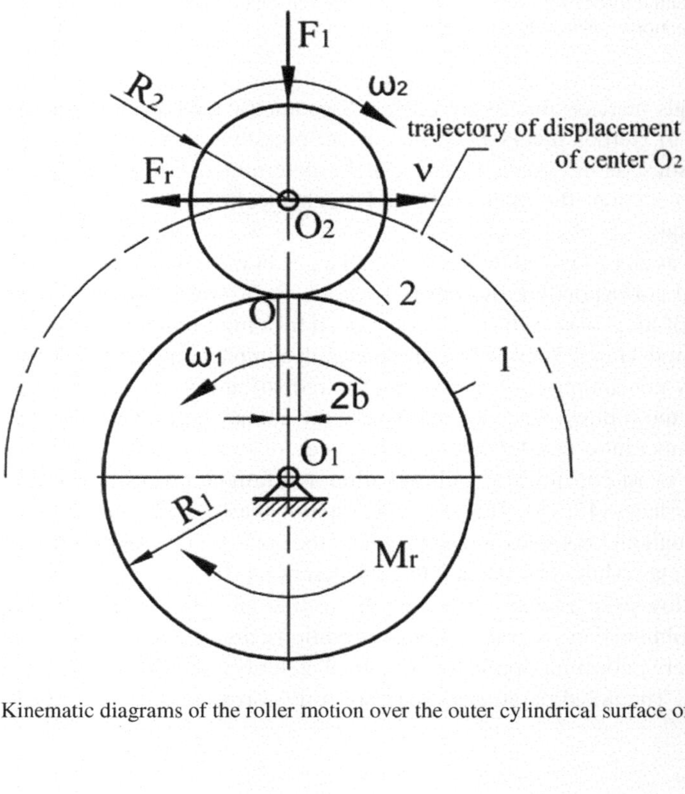

Fig. 25.4 Kinematic diagrams of the roller motion over the outer cylindrical surface of the tyre

The rolling friction moment M_r intended for determination by the experiments is a fundamental parameter, which consists of the sum of the rolling resistance force F_r moments occurring during the rolling of the roller 2 around the wheel 1:

$$M_r = F_r(R_1 + R_2) \tag{25.1}$$

According to [13, 14], the pressure at the point of contact of the tyre and support roller is determined by the formula:

$$p = \frac{2F_1}{\pi bl}, \tag{25.2}$$

where l is the length of the contact area; b is half of the contact area width:

$$b = \sqrt{\frac{2F_1}{\pi l} \cdot \frac{\eta_1 + \eta_2}{\frac{1}{D_1} + \frac{1}{D_2}}} \tag{25.3}$$

where $\eta_i = \frac{1-\upsilon_i^2}{E_i}$, $(i = 1, 2)$; υ_i and E_i—the Poisson's ratios and Young's modulus of tyre materials and support roller; $D_1 = 2R_1$ and $D_2 = 2R_2$ are respectively tyre and support roller diameters.

Next, it is necessary to take into account the contact parameters to determine the rolling friction moment. For this case, the rolling process balance condition is considered, taking into account the contact scheme, which has a general view:

$$F_1 b = F_r R \tag{25.4}$$

where $R = \frac{R_1 \cdot R_2}{R_1 + R_2}$ is the equivalent radius of curvature of the contact surfaces.

From Formula (25.4) we obtain the correlation:

$$f_r = \frac{b}{R} = \frac{F_r}{F_1} \tag{25.5}$$

which is interpreted as a dimensionless of rolling friction coefficient, as well as a formula for the force rolling resistance calculation:

$$F_r = f_r F_1 \tag{25.6}$$

The calculation of the rolling friction moment which is taking into account the actual operating conditions has fundamental importance for practice. Therefore, the rolling friction moment M_r the formula will have the form taking into account Formulas (25.2) and (25.6):

$$M_r(p) = f_r(p)pl\sqrt{\pi Rq(\eta_1 + \eta_2)}(R_1 + R_2) \tag{25.7}$$

where $f_r(p)$ is the rolling friction coefficient, which is determined experimentally taking into account the rolling conditions and the effective contact pressure p; $F_1 = pl\sqrt{\pi Rq(\eta_1 + \eta_2)}$—radial force; $q = F_1/l$—linear contact load.

Formula (25.7) can be used for determining the rolling friction moments for different sizes and loads, based on the available values of the contact pressure p, the length of the contact elements l, the radius R and the linear contact load q. In this case, it is necessary to use the relation $f_r(p)$, obtained for known rolling contact parameters and surface condition, namely, lubricated with a certain oil viscosity. For this purpose, it is necessary experimentally to determine the value of the rolling friction moment M_r at the different parameters of the rolling speed v and contact pressure p.

25.3 Experimental Stand

The studies of rolling friction coefficients are performed on an experimental stand (Fig. 25.5), which imitates the process of the tyre rolling on the support rollers, by the analogy of the pneumatic tyres rolling friction study [22]. The experimental stand is based on three similarity parameters (geometric, kinematic and dynamic) for to the operating equipment.

The geometric similarity is established by the condition of equality of the correlations of corresponding diameters on the operating assembly D_{real} and on the experimental stand D_{model}, as well as contact lengths l_{real} and l_{model}:

$$K_1 = \frac{D_{real}}{D_{model}} \equiv \frac{l_{real}}{l_{model}} \tag{25.8}$$

(a) (b)

Fig. 25.5 Testing stand for the rolling friction research: **a** diagram, **b** general view

The kinematic similarity criterion is established based on equality of the rolling speeds on the full-scale kiln υ_{real} and on the experimental model υ_{model}:

$$\upsilon_{\text{real}} = \upsilon_{\text{model}}, \quad \text{or } \omega_{\text{real}}\frac{D_{\text{real}}}{2} = \omega_{\text{model}}\frac{D_{\text{model}}}{2} \tag{25.9}$$

As a result, the rotation elements frequency on the model n_{model} is determined by the formula:

$$n_{\text{model}} = K_1 n_{\text{real}} \tag{25.10}$$

Dynamic similarity characterizes the proportionality of the forces and equality of the contact pressures, which takes place in the support unit of the rotary kiln p_{real} and of the model p_{model}. Using the condition:

$$p_{\text{real}} = p_{\text{model}} \tag{25.11}$$

and Formulas (25.2) and (25.3) it is obtained the relation between the forces on the model F_{model} and the full-scale kiln F_{real}:

$$F_{\text{model}} = \frac{F_{\text{real}}}{K_1^2} \tag{25.12}$$

The most common rotary kiln Ø5 × 185 m was chosen as a prototype. The geometric scale for modelling the process of tyres rolling friction on the support rollers is selected as $K_1 = 10$. The experimental model influential force and kinematic characteristics are determined using Formulas (25.10) and (25.12): $n_{\text{model}} = 10 n_{\text{real}}$, $F_{\text{model}} = 0.01 F_{\text{real}}$. Thus, during the stand designing, the rolling conditions with the actual power and kinematic parameters are taken into account.

In the designed experimental stand the rotation from the DC motor 1 ($P = 6.5$ kW, $n = 1450$ rpm) is transmitted through a conical–cylindrical gearbox 2 with a gear-ratio $i = 40$ and the plate clutch 3 to the shaft 7. The design of the installation includes the wheel 15 mounted on the shaft 7, which is assumed as the tyre. The working force F_1 is provided by a variable load Q 13, which rises at the free end of the beam 11. The counterweight 14 is rise through the unit 12 and proposed to remove the preload formed by the own weight of the boom. The force parameter is determined by the formula:

$$F_1 = Q\frac{l_1}{l_2} = 14.9Q \tag{25.13}$$

where Q is the weight of the variable load 13.

The rolling friction moment is investigated using the constant wheel 15 and a replaceable roller 8 (Fig. 25.6a). The drive wheel 15 with a dimension Ø640 × 100 mm

(a)

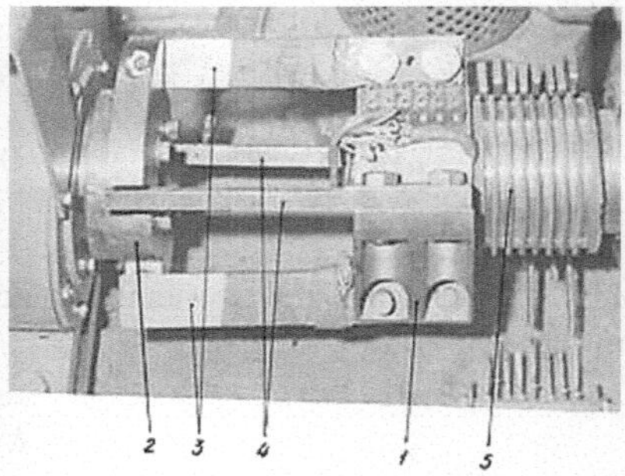

(b)

Fig. 25.6 General view of testing stand: **a** rolling pair; **b** the plate clutch in the driving gear

is made of steel 50 L $\left(E_1 = 2.19 \times 10^5 \text{ MPa}, \quad \upsilon_1 = 0.288\right)$with a working surface hardness of 190–200 HB, yield strength 340 MPa, tensile strength 580 MPa. Replaceable roller 8 with dimensions Ø180 × 100 mm is made of steel St3 $\left(E_2 = 2.05 \times 10^5 \text{ MPa}, \quad \upsilon_2 = 0.25\right)$ with a working surface hardness of 120–190 HB, yield strength 235 MPa, tensile strength 490 MPa. In the real kilns, the carbon steel grades are used for tyre and rollers respectively for the sake of low cost, although different methods are available of contact surfaces hardening and increasing the wear resistance [23].

The number of load cycles of the experimental models is measured by a counter 5 connected with a flexible shaft 4 to a low-speed gearbox shaft. Working surfaces lubrication during the rolling friction research is carried out by the wheel 15 immersing to the bath 6 with oil.

The plate clutch (Fig. 25.6b), which was mounted between the gearbox output shaft and the shaft 7 with the wheel 15, is used for the friction moment measuring. The designed plate clutch consists of driving and driven half-coupling 1 and 2, working and safety plates 3 and 4. Safety plates are served to prevent the breakage of working plates at the sudden stops or the overloads. They do not transfer the loads under the normal conditions, because they are installed with a gap concerning the driven half-coupling.

One of two working plates made of 40Cr steel with a hardness of 45HRC is installed on the driving half of coupling. On these plates, the strain gauges are glued on both sides in the longitudinal directions and they are switched on to the measuring bridge opposite diagonals. The plate bends due to the loading. As a result, the strain gauge gives a signal proportionally to the applied load. The speed of both rollers rotation is measured by the contact sensors 16. The rolling friction moment is recorded via the strain gauge amplifier and the ampere metre into the computer.

25.4 Results of Experiments

The experiments were performed under conditions that imitate the operation of the rotary kiln support unit. The load in the contact F_1 and the applied weight Q were determined based on the contact stress equality, which takes place between the tyre and the roller on the full-scale equipment and the experimental stand test rollers. The experiments were performed step-by-step with a speed $v = 0.134-1.34$ m/s and the load $F_1 = 7.2-42.5$ kN, which corresponds to a contact pressure changes by (25.4) within $p = 193-469$ MPa.

25.4.1 Rolling Moment

The actual rolling friction moment M_r was defined as a difference between the moment measured by the plate clutch and the rolling bearings rolling friction moment measured beforehand. The measured rolling friction moment M_r (Fig. 25.7) due to the rolling speed v firstly decreases to a certain minimum value and then rises due to the increasing speed.

The rolling friction coefficient f_r was determined from Formula (25.9) taking into account the applied radial load F_1:

$$f_r = \frac{M_r}{F_1(R_1 + R_2)} \tag{25.14}$$

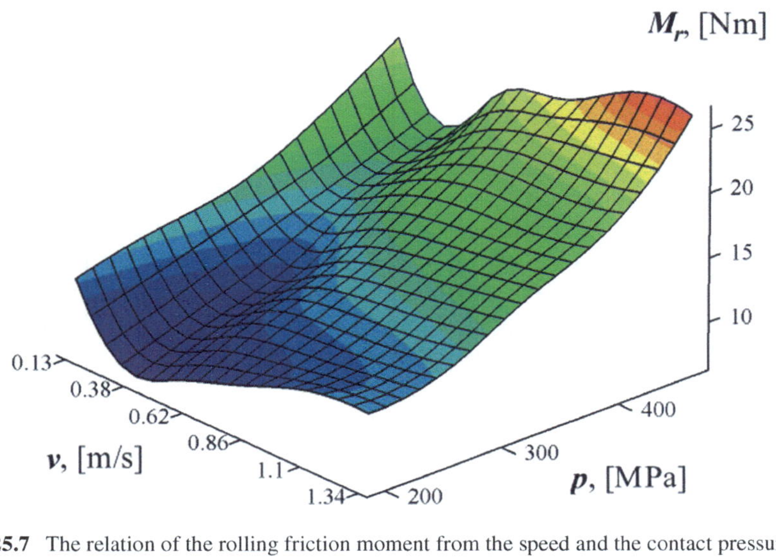

Fig. 25.7 The relation of the rolling friction moment from the speed and the contact pressure

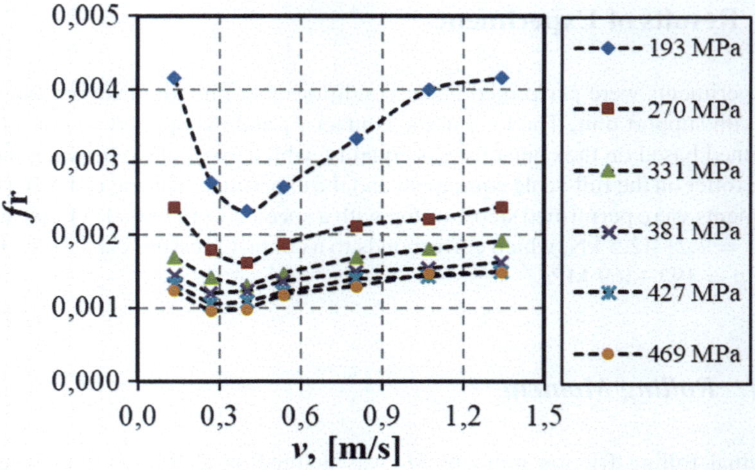

Fig. 25.8 The dependences of the rolling friction coefficient on the circular speed at different contact pressures

25.4.2 Rolling Speed

The rolling friction coefficient f_r of the speed v dependence at different radial loads F_1 has an extremum point with a minimum value (see Fig. 25.8). In the speed range of $v = 0.4-1.34$ m/s the rolling friction coefficient f_r rises, because of the contact force F_1 increase. At the same time, a larger range of the rolling resistance coefficient changes is observed at lower contact pressures.

25.4.3 Effect of Contact Pressure

The rolling friction coefficient f_r decreases within a range $f_r = 0.004-0.001$ by the power function with the load F_1 increasing and accordingly the contact pressure p (see Fig. 25.9). At the same time, the deviations of the rolling friction coefficient are much smaller at the higher contact pressures. Due to that fact, if the load increases, the rolling friction coefficient stabilizes up to the initial values during the speed changes that is why it can be assumed that it does not depend on the speed at the calculating stage. The contact load effect is determinant, which is figured in the calculation Formula (25.7).

Based on the rolling friction coefficient experimental data, the regression formula is obtained by means the values averaging at the intermediate pressure points:

$$f_r(p) = 27.977p^{-1.689} \tag{25.15}$$

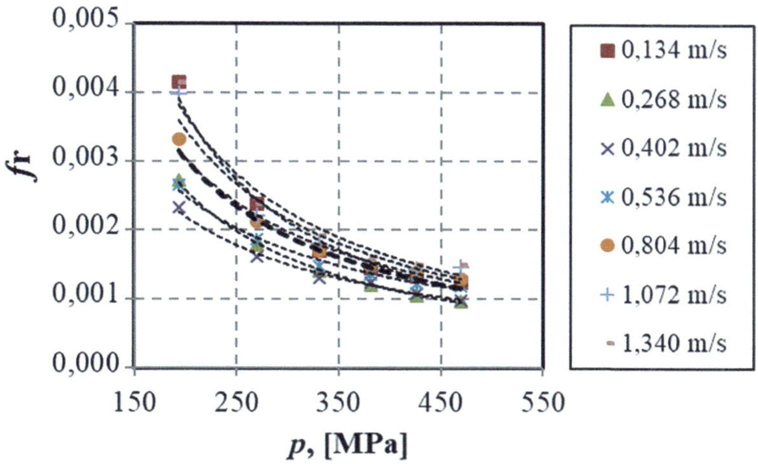

Fig. 25.9 Rolling friction coefficient dependence on the working force at the different velocities

The obtained relation approximates the dependence of rolling friction coefficient within a contact pressure range $p = 193–469$ MPa.

25.4.4 Effect of Oil Viscosity and Abrasive

The effect of surfaces lubrication and rollers wear is investigated at the contact pressure $p = 469$ MPa. The influence on the cylinder rollers lubrication viscosity on rolling friction coefficient is shown in Fig. 25.10. The presence of low-viscosity oil (0.12–1.57 stokes) on the rolling surfaces practically does not change the rolling friction coefficient at all rolling speeds. At the same time, if the oil with a higher viscosity (4 stokes) is used, that increases the rolling friction coefficient by 1.1–1.5 times in comparison with no lubricated surfaces.

In general, the rolling friction moment is sharply increasing at the higher rolling speeds. This phenomenon can be explained based on the hydrodynamic theory of lubrication law by the contact interaction of two surfaces, according to which the force of viscous shear in the oil layer must be proportional to the oil viscosity and the rolling speed. The presence of abrasives (sand, cement) on the rolling surfaces leads to an increase in the rolling friction coefficient by 1.2–1.5 times in comparison with the dry cleaned surfaces.

The results of the research are presented in Table 25.1 where the main regularities of the rolling friction coefficient changes are established. In particular, due to the three-fold rolling speed increasing, the rolling friction coefficient increases by approximately 50% for the whole load range, both for dry and lubricated rolling surfaces.

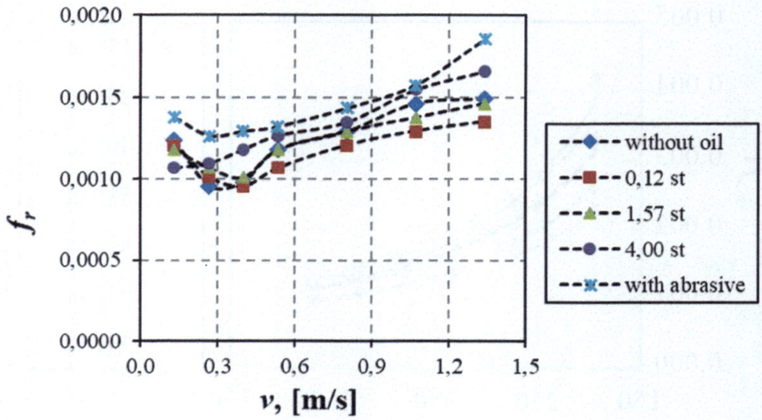

Fig. 25.10 Dependence of the rolling friction coefficient f_r on the rolling speed v under the different conditions of lubrication at the contact pressure $p = 469$MPa

Table 25.1 Summary values of rolling friction coefficients $f_r \times 10^{-3}$

Rolling speed, [m/s]		Dry surfaces			Lubricated surfaces		
		0.4	0.8	1.34	0.4	0.8	1.34
Contact pressure, [MPa]	270	1.62	2.13	2.38	2.56	3.80	4.05
	331	1.30	1.69	1.92	2.27	3.03	3.25
	381	1.24	1.49	1.62	2.15	2.61	2.69
	409	1.09	1.39	1.49	1.61	2.09	2.10
	469	0.98	1.29	1.49	1.01	1.29	1.46

The lubrication influence on the rolling friction coefficient (up to 70%) is greater at the lower loads and is observed in all ranges of rolling speeds. If the load increases by 3 times, the rolling friction coefficient reduces by 40% for dry surfaces and more than 60% for lubricated surfaces.

25.4.5 Effect of Loading Cycles

During the research, the rolling duration effect is observed of the numbers N of the load cycles. Rolling torque decreases over time almost by 2 times to a certain value, then remaining almost constant (Fig. 25.11).

To obtain a clearer understanding of the number of load cycles effects on the tyre-roller pair capacity, the wear of the rollers working surfaces is investigated. The value of the linear wear is determined as the difference between the diameters of the samples before and after the tests. For this purpose, indicators are installed on special corbels, which could move radially and measure the changes in diameter at

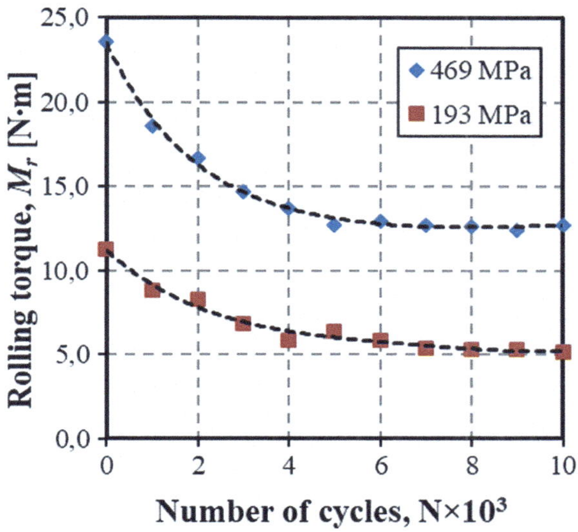

Fig. 25.11 The rolling friction moment changes from the numbers of load cycles

certain points (Fig. 25.6a). The diagrams of changes in absolute radial wear W_a (μm) and specific wear rate W_r, (mm^3/N m) of a replaceable cylindrical roller made of steel 35 grade with the number of load cycles N during the experiments are shown in Fig. 25.12.

The absolute wear W_a at both low and high contact stresses rises sharply during the run-in period up to the numbers of cycles $N = 10 \times 10^3$, then the intensity of wear

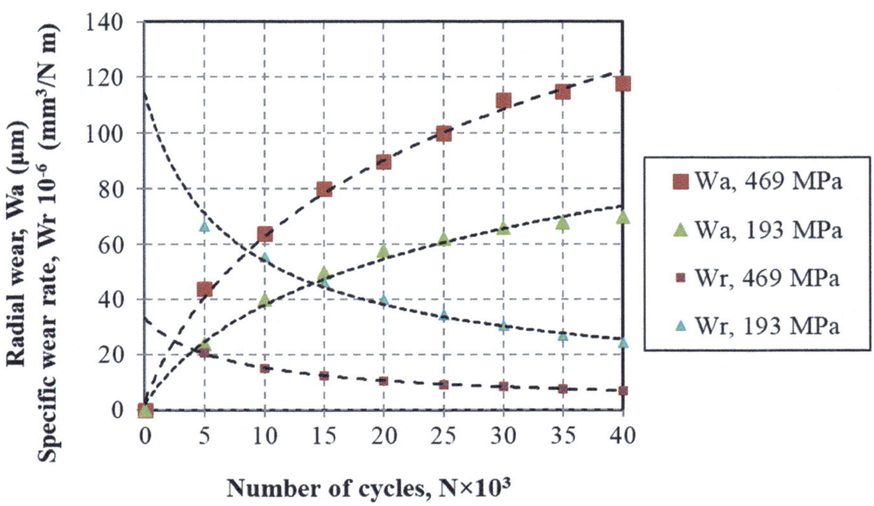

Fig. 25.12 The dependence of wear on a cylindrical driven roller by the number of test rotations

decreases. Specific wear rate W_r, which is a wear volume normalized by the numbers of cycles (length of rolling) and applied radial load, similarly decreases during the first load cycles and then stabilizes. It takes place due to plastic deformations accompanied by the hardening of the working surfaces.

The testing process resulted in a 20–30% increase of rollers surface hardness after all experiments. A detailed rollers material microstructure analysis showed certain changes in under surface layers where plastic deformation takes place.

25.4.6 Case Study of Friction Losses Calculation

The obtained dependences allow determining the predicted power losses due to rolling friction in the support units of the real rotating assemblies. In particular, for the basic rotary kiln Ø5 m × 185 m, the design characteristics are $D_1 = 6.4$ m, $D_2 = 1.8$ m, $l = 1$ m, the angle between the vertical axis and rollers $\alpha = 30°$ (see Fig. 25.1).

The radial load on the support unit is about $F_1 = 0.72-3.53$ MN and the contact pressure between the roller and the tyre under such conditions according to Formula (25.2) and (25.3) is within a range $p = 193-469$ MPa. The rolling friction coefficient by the (25.15) is $f_r = 0.0039-0.0013$. Taking into account the f_r values and Formula (25.7), the rolling friction moment is $M_r = 11.4-14.6$ kN m.

The rolling friction power losses with two rollers in one support of rotary kiln is $P = 2M_r\omega_1 = 6.3-8.2$ kW at the rolling speed $v = 0.877$ m/s. If the number of the support units is 7, then the total frictional losses will be $P_\Sigma = 44.1-57.4$ kW or 9.8–12.8% of the drive power 450 kW.

25.5 Conclusions

Based on the experimental results it is recommended to assume in design the rolling friction coefficient $f_r = 0.0039$–0.0013 at the contact pressures $p = 193$– 469 MPa where the lower pressure corresponds to the higher values of coefficient and vice versa.

Explicit non-linearity of rolling friction coefficient is observed in both speed and contact pressure relations with a minimum value for rolling speed about 0.4 m/s.

Contact lubrication with moderate values of viscosity (0.12–1.57 St) has minimal effect on rolling friction coefficient, while high viscosity oil (4 St) similar to an abrasive presence in a dry contact has notably increased the resistance to rolling.

Wear of contacting rollers has a non-linear relationship with the number of rotations (cycles) and proportionally increases with the contact pressure.

The obtained results can generally be used in the design calculations of rolling frictional power losses and wear prediction in the maintenance and repairs planning of the continuously working rotational heavy machines with roller supporting units, in particular, the drum mills, kilns, etc.

Acknowledgements This activity has received partial funding from the European Institute of Innovation and Technology (EIT), a body of the European Union, under the Horizon 2020, the EU Framework Programme for Research and Innovation. This work is supported by EIT RawMaterials GmbH under Framework Partnership Agreements No. 18253 (OPMO. Operation monitoring of mineral crushing machinery).

References

1. Van Dyk DJ, Pretorius L (1995) Analysis of dynamic effects in a rotary kiln system used for iron production. R&D J 11(1):12–20
2. Krot PV (2010) Dynamics and diagnostics of the rolling mills drivelines with non-smooth stiffness characteristics. In: Proceedings—3rd international conference of nonlinear dynamic (ND-KhPI2010), Kharkiv, Ukraine, pp 115–120
3. Cherepanov GP (2014) Theory of rolling: solution of the coulomb problem. J Appl Mech Tech Phys 55:182–189
4. Zhou X, Liu Y-L, Zhao X-Q et al (2002) Mechanical model and contact stress emulational analysis of rotary kiln's tyre. J Cent South Univ Technol 33(5):526–529
5. Yang XY, Xiao YG, Lei XM et al (2013) Contact pressure of loose-fitted tyre under intermittent contact. Adv Mater Res 815–817:1015–1018
6. Krot PV (2008) Methods and instrumentation for measuring wear in drivelines of rolling mills. Metal Proc and Equip 2:45–53
7. Żak G, Obuchowski J, Wyłomańska A, Zimroz R (2014) Novel 2D representation of vibration for local damage detection. Mining Sci 21:105–113
8. Putnoki AY, Klevtsov OM, Ermolenko AA et al (2003) Evaluation of operation of equipment at the rolling mill. Stal' 10:56–58
9. Peppin C, Carlson C (2020) Causes of tire and trunnion wear. https://feeco.com/causes-of-tire-trunnion-wear. Last accessed 30 Nov 2020
10. Kuzio IV, Dzyubik LV (2007) The influence of the position of the geometric axis on the strength of rotating units. Bull Nat Univ Lviv Polytechnica 688:53–57
11. Xiao Y, Li X, Chen X (2008) General solution to kiln support reactions and multi-objective fuzzy optimization of kiln axis alignment. Struct Multidiscip Optim 36:319–327
12. Li X-J, Shen Y-P, Wang Y-Q et al (2006) The contact finite element analysis of support structure of large-scale rotary kiln with multi-supporting. Eng Mech 23:109–113
13. Jurkiewicz A, Pyryev Y (2011) Compression of two rollers in sheet-fed offset printing machine. Acta Mech Autom 5(4):58–61
14. Greenwood JA, Minshall H, Tabor D (1961) Hysteresis losses in rolling and sliding friction. Proc R Soc London. Ser A Math Phys Sci 259(1299):480–507
15. Žiga A, Kačmarčik J (2017) Stress state in rotary kiln support rollers. Mašinstvo 1(14):3–10
16. Tadić B, Kocovic V, Matejić M et al (2016) Static coefficient of rolling friction at high contact temperatures and various contact pressure. Tribol Ind 38(1):83–89
17. Deshpande V, Dhekhane A (2014) Contribution to kiln tyre contact stress analysis. Int J Innov Res Sci Eng Technol 3(2):9500–9504
18. Tharoon T (2016) Analysis of rotary kiln support roller by using analytical method and FEA software. Int J Res Appl Sci Eng Technol 4(12):603–611
19. Wang H, Hu Y, Gao F et al (2015) Nominal friction coefficient in spread formulas based on lead rolling experiments. Trans Nonferrous Met Soc China 25(8):2693–2700

20. Scaraggi M, Persson B (2014) Rolling friction: comparison of analytical theory with exact numerical results. Tribol Lett 55:15–21
21. Shen Y, Wang S, Li X et al (2010) Multiaxial fatigue life prediction of kiln roller under axis line deflection. Appl Math Mech Engl Ed 31:205–214
22. Wiegand BP (2016) Estimation of the rolling resistance of tires. SAE Tech Paper, SAE International, Warrendale, PA
23. Krot P, Bobyr S, Dedik M (2017) Simulation of backup rolls quenching with experimental study of deep cryogenic treatment. Int J Microstruct Mater Prop 12(3/4):259–275

Chapter 26
Safe Operation of Underground Mining Vehicles Based on Cyclic Fatigue Monitoring of Powertrains

Pavlo Krot, Radoslaw Zimroz, Pawel Sliwinski, and Norbert Gomolla

Abstract This research is dedicated to the problem of automatic monitoring of the operational loading of heavy vehicles for minerals transportation in harsh conditions of underground mines. The torsional vibrations excited in the powertrains by the driver's control actions (running, loading, reverse) are investigated on the multi-body nonlinear dynamical model, which accounts diesel engine torque characteristics, changes in transmission structure due to gears switching, lock-up activation in the hydraulic torque converter and angular clearances in the Cardan shafts couplings. The slowly sampled real signals of the on-board monitoring system are combined with the fast modelled transient responses in the range of torsional natural modes. The obtained results of simulations allow observing and optimisation of the most severe working modes to reduce amplitudes of torsional vibrations, which cause the abrupt failures in the vehicles. Besides, the database is under development for fatigue cycles accumulation and remained useful life prediction of powertrains elements for diagnostics procedures and maintenance planning.

Keywords Load-haul-dump (LHD) vehicles · Torsional vibration · Nonlinear model · Cyclic fatigue

P. Krot (✉) · R. Zimroz
Faculty of Geoengineering, Mining and Geology, Wroclaw University of Science and Technology, 50-370 Wroclaw, Poland
e-mail: pavlo.krot@pwr.edu.pl

R. Zimroz
e-mail: radoslaw.zimroz@pwr.edu.pl

P. Sliwinski
KGHM Polska Miedź S.A, 59-301 Lubin, Poland
e-mail: pawel.sliwinski@kghm.com

N. Gomolla
DMT GmbH & Co. KG, 45307 Essen, Germany
e-mail: norbert.gomolla@dmt-group.com

G. Lesiuk et al. (eds.), *Structural Integrity and Fatigue Failure Analysis*,
Structural Integrity 25, https://doi.org/10.1007/978-3-030-91847-7_26

26.1 Introduction

In the underground mines, the heavy load-haul-dumpers (LHD) and hauling trucks (HT) are operated in harsh conditions and subjected to severe dynamic loads due to hard regimes of powertrains loading and intensive steering by operators [1]. To prevent abrupt failures and optimise maintenance process, the monitoring systems are installed on board of vehicles [2, 3]. The obtained data allow determining operational cycles of machines by different signals [4, 5], but the high-frequency quickly damped transient torsional vibrations of large amplitudes are beyond of observation. Their experimental investigation in the complicated transmissions with a split-path structure [6] or switchable gears [7] in heavy industrial machines requires the installation of strain gauges and special telemetry torque metres on the shafts [8] with subsequent identification of natural oscillations modes and nonlinear behaviour [9, 10]. The analysis of torsional vibrations allows improving machine control [11].

The experimental research of automotive drivetrain on the laboratory testing rig showed response up to 200 Hz, which is also predicted by the 5-degree-of-freedom lumped parameters model [12]. The vibration properties of the powertrain system have a great influence on the cyclic fatigue damages due to possible resonances at different ranges of shafts rotations and gears meshing frequencies, while the mean engine torque is the main factor causing the fatigue damage of gear parts [13].

To estimate a residual (remaining) useful life (RUL), the different prognosis models are under development to accurately predict the crack growth of powertrain components and internal combustion engines, e.g. cracked cylinder heads, intake valves [14], based on both physics-based modelling and data-driven approaches [15].

Nevertheless, permanent torsional vibration monitoring systems are not characteristic for mining machines due to harsh conditions and issues of permanent power supply on the rotating shafts. Methods of electrical motors currents monitoring for machines operation analysis [16] will probably be available in case of wide implementation of battery-powered underground vehicles. For a while, the mathematical modelling of vehicles is the main method to study dynamic behaviour under different working conditions. However, some issues complicate the simulation. These are the nonlinear characteristics of drivetrain elements, e.g. angular clearances [17], which are difficult to measure by known methods [18], or hydraulic torque converter [19] with lock-up, or manual shifting of gears, which change the structure of drivetrain. The lock-up, besides the influence on the torsional dynamics, can decrease fuel consumption at higher speeds [20] and affect some other operational parameters of heavy underground vehicles.

Since the underground vehicles have no spring suspension units because of height restrictions, one of the most important elements influencing their dynamics and performance are the tyres. The large-diameter wheels, having huge weight, with corresponding inertial moments affect the natural modes of torsional vibrations in the powertrains. Besides, tyre protector configuration and its intensive wear directly influence on torsional dynamics when the vehicle is accelerating and wheels are subjected to stick–slip contact interaction with the uneven or watered road surface

[21]. Since the underground heavy vehicles have no spring suspension, tyres stiffness and damping properties determine operator's comfort under the impacts of large pieces of bulk material and machine safety [22] in case of tyres failure.

26.2 Underground Mining Vehicles

The design of underground vehicles (see Fig. 26.1) is adopted to operate in the low tunnels (less than 2 m) with frequent reversal motions and sharp turns in a confined space, as well as running on the usually uneven and watered road surface with slopes. They have an articulated body and four driven wheels. To counterbalance the weight of carried bulk material, front or back placement of diesel engines is opposite to the bucket depending on vehicle function.

Design of powertrains in such vehicles is similar for a variety of main producers (see Fig. 26.2). Powertrains include diesel engine combined with hydraulic torque converter (TC) and manual gearbox, several Cardan shafts and differential type axles. TC includes the lock-up function so that to increase power transfer efficiency at high speeds. From the other side, TC efficiently reduces impacts on transmission at the moments when the machine starts motion under load.

The on-board monitoring systems collect a wide variety of digital signals from different sensors and electronic control units (ECU) via the internal CAN bus, which connects different subsystems (see Fig. 26.3). Based on these signals, the operator's assistance modules and remote control systems are acting. Collected data is transferred via Wi-Fi to the server once per working shift. Unfortunately, the majority of signals are sampled with above the 1 s time interval because of certain technical restrictions.

Fig. 26.1 Underground mining vehicles: load-haul-dumper (**a**); hauling truck (**b**)

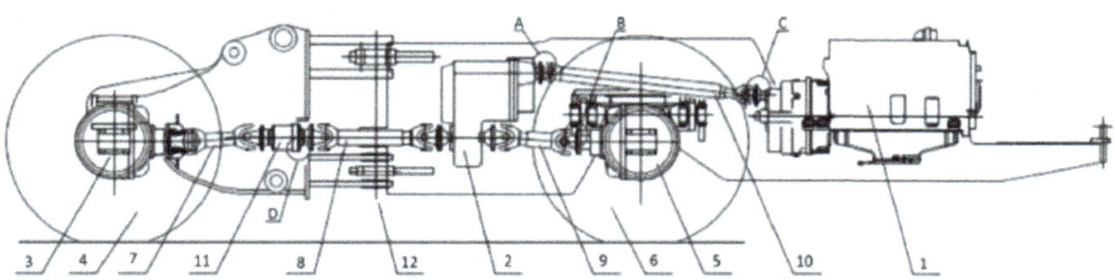

Fig. 26.2 Design of the underground mining vehicle: 1—diesel engine with hydraulic TC; 2—gearbox; 3, 4—front axle with a limited-slip differential, brakes and wheels; 5, 6—rear axle with brakes and wheels; 7, 8, 9, 10—Cardan shafts; 11—intermediate support; 12—articulation joints; A, B, C, D—couplings with potential angular clearances

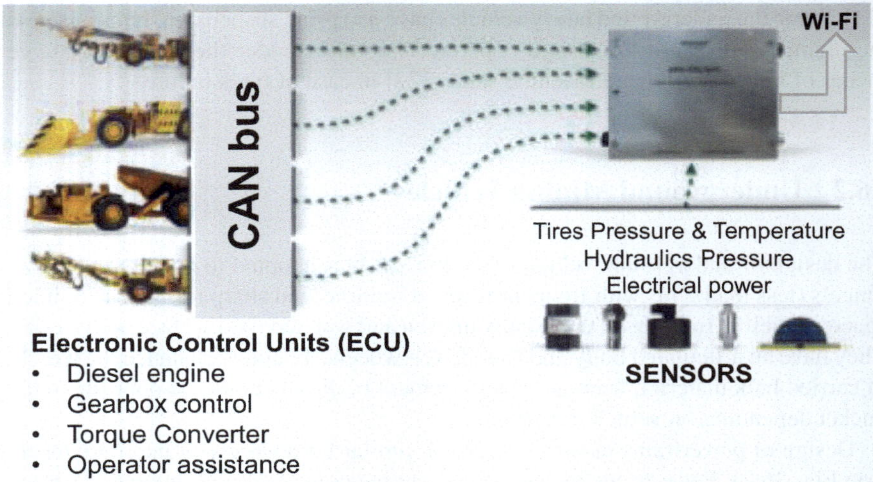

Fig. 26.3 On-board monitoring systems of underground vehicles

Table 26.1 Parameters for dynamical model identification and simulation

No.	Parameters	Description	Units
1	ENGTPS	Acceleration pedal position	%
2	SELGEAR	Gear selection	– 4.0.0.0.4
3	TRNBPS	Brake pedal position	%
4	BRAKEP	Brake pressure	kPa
5	TRNLUP	State of lock-up	0/1
6	SPEED	Vehicle speed	km/h
7	ENGRPM	Engine rotation speed	Rpm
8	ENGTORQ	Engine output torque	%

In the proposed approach, time series from the on-board monitoring system (see Table 26.1) are used as input parameters for dynamical model simulation.

26.3 Dynamical Model

When designing the dynamical model, certain issues encountered in the practice of underground vehicles operation should be taken into account, e.g. zero values of monitoring data, missing data due to failures of sensors and cables, different parameters of ECU tuning (engine, gearbox), variable road conditions (water,

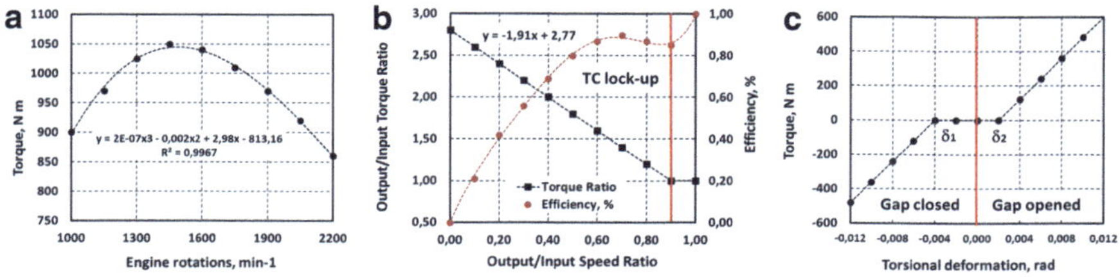

Fig. 26.4 Nonlinear characteristics of elements in transmission: **a** DEUTZ engine function of torque by rotation speed; **b** torque converter with lock-up; **c** Cardan shafts with backlash

gradient, unevenness) and rocks properties, specific driving manner of each operator. Therefore, the dynamical model of an underground vehicle is based on certain assumptions:

- Inertial moments of the engine crankshaft, pistons, rods and flywheel are constant.
- Engine torque is non-linearly related to the speed of rotation (Fig. 26.4a).
- Speed of engine rotation is linearly proportional to acceleration pedal position.
- Torsional stiffness of wheels tyres is constant and not dependent on deformation.
- Braking forces on all wheels are equal and linearly depends on the pedal position.
- Torque converter output is linearly related to the speed ratio of shafts (Fig. 26.4b).
- Stiffness of Cardan shafts is nonlinear (Fig. 26.4c) and independent on inclination.
- Interaction of torsional vibration and spatial motions is neglected.

The calculation scheme of the powertrain model with a scheme of a torque converter (TC) is shown in Fig. 26.5.

The systems of differential equations of the dynamical model in absolute angles of rotation and relative angles of shafts torsional deformation are as follows:

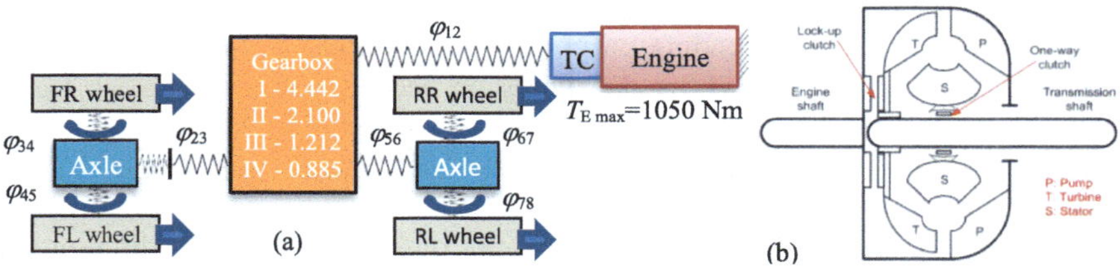

Fig. 26.5 Calculation scheme of torsional vibrations (**a**); TC scheme (**b**)

$$\begin{cases} J_1\ddot{\varphi}_1 = T_E(n_E,t) - T_{\text{TC}} \\ J_2\ddot{\varphi}_2 = T_{\text{TC}} \cdot i_G - T_{\text{GF}} - T_{\text{GR}} \\ J_3\ddot{\varphi}_3 = T_{\text{GF}} - T_{\text{AFR}} \cdot i_{\text{DF}} - T_{\text{AFL}} \cdot (1 - i_{\text{DF}}) \\ J_4\ddot{\varphi}_4 = T_{\text{AFR}} \cdot i_{\text{DF}} - T_{\text{WFR}} - T_B(t) \\ J_5\ddot{\varphi}_5 = T_{\text{AFL}} \cdot (1 - i_{\text{DF}}) - T_{\text{WFL}} - T_B(t) \\ J_6\ddot{\varphi}_6 = T_{\text{GR}} - T_{\text{ARR}} \cdot i_{\text{DR}} - T_{\text{ARL}} \cdot (1 - i_{\text{DR}}) \\ J_7\ddot{\varphi}_7 = T_{\text{ARR}} \cdot i_{\text{DR}} - T_{\text{WRR}} - T_B(t) \\ J_8\ddot{\varphi}_8 = T_{\text{ARL}} \cdot (1 - i_{\text{DR}}) - T_{\text{WRL}} - T_B(t) \end{cases}$$

$$\begin{cases} \ddot{\varphi}_{12} = (T_E(n_E,t) - T_{\text{TC}})/J_1 - (T_{\text{TC}} - T_{\text{GF}} - T_{\text{GR}})/J_2 \\ \ddot{\varphi}_{23} = (T_{\text{TC}} \cdot i_G - T_{\text{GF}} - T_{\text{GR}})/J_2 - (T_{\text{GF}} - T_{\text{AFR}} \cdot i_{\text{DF}} - T_{\text{AFL}} \cdot (1 - i_{\text{DF}}))/J_3 \\ \ddot{\varphi}_{34} = (T_{\text{GF}} - T_{\text{AFR}} \cdot i_{\text{DF}} - T_{\text{AFL}} \cdot (1 - i_{\text{DF}}))/J_3 \\ \quad -(T_{\text{AFR}} \cdot i_{\text{DF}} - T_{\text{WFR}} - T_B(t))/J_4 \\ \ddot{\varphi}_{45} = (T_{\text{GF}} - T_{\text{AFR}} \cdot i_{\text{DF}} - T_{\text{AFL}} \cdot (1 - i_{\text{DF}}))/J_3 \\ \quad -(T_{\text{AFL}} \cdot (1 - i_{\text{DF}}) - T_{\text{WFL}} - T_B(t))/J_5 \\ \ddot{\varphi}_{56} = (T_{\text{TC}} - T_{\text{GF}} - T_{\text{GR}})/J_2 - (T_{\text{GF}} - T_{\text{ARR}} \cdot i_{\text{DR}} - T_{\text{ARL}} \cdot (1 - i_{\text{DR}}))/J_6 \\ \ddot{\varphi}_{67} = (T_{\text{GF}} - T_{\text{ARR}} \cdot i_{\text{DR}} - T_{\text{ARL}} \cdot (1 - i_{\text{DR}}))/J_6 \\ \quad -(T_{\text{ARR}} \cdot i_{\text{DR}} - T_{\text{WRR}} - T_B(t))/J_7 \\ \ddot{\varphi}_{78} = (T_{\text{GF}} - T_{\text{ARR}} \cdot i_{\text{DR}} - T_{\text{ARL}} \cdot (1 - i_{\text{DR}}))/J_6 \\ \quad -(T_{\text{ARL}} \cdot (1 - i_{\text{DR}}) - T_{\text{WRL}} - T_B(t))/J_8 \end{cases} \tag{26.1}$$

where φ_{i}, φ_{ij}—angles of rotation and shafts deformation; J_{i}—moments of inertia; $T_{\text{E}} = f(n_{\text{E}}, t)$—engine torque, depending on n_{E}—rotations; $T_{\text{TC}} = T_{\text{E}}(n_{\text{E}}, t) f_{\text{TC}}(n_{\text{E}}, n_{\text{G}})$—TC output torque, depending on input n_{E} and output n_{G} speed ratio; $T_{\text{B}} = f(n_{\text{B}})$—braking torque on a wheel, depending on pedal position n_{B}; T_{WFR}, T_{WFL}, T_{WRR}, T_{WRL}—rolling resistance torques on wheels, depending on road conditions and tyre wear; T_{GF}, T_{GR}—intermediate shafts torques; T_{AFR}, T_{AFL}, T_{ARR}, T_{ARL}—torques in front and rear axles; i_{G}—a current ratio of the manual gearbox; i_{A}—a constant ratio of the axle; i_{DF}, i_{DR}—variable ratios of front and rear axle differential, depending on vehicle steering and road conditions.

Viscous damping and angular clearances in the elastic joints are introduced via corresponding coefficients and functions. The resonance regimes from periodical excitation of a diesel engine are possible for simulation concerning output shaft speed of rotation, but not considered in this research. The overall torsional dynamics greatly depends on angular clearances in couplings under non-stationary loading conditions. This effect is accounted with the estimated wear intensity of every shaft coupling by the service time. Additionally, methods of angular gaps diagnostics can be applied in vehicles powertrains to estimate real values of clearances.

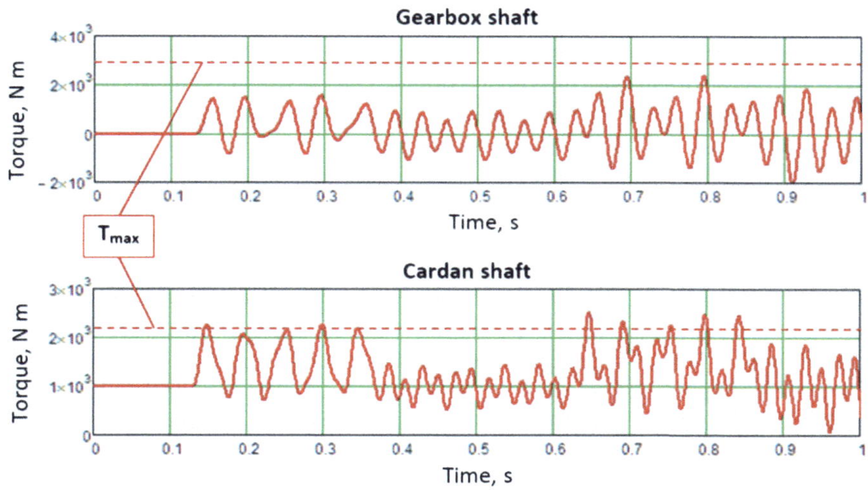

Fig. 26.6 Transient responses to impacts in powertrain: gearbox and Cardan shafts

The examples of torsional vibrations simulation are represented in Fig. 26.6, where gearbox shaft and front intermediate Cardan shaft torques are given with their endurance limiting torques T_{max}. These values are recalculated from the stress for corresponding steel grades taken from the powertrain components specification. The amplitude and composition of natural modes in transient signals depend on frequency response characteristics in every elastic shaft.

Further, the calculation procedure is implemented for fatigue cycles accumulation in the database for the powertrain elements depending on working hours of vehicles. The example of information on remained useful life (RUL) of certain component (Cardan shaft No. 2) retrieved from the database is shown in Fig. 26.7. The S–N diagrams (Wohler curve) are used to determine cyclic fatigue in elements. Goodman's

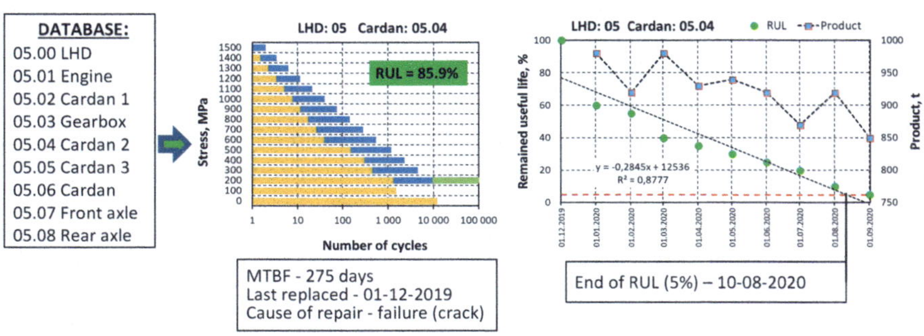

Fig. 26.7 Remained useful life (RUL) calculation based on data from the monitoring system

law is used to process combined asymmetrical cycles of loading and transform them into equivalent reversal stress cycles.

The prediction of the date for a predetermined level of RUL (5–10%) achievement is based on standard Miner's rule of damage accumulation. The convenient representation of RUL (see Fig. 26.7) with appropriate working load in terms of transported tonnage of materials helps to schedule maintenance and production together. Other useful information such as mean time between failures (MTBF), dates and reasons of replacement allows analysing the reliability of elements and the vehicle in general.

Nonlinear stiffness (gaps) in different elements of powertrain resulted in more scattered torsional loads and a correspondingly wider range of accumulated fatigue cycles depending on regimes of vehicle operation (material loading, acceleration, quick stopping). The most dangerous regime for the powertrain of LHD vehicles is the intensive digging into a hill of blasted bulk material. Unavoidable slipping of wheels causes sharp rising and drops of torques in transmission. Torque converter is able to dampen some impacts to a certain extent. At the same time, the reverses of LHD with additional weight in the bucket also causes torsional vibrations of large amplitudes.

Full-scale vehicle simulation requires additional signals from the monitoring system—articulation angle steering (up to $\pm$ 40°). Then, more real ratios i_{DF}, i_{DR} in front and rear axle differentials can be accounted in torsional vibrations calculations.

26.4 Conclusions

The nonlinear dynamical model of the articulated underground vehicle is developed, which allows determining the most severe working regimes from the viewpoint of peak torsional loads in elements of transmission.

Including such dynamical models into on-board monitoring systems gives a possibility to calculate and accumulate in the database fast cycles of high amplitude by the slowly sampled data from the on-board monitoring systems that greatly increase the RUL prediction accuracy of elements and safe operation without abrupt failures.

In case of thorough registration of maintenance actions and working hours of vehicles, the developed approach of model-based fatigue monitoring can be applied for maintenance actions and parts replacement planning.

Also, using dynamical models helps in the assessment of each operator manner of control and improvement of working regimes from the reliability point of view.

Acknowledgements This activity has received funding from the European Institute of Innovation and Technology (EIT), a body of the European Union, under the Horizon 2020, the EU Framework Programme for Research and Innovation. This work is supported by EIT Raw Materials GmbH under Framework Partnership Agreement No. 17031 (MaMMa. Maintained Mine and Machine).

References

1. Trushin N, Antsev A (2020) Combined torque converter for mining machines. IOP Conf Ser Earth Environ Sci 459042078
2. Sliwinski P, Kaniewski T, Hebda-Sobkowicz J et al (2019) Analysis of dynamic external loads to haul truck machine subsystems during operation in a deep underground mine. In: Proceedings 39th international symposium on application of computers and operations research in the mineral industry (APCOM-2019), Taylor and Francis, pp 515–524
3. Zimroz R, Wodecki J, Król R et al (2013) Self-propelled mining machine monitoring system—data validation, processing and analysis. In: Drebenstedt C, Singhal R (eds) Mine planning and equipment selection. Proceedings of the 22nd MPES conference, Springer, Cham, pp 1285–1294
4. Krot P, Sliwinski P, Zimroz R et al (2020) The identification of operational cycles in the monitoring systems of underground vehicles. Measure J Int Measure Confederation 151:107111
5. Stefaniak P, Zimroz R, Obuchowski J et al (2015) An effectiveness indicator for a mining loader based on the pressure signal measured at a bucket's hydraulic cylinder. Proc Earth Planet Sci 15:797–805
6. Krot P (2019) Dynamical processes in a multi-motor gear drive of heavy slabbing mill. J Vibroeng 8:2064–2081
7. Krot PV (2010) Dynamical model of geared drive with changeable structure of high-speed wire rolling block. Vestnik NTU KhPI Prob Mech Drive 27:96–109
8. Krot PV (2008) Telemetering systems for monitoring dynamic loads in drive lines of rolling mills. Vibr Mach Measure Reduction Prot 1:46–53
9. Yoshida K, Takamatsu H, Matsumoto S (2016) Nonlinear identification of torsional driveshaft vibrations in a full-scale automotive vehicle during acceleration. Nonlinear Dyn 86:711–721
10. Krot PV, Korennoy VV (2016) Nonlinear effects in rolling mills dynamics. In: Proceeding of the 5th international conference on nonlinear dynamics ND-KhPI2016, Sep 27–30, 2016, Kharkiv, Ukraine, pp 117–124
11. Wang Y, Zhu G, Zhang F (2013) Powertrain control parameter optimisation using HIL simulations of a heavy-duty vehicle. Int J Powertrains (IJPT) 2(1):24
12. Galvagno E, Velardocchia M, Vigliani A (2016) Torsional oscillations in automotive transmissions: experimental analysis and modelling. Shock Vib 2016:5721960
13. Bai J, Wu X, Gao F et al (2017) Analysis of powertrain loading dynamic characteristics and the effects on fatigue damage. Appl Sci 7(10):1027
14. Azadi M (2016) Failure analysis and prevention in powertrain systems. In: Hamdy AS, Aliofkhazraei MM (eds) Handbook of materials failure analysis with case studies from the aerospace and automotive industries, Butterworth-Heinemann, pp 471–492
15. Neerukatti RK, Liu KC, Kovvali N, Chattopadhyay A (2014) Fatigue life prediction using hybrid prognosis for structural health monitoring. J Aerosp Inf Syst 11(4):211–232
16. Putnoki AY, Klevtsov OM, Ermolenko AA et al (2003) Evaluation of operation of equipment at the rolling mill. Stal' 10:56–58
17. Krot PV (2010) Dynamics and diagnostics of the rolling mills drivelines with non-smooth stiffness characteristics. In: Proceeding 3rd international conference of nonlinear dynamics (ND-KhPI2010), Kharkiv, Ukraine, pp 115–120
18. Krot PV (2003) Methods and instrumentation for measuring wear in drivelines of rolling mills. Metall Process Equip 2:45–53
19. Asl HA, Azad NL, McPhee J (2012) Modeling torque converter characteristics in automatic drivelines: lock-up clutch and engine braking simulation. In: Proceedings ASME international design engineering technical conference, computers and information, vol 6, Chicago, Illinois, USA, pp 359–367
20. Kaniewski T, Sliwinski P, Hebda-Sobkowicz J et al (2019) Comprehensive, experimental verification of the effects of the lock-up function implementation in LHD haul trucks in the deep

underground mine. In: Proceeding of the 39th international symposium on application of computer and operations research in the mineral industry (APCOM-2019), Taylor and Francis, pp 506–514
21. Gipser M (2007) FTire—the tire simulation model for all applications related to vehicle dynamics. Veh Syst Dyn 45(S):139–151
22. Verbas VV, Naumenko AP et al (2010) To the formation of a calculation-experimental model of a safe wheel mover. Issues Chem Chem Technol 3:89–94

Chapter 27
Optimization of Wastes Compaction Parameters in Case of Gradual Wear of the Briquetting Press Rolls

Kostiantyn Baiul, Sergii Vashchenko, Aleksander Khudyakov, Pavlo Krot, and Nataliia Solodka

Abstract This paper represents the study of the briquetting presses operation under conditions of rolls gradual wear and diversity of bulk raw materials with varying fractions and physical–mechanical properties. The model of the contact wear is developed to predict product quality (briquettes density) and energy-power parameters of the compaction process. The experimentally established wear patterns of the replaceable tires have a good correlation with predicted values. Based on the model simulations, the optimized modes of briquetting process are recommended at different stages of rolls wear, which are tested in a production plant. This approach provided tires service up to 11 mm of wear instead of 5–6 mm with satisfying the required quality of briquettes made of silicomanganese with an organic binder and ensured the limitations on available mechanical drive loading and electric motor power. The results of the research are the basis for the development of monitoring, controlling systems, and maintenance optimization, taking into account the gradual wear of the tires.

Keywords Roller briquetting press · Tire profile wear · Briquette quality

K. Baiul · S. Vashchenko · A. Khudyakov
Z.I. Nekrasov Iron and Steel Institute of National Academy of Sciences of Ukraine, 1 Acad. Starodubov Sq, Dnipro 49005, Ukraine

P. Krot (✉)
Faculty of Geoengineering, Mining and Geology, Wroclaw University of Science and Technology, 15 Na Grobli st, 50-421 Wroclaw, Poland
e-mail: pavlo.krot@pwr.edu.pl

N. Solodka
Information System Department, State Higher Educational Institution, Ukrainian State University of Chemical Technology", 8 Gagarin Ave, Dnipro 49005, Ukraine
e-mail: solodka_n_o@ukr.net

G. Lesiuk et al. (eds.), *Structural Integrity and Fatigue Failure Analysis*, Structural Integrity 25, https://doi.org/10.1007/978-3-030-91847-7_27

27.1 Introduction

In the modern metallurgical, mining, and chemical enterprises, roller presses (Fig. 27.1a) are used to produce briquettes from fine wastes of various bulk raw materials that significantly differ by physical and mechanical properties [1–6]. The need to expand the operational capabilities of the presses is caused by the increased volumes of small-fraction raw materials.

Roller presses are equipped with the two rolls and replaceable pressing tires (Fig. 27.1b, c) which are subjected to high contact stresses, intensive wear and are quite expensive parts made of alloy steels. Circular tires production includes their case carburization and possible subsequent deep cryogenic treatment [7]. The gradual wear of tires profile greatly influences the product quality and above the certain level of deterioration makes it impossible the separated briquettes formation due to a decreased density as the main parameter of product quality (Fig. 27.2).

To achieve the good quality of briquettes, the gradual tire wear requires to increase of the pressing force and drive torque in the compaction process control. Figure 27.3 shows an experimentally measured typical wear pattern on the working surface of a briquette press roll.

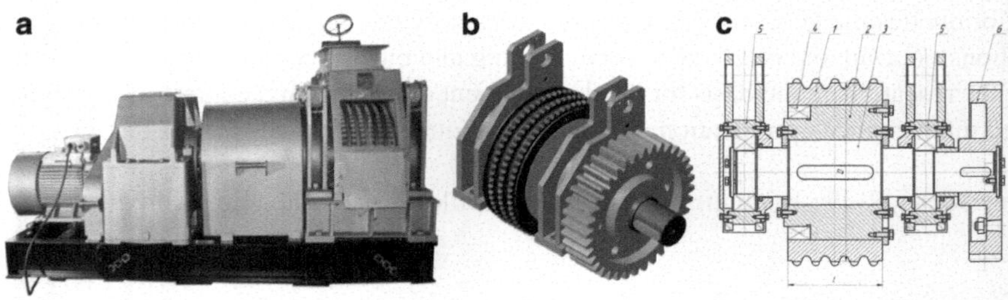

Fig. 27.1 Briquetting roller press: general view (**a**); design of roll (**b**); components (**c**): 1—gear-type replaceable pressing tire; 2—hub; 3—driving shaft; 4—clamping ring; 5—bearing supports; 6—synchronizing gear

(a) (b) (c)

Fig. 27.2 The worn-out tire (**a**); briquette of good quality (**b**) and produced on worn tires (**c**)

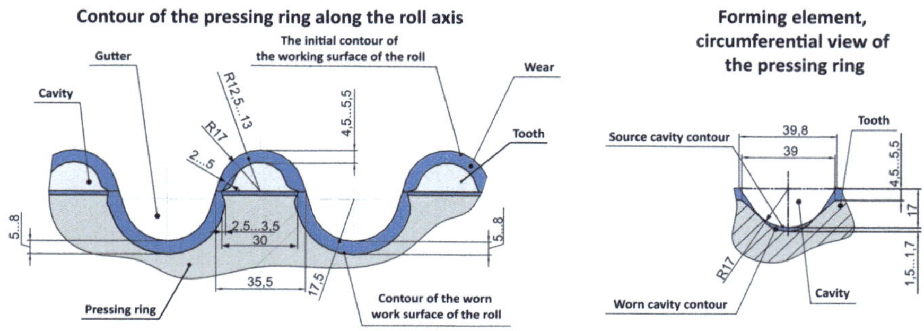

Fig. 27.3 The experimental pattern of the contact surfaces wear on the tires

In the theory of bulk material compaction and the practice of rollers presses design, the major efforts are focused on wear forecasting and methods for protecting and restoring parts of industrial equipment, using more wear-resistant alloy steels and ceramic composites. The same applies to the press rolls [8] where attention is also paid to the configuration and calibration [9–11] because compacted materials have elastic after-effect [12]. The closest similar example of a working tool is the rolling mills' rolls [13–15] where some models are developed for rolls wear prediction and process stability assessment by the electrical drives current [16].

However, at present, there is very little data on the operating modes of roller presses, parameters of the operation and periods of replaceable ring tires maintenance in real production conditions. These data are usually short and not systematized and methods of wear diagnostics in the drivelines of rolling mills [17] are not sufficient here. There is no holistic methodology developed to predict the wear of the briquetting press tires.

Therefore, in this paper authors proposed a model-based approach for rational operating parameters control at different stages of the working surfaces wear allowing in time maintenance of roller presses without decreasing the output product quality.

27.2 Methodology

27.2.1 The Constitutive Model of the Rolls Wear

The wear of the roller press tires is caused by the interaction with solid particles of the material within the repeated cycles of elastic and plastic deformation of the micro volumes at the contacts. The material feeding into a roller press is shown in Fig. 27.4a.

Constructing an analytical model of the compaction process using tooth-and-groove forming elements is more complicated in comparison with lenticular,

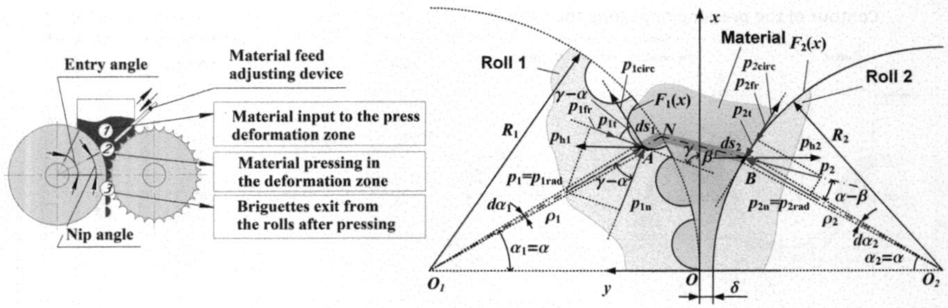

Fig. 27.4 The material feeding with its compaction stages (**a**) and rolls with material interaction scheme (**b**)

pincushion, and similar configurations. The scheme of an analytical model for determining the power and wear in the deformation zone is shown in Fig. 27.4b.

The press rolls rotate at the same angular velocities, i.e., $\omega_1 = \omega_2 = \omega$, but the forming elements are configured so that radii R_1 and R_2 are not equal. The α_1 and α_2 are the current rotation angles of the rolls and they are equal. To cover the entire contact area of forming elements, the deformation zone is represented as the series of bands. This approach accounts for the geometry of forming elements in stress–strain studies, determining the direction and magnitude of exerted forces acting on the rolls. The complete description of the created model is given in work [18] and forming elements design considerations in [19].

The compressive resistance of the treated material is described as:

$$p = a \cdot K_{\mathrm{c}}^{b}, \tag{27.1}$$

where p is compaction pressure; K_{c} is the compressibility factor; a and b are experimentally determined coefficients.

The torque on rolls is determined as the result of partial forces:

$$M_{1,2} = h_1 \sum_{i=1}^{n} dM_i \tag{27.2}$$

$$M = M_1 z_1 + M_2 z_2 \tag{27.3}$$

Pressing force:

$$P_1 = h_1 \sum_{i=1}^{n} dP_i \tag{27.4}$$

$$P = P_1 (z_1 + z_2)/2, \tag{27.5}$$

where z_1 and z_2 are the total numbers of toothed rows and grooves on both rolls; h_1 is the reduced width of one element; i—an index of bands over the contact zone.

Wear rate and specific wear rates are determined as:

$$I_h = \frac{dh}{dS} = \frac{dW}{AdS}; \quad i_h = \frac{V_D}{nA_r d_c}; \quad I_h = i_h \frac{p_a}{p_r}, \tag{27.6}$$

where W—wear, S—friction path; A—friction contact area; V_D—deformed volume; n—cycles of frictional interaction; d_c—diameter of the contact spot; p_a, p_r—contact pressures. Based on the developed model of the compaction process, the tires wear is simulated for k_1–k_5 stages of profile deterioration. The calculated radial and tangential contact forces, as well as wear patterns of profiling elements, are shown in Fig. 27.5.

The theoretical dependences of controlled and technological parameters on wear grade of rolls tires are represented in Fig. 27.6. The permissible limits of briquettes density are 3.8–4.9 g/cm^3 must be provided throughout the entire period of one set of rolls operation to obtain briquettes of the required quality. At the same time, the power parameters of the briquetting process (force and torque or electrical motor current)

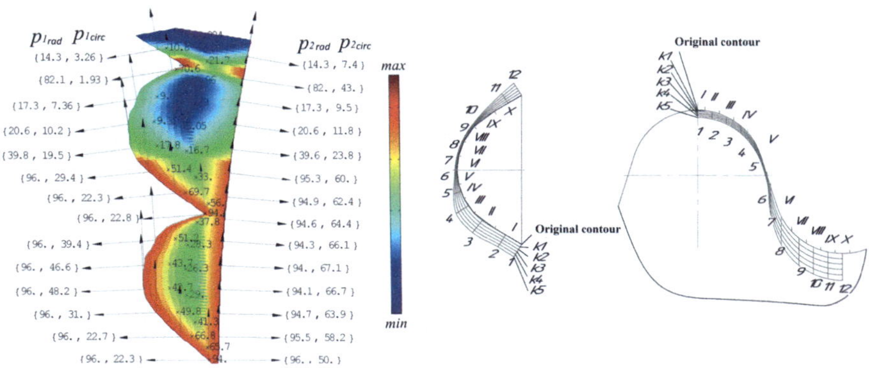

Fig. 27.5 Rolls wear simulations; forces (**a**); transverse (**b**) and longitudinal (**c**) wear patterns

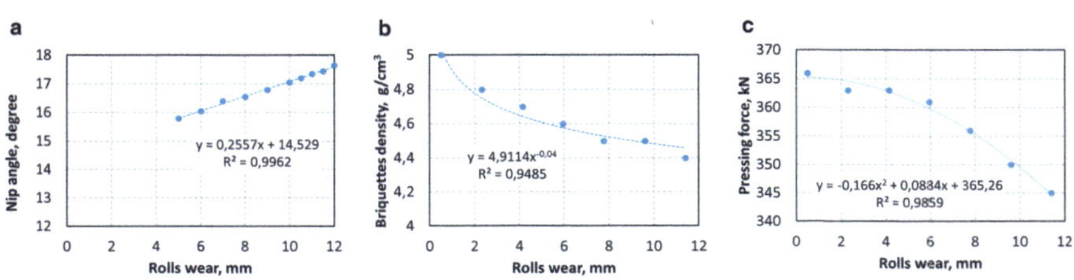

Fig. 27.6 Results of roller press rolls wear simulations: nip angle (**a**); briquettes density (**b**); pressing force (**c**)

should not exceed the technical characteristics of the press to avoid overloading or failure of the certain type of pressing equipment. Besides, in real production conditions, there are fluctuations in the properties of the mixture.

Therefore, at different stages of tires wear, it is important to adjust the process parameters, taking into account the specific properties of the mixture (bulk material fractions, moisture and binder content), to ensure the production of high-quality briquettes, the normal operation of the pressing equipment and maximum service life of the pressing tires.

27.3 Industrial Trials

27.3.1 The Operational Parameters of the Press

The industrial trials are conducted on the industrial roller press which main parameters are given in Table 27.1. The set of new tires have the dimensions of forming elements 63.6 × 60.0 × 27.0 mm. A compacted mixture is the silicomanganese particles of −5 … −3 mm and organic binder. The density of produced briquettes is determined by the direct measurements and regulated by producer specifications.

Although the chosen mixture of material is based on silicomanganese, the other bulk materials (lump chamotte, silicon carbide, coal sludge) are also available [20, 21] for compaction in this press.

Since the hardened layer of gear-type profile tires is approximately 3–6 mm after case carburising, it determines the service time of rolls. Maintenance staff uses weld fusion technology to restore hardened layer that anyway required press downtime.

Table 27.1 Parameters of the investigated briquetting press[a]

Parameters	Description
Forming elements type	Gear-type
Nominal cavity size, mm	63.6 × 60.0 × 27.0 mm ($50\ cm^3$)
Tire material	Steel 40Cr
Tire hardened layer	58–63 HRC (3 mm)
Rolls clamping and safety	Hydraulic cylinders
Nominal pressing force, kN	1000
Nominal pressing torque, kN m	35
Electric motor power, kW	30
Electric motor current, A	57
Rolls rotation speed, min^{-1}	3–9
Ratio of gearbox and chain	160 × 1.5 = 240

[a] Briquetting press is designed and produced in the iron and steel institute NAS of Ukraine

27.3.2 Process Parameters Optimisation

As of now, there is no open-access database for practical use on the intensity of the briquetting presses rolls wear at the different working conditions and compacted materials. Therefore, the proposed approach intends to use the results of theoretical calculations and practical data obtained when ferroalloys screenings have been compacted for the optimization of the roller press operating in an industrial environment.

The compaction process and roller press operation include the following parameters: ρ_{bulk}—input bulk density; W—input humidity; F_r—fraction size of the particles; B—binding organic compound percentage; δ—wear of rolls; α_{pr}—pressing angle; L_{reg}—position of the input flow regulator; K_y—compaction ratio; ρ_{br}—briquettes density; P—pressing force; M—pressing rolls torque; I_{m}—drive motor current; n_{roll}—roll speed; Q—press productivity. The overall scheme of technological parameters relations is shown in Fig. 27.7. The investigated roller press is not equipped with a monitoring system and the electric motor current was the only available for analysis parameter. Nevertheless, the real data on stages of tires spatial wear is recorded and systematized at different periods of press operation.

To stabilize the density of briquettes under conditions of the gradual wear of rolls, the input flow nip angle is controlled by the material feeding device (Fig. 27.4a). Also, the rolls wear compensation can be achieved by adjusting the properties of bulk material (moisture, fractions) as well as the amount of the organic binder.

Table 27.2 shows the original and adjusted parameters, which are based on predicted wear of pressing tires. This table contains the most rational values of the predicted options for the implementation in a certain case.

The working regimes of the press are tested in an industrial plant through the whole period of one set of rolls operation (11 months) and provided the desired

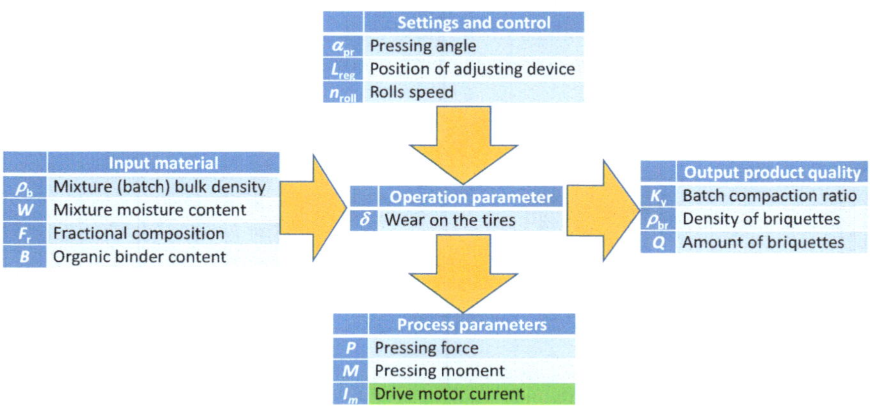

Fig. 27.7 The scheme of bulk material compaction process optimization

Table 27.2 Technological parameters of the briquetting process

	Mixture (bulk) properties				Technological parameters				
Options	ρ_{bulk}, g/cm^3	W, %	B %	F_r, mm	α_{pr}, deg	K_y	r_{br}, g/cm^3	P, kN	M, kN m
Original	2.1	3.5	3.5	−6	14.5	2.30	4.9	379	36
Adjusted			1.5	−3	15.5	2.28	5.0	366	35

density of briquettes up to 11 mm of tires wear. The experimentally determined wear patterns are correlated with the values calculated by the developed model.

The actual performance parameters of the press are reflected in Fig. 27.8. The actual indicators of the briquetting process are slightly different from the recommended ones. This is due to the instability of the physical and mechanical properties of the silicomanganese screenings, primarily due to the change in the percentage of large and small fractions of particles in the compacted mixture. However, based on the recommended process parameters, it was possible to ensure the characteristics of the briquettes and the operating modes of the press within acceptable limits during the entire cycle of operation of the pressing rings.

The density of briquettes remained within the predicted by model limits until the pressing rings wear of about 11 mm. This is almost twice greater than usually admitted value for rolls repair or tires restoration or replacement.

The involved in industrial research press is not equipped with the appropriate control and measuring equipment. However, the values of the rolls torque calculated by the electric motor current for the entire operation period of the pressing rings did not exceed the nominal value on the rolls (93.6 kN m) and slightly greater than gearbox nominal value (35 kN m). The pressing force (recalculated from the drive torque) is higher than predicted values but still less than the allowable level (1000 kN).

It was possible to record only the total amount of briquettes produced over the entire period of operation of one set of tires—2500 tons. However, it is desirable to fix also the intermediate values of the produced amount of briquettes at various stages of tire wear that allows correlating with produced tonnage and properties of a material.

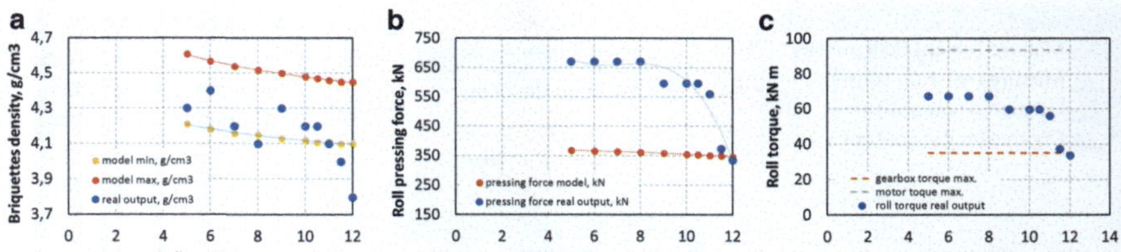

Fig. 27.8 Results of industrial trials under conditions of roller press tires gradual wear: briquettes density (**a**); rolls pressing force (**b**); drive torque (**c**)

27.4 Conclusions

The model-based approach is developed for the optimization of briquetting press regimes to obtain required quality with account predicted gradual wear of tires and limitations on strength capacity of gear drive and electric motor power.

The developed procedure is verified on the example of a set of gear-type tires used in production within almost one year. Implementation of the proposed approach allowed to control output product density within desired limits up to 11 mm of wear that is much greater than the hardened layer (3–6 mm) and usually admitted by the maintenance rules.

Besides the output product quality, the proposed methodology ensures reliable operation of roller presses without failures caused by overloading. The results of this research form the basis for the condition monitoring and maintaining the mechanical equipment of roller presses in the real production plants taking into account the gradual wear of the pressing rolls tires.

Further enhancement of roller press maintenance and briquetting technology requires a wider set of online monitored parameters and their historical data analysis.

Acknowledgements This activity has received partial funding from the European Institute of Innovation and Technology (EIT), a body of the European Union, under the Horizon 2020, the EU Framework Programme for Research and Innovation. This work is supported by EIT Raw Materials GmbH under Framework Partnership Agreements No. 18253 (OPMO. Operation monitoring of mineral crushing machinery).

References

1. Barnett T (2010) Roll-press briquetting: compacting fines to reduce waste-handling costs. Powder and Bulk Eng 24(10):1–6
2. Baiul KV (2019) Synthesis of roller press rational design for composite solid fuel production. Probl Reg Energ 2(43):103–116
3. Bembenek M, Zięba A, Kopyściański M et al (2020) Analysis of the impact of the consolidated material on the morphology of briquettes produced in a roller press. J Mater Eng Perform 29:3792–3799
4. Vashchenko SV, Khudyakov AY, Baiul KV et al (2018) Selecting the batch composition in briquetting. Steel Transl 48:509–512
5. Bygrave J (2011) Briquetting of steel industry waste streams for value added recycling. In: 32nd IBA biennial conference, New Orleans, LA, pp 1–13
6. Khudyakov AY, Vashchenko SV (2020) Analysis of known dependences and the construction of new compaction equations for the fine-fraction materials of the mining and metallurgical complex. Refract Ind Ceram 60:618–626
7. Krot P, Bobyr S, Dedik M (2017) Simulation of backup rolls quenching with experimental study of deep cryogenic treatment. Int J Microstruct Mater Prop 12(3/4):259–275
8. Babailov NA, Polyanskii LI, Loginov YN (2016) Briquetting metallurgical lime screenings and parameters making it possible to improve process efficiency. Metallurgist 60:576–580
9. Roman TD, Hryniewicz M, Komarek RK (2007) Testing the relative wear characteristics of different alloys for roller press forming elements. In: Proceedings of 30th Biennial Conference of the Institute for Briquetting and Agglomeration, vol 30, Savannah, GA, pp 130–139

10. Bembenek M (2017) The issue of selection: shape and volume of briquettes agglomerated in a roller press. J KONES: Powertrain Trans 24(2):31–38
11. Roman TD, Zavaliangos A (2008) Optimizing of roller press forming cavity profile using finite element simulation. In: Proceeding of XXIV international mineral processing congress, vol 3, Beijing, China, Sep. 24–28, pp 3175–3181
12. Baiul K, Khudyakov A, Vashchenko S et al (2020) The experimental study of compaction parameters and elastic after-effect of fine fraction raw materials. Mining Sci 27:7–18
13. Kim DH, Lee Y, Yoo SJ et al (2003) Prediction of the wear profile of a roll groove in rod rolling using an incremental form of wear model. Proc Inst Mech Eng Part B J Eng Manuf 217(1):111–126
14. Turk R, Fajfar P, Robič R et al (2002) Prediction of hot strip mill roll wear. Metalurgija 41(1):47–51
15. Galakhar AS, Daniel WJT, Meehan PA (2009) Prediction of roll profile wear in the cold roll forming process. Key Eng Mater 410–411:643–660
16. Putnoki AY, Klevtsov OM, Ermolenko AA et al (2003) Evaluation of operation of equipment at the rolling mill. Stal' 10:56–58
17. Krot PV (2003) Methods and instrumentation for measuring wear in drivelines of rolling mills. Metall Process Equip 2:45–53
18. Baiul KV (2012) Effect of the geometrical parameters of roll press forming elements on the briquetting process: analytical study. Powder Metal Metal Ceram 51:157–164
19. Baiul KV, Solodka NA, Khudyakov AY et al (2020) Selection of rational surface configuration for roller press tires. Powder Metall Met Ceram 59(1–2):9–21
20. Khudyakov AY, Vashchenko SV, Baiul KV et al (2018) Kaolin raw material briquetting for lump chamotte production. Refract Ind Ceram 59(4):333–337
21. Semenov YS, Gorupakha VV, Kuznetsov AM et al (2020) Experience of using manganese-containing materials in blast-furnace charge. Metallurgist 63(9–10):1013–1023

Chapter 28
Multiple Linear Regression Parameters for Generating Fatigue-Based Entropy Characteristics of Magnesium Alloy

M. A. Fauthan, S. Abdullah, M. F. Abdullah, S. S. K. Singh, and I. F. Mohamed

Abstract This paper presents the development of a multiple linear regression based on stress ratio and applied load approach, assessed via thermodynamic entropy generation. Entropy generation as an early detection method that enables researchers to discover temperature evolution caused by dissipated plastic energy if the material experienced fatigue load. This relationship was developed through a complete entropy generation prediction using statistical approach, whereby a constant amplitude loading was applied to evaluate the fatigue life. By conducting compact tension tests, different stress ratios of 0.1, 0.4, and 0.7 were applied to the specimen. During the tests, temperature change was also observed. The highest entropy generation was 2.922 MJ m^{-3} K^{-1} when 2600 N load with a stress ratio of 0.7 was applied to the specimen. The assumptions of the models were considered through graphical residual analysis. As a result, the predicted regression model based on the applied load and stress ratio corresponded with the results of the experiment, deviating at only 4.6% from the actual experiment.

Keywords Entropy · Multiple linear regression · Stress ratio

28.1 Introduction

Magnesium alloys have many useful features such as high strength-to-weight ratio and great damping, becoming the better choice in structural application. As a consequence of applying cyclic mechanical load or cyclic deformation, comes fatigue as a gradual reduction in material strength. Without any noticeable warning, fatigue failure will happen unexpectedly and sometimes catastrophically. Fatigue is typically

---

M. A. Fauthan · S. Abdullah (✉) · I. F. Mohamed
Department of Mechanical and Manufacturing Engineering, Faculty of Engineering and Built Environment, Universiti Kebangsaan Malaysia, Bangi, Malaysia
e-mail: shahrum@ukm.edu.my

M. F. Abdullah · S. S. K. Singh
Department of Mechanical Engineering, Faculty of Engineering, Universiti Pertahanan Nasional Malaysia, Kuala Lumpur, Malaysia

G. Lesiuk et al. (eds.), *Structural Integrity and Fatigue Failure Analysis*, Structural Integrity 25, https://doi.org/10.1007/978-3-030-91847-7_28

the main problem hampering total efficiency, prompting investigation on magnesium alloys fatigue behavior.

Recent studies in infrared thermography for non-destructive evaluation have provided new research in studying crack propagation subjected to cyclic loading [1]. Furthermore, fatigue is probabilistic in nature and involves, at the same time, a multimode process [2]. In a conventional test, many unknown input parameters are required. Therefore, dissipation of energy is introduced to predict the life span of materials [3]. This relationship can also be described through the introduction of multiple linear regression (MLR). The MLR is used to determine the value of a variable based on two or more of other variables' values [4].

Based on previous studies, temperature evolution is used to predict any failures upfront [5]. Furthermore, entropy generated is also associated with damage. If the total generation of entropy can be approached through regression, then fatigue life can be predicted. Hence, this paper was aimed at describing the MLR relationship in order to predict total entropy generation of magnesium alloy, AZ31B.

28.2 Literature Background

28.2.1 Fatigue Crack Growth Assessment Toward Entropy Generation

The thermodynamic approach to assess the fatigue behavior of materials requires system definition that abides to the thermodynamics laws. For practical purposes, it is considered that the material subjected to the analysis is a closed system since there is no mass inter-change with the environment, but there is heat transfer through the boundaries. According to the first law of thermodynamics, they are expressed by:

$$W = Q + U \tag{28.1}$$

where W is the mechanical energy, Q is dissipated energy, and U is internal energy. The benefit of using thermodynamic forces and flows is that the entropy production can be explained in term of experimentally measurable quantities.

28.2.2 Entropy Generation Method

In fatigue, the dissipative process $p = p(\zeta)$ depends on a time-dependent phenomenological variable ζ. In general, the change in system's entropy, dS through a form of modification is connected to δQ by:

$$dS = {}^{\delta Q}/_{T} \quad (28.2)$$

where T is the temperature.

For the second law of thermodynamics in solids with internal friction, the relationship between plastic deformation and the thermal dissipation is given by the following relationship [6]:

$$\dot{\gamma} = w_p / T - J_q \cdot \mathrm{grad}T / T^2 \quad (28.3)$$

In this instance, $\dot{\gamma}$ signifies the rate at which entropy is manufactured ($\dot{\gamma} \geq 0$), J_q is the heat flux, T the temperature of the surface, and w_p the recurring plastic energy mass per unit which comes about form the calculation from Morrow's estimate.

$$w_p = AN_f^{\alpha} \quad (28.4)$$

where N_f represents the number of cycles to failure, and the constants A and α the material specifications, which can be calculated from the following relationship:

$$A = 2^{2+b+c} \varepsilon'_f \sigma'_f \left(\frac{c - b}{c + b} \right) (N_f)^{b+c} \quad (28.5)$$

whereby $\varepsilon^{'}_{f}$, and $\alpha^{'}_{f}$ are cyclic ductility and fatigue strength coefficient, respectively. Then, the term of b and c is the fatigue strength exponent and the fatigue ductility exponent, respectively.

28.2.3 Multiple Linear Regression (MLR)

The MLR was chosen to develop a relationship between entropy and the applied load as well as stress ratio to ensure the linear relation between dependent and independent variables. The MLR is able to immediately predict the dependent variable by best matching the observational data, eliminating the need for repeated research with commercial software. The general multiple regression model, given and observation was defined to be:

$$y_1 = \alpha + \beta_1 x_{i1} + \beta_2 x_{i2} + \cdots + \beta_n x_{in} + \varepsilon_i \quad (28.6)$$

The MLR can deliver instant forecasts of the reliant variable by fitting the very best line of observed information, reducing the requirement for repetitive analyses utilizing business software application.

28.3 Experimental Procedure

The applied approach in this study began with determining fatigue crack development, in addition to the evolution of temperature. This study made use of the commercial AZ31B magnesium alloy. The AZ31B is a wrought magnesium alloy with both notable room-temperature ductility and strength including weldability and corrosion resistance properties.

To study the fatigue crack growth (FCG), compact tension (CT) test specimens were prepared following the recommendation of ASTM E647 as illustrated in Fig. 28.1. The CNC milling machine and electrical discharge wire-cutting device were used in cutting the sample to the required dimensions. A uniaxial servo-hydraulic at a load capacity of 100kN was used to perform all the FCG experiments as in Fig. 28.2. During the fatigue process, a non-contact infrared sensor was utilized to monitor and trace the surface material's temperature. It is essential to use the infrared sensor as it is sensitive to small changes in temperature caused by deformation. These small changes may not be captured by conventional sensors such as a thermocouple.

All the specimens were then tested using a constant amplitude sinusoidal loading with a load of 2600 N and 2800 N and load ratio (ratio of minimum to maximum load) of $R = 0.1$, 0.4, and 0.7, respectively, at a constant 10 Hz frequency. During the test, the temperature trending of the specimen was detected using the infrared sensor set at 50 mm gap between the sensor and specimen.

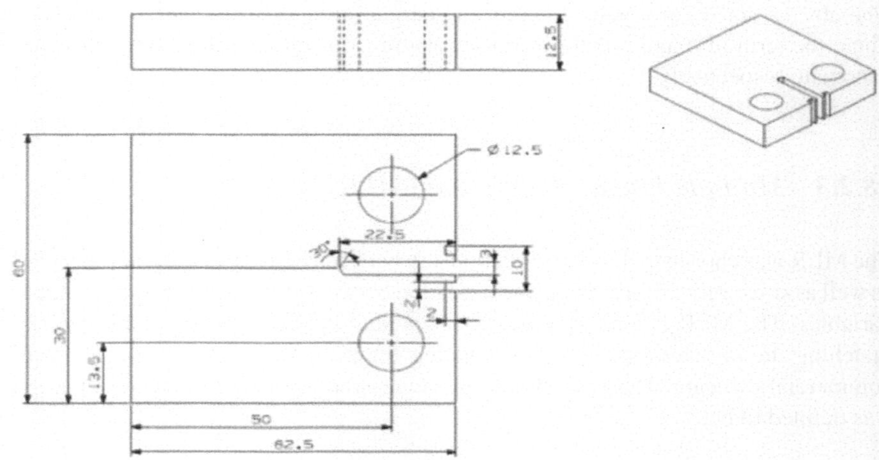

Fig. 28.1 Geometry of the specimen with dimension in mm

Fig. 28.2 The setup of the IR sensor and the specimen

28.4 Results and Discussion

The constant amplitude was used during the fatigue crack growth test to regulate the entropy generation for the applied load and stress ratio. The test has essentially been done to predict the total number of cycles before it succumbs to fatigue loading failure. The surface temperature of the sample was then measured to discover its temperature changes. The experiments showed a significant outcome that the temperature of the surface could be related to the cycle, leading up to their failure. For the first step, the fatigue crack was observed from the experiment. Figure 28.3 shows the crack length during different stress ratios at 2600 N, where the stress ratios are 0.1, 0.4, and 0.7.

The evolution of the crack temperature throughout FCG investigation is displayed in Fig. 28.4. It shows the change of values going through three different phases. At the beginning of the FCG test, the sudden movement and the disruptions of the grains will cause the rise in the surface temperature, concerning 10% of the material's lifecycle.

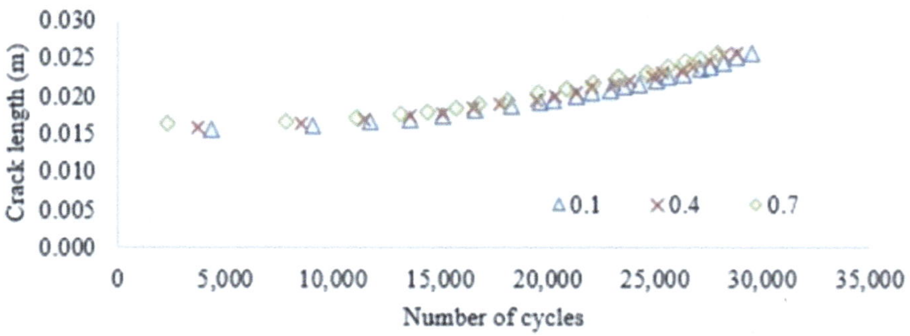

Fig. 28.3 The fatigue crack growth rates at stress ratio of 0.1, 0.4, and 0.7 for load 2600 N

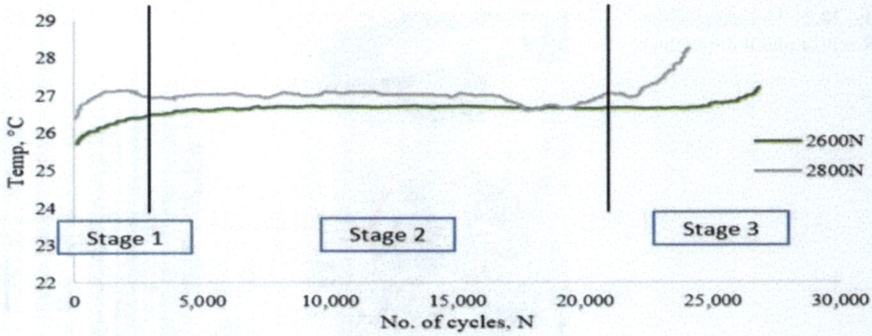

Fig. 28.4 The temperature evolution measured during fatigue crack growth tests under 2600 N and 2800 N for 0.1 stress ratio

The phenomenon of intrusion and extrusion also occurred during this stage. After that, the temperature becomes stable due to the volume of heat generated equals to the heat released to the surrounding. However, at the end of the test, the temperature starts to rapidly increase as more extensive plastic deformation occurs compared to the second stage.

As the crack growth increased, the total entropy was calculated until the specimen utterly fractured. The total entropy generation at applied load of 2600 N was 3.424, 3.101, and 2.922 MJ m^{-3} K^{-1} for stress ratios of 0.1, 0.4, and 0.7, respectively. According to Fig. 28.5, the total entropy generation increased as a higher stress ratio was applied. This was due to the distribution of a higher energy per unit volume, which led to failure. It indicates that at a higher entropy generation, the specimen should have a lower fatigue life.

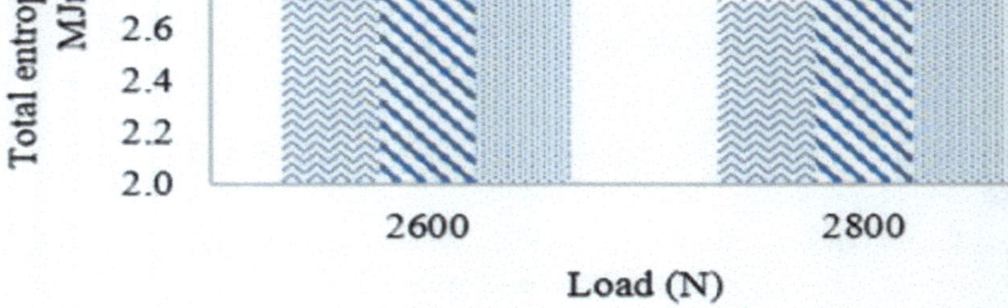

Fig. 28.5 Total entropy generation with different loads and stress ratios

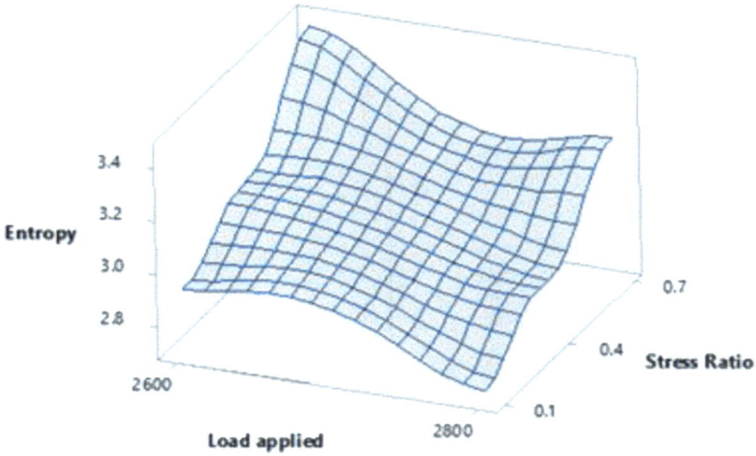

Fig. 28.6 The response surface for entropy MLR-based entropy model

The regression model can be merely described as indirect regression examination with just one explanatory variable, utilizing a two-dimensional plot of the reliant variable as a function of the independent variable. From the data set, a straight line can be obtained to represent the regression model. Moreover, R^2 value determines fit consistency. Nonetheless, there are many explanatory variables associated with MLR analysis and, as a result, the presumptions of linearity, homoscedasticity and normality must be tested to confirm that the MLR-based entropy models obtained in this research can be generated with valid inferences. Overall, the response surface plot in Fig. 28.6 shows that the entropy varies inversely with the load applied and linearly with the stress ratio (independent variables), which validates the MLR-based entropy model's linearity.

Next, the assumptions of the MLR model were assessed. The four different conditions that need to be evaluated for multiple regression to give a valid result are the linear function, independent function, normal distribution and equal variance. The results are shown in Fig. 28.7. Hence, all of the MLR-based entropy models justify the requirement that most of the error terms are generally dispersed.

When the goodness of fit, homoscedasticity, normality, and also linearity of the MLR-based entropy model had been examined, the model was verified by comparing the entropy values of the models with those observed from the experiment, as presented in Fig. 28.8. From that figure, the anticipated entropy by the MLR-based entropy model shows noble conformity with the entropy values from the experiments, with the R^2 value of 0.9760. The R^2 value is more significant than 0.9000, suggesting dependability of the model in predicting the entropy of the specimen.

Then, the MLR model is applied in producing a meaningful entropy prediction model. The datasets comprising the entropy generation values of the CT specimens, stress ratio (R), and load applied (P), as shown in Eq. (28.7), were used to establish

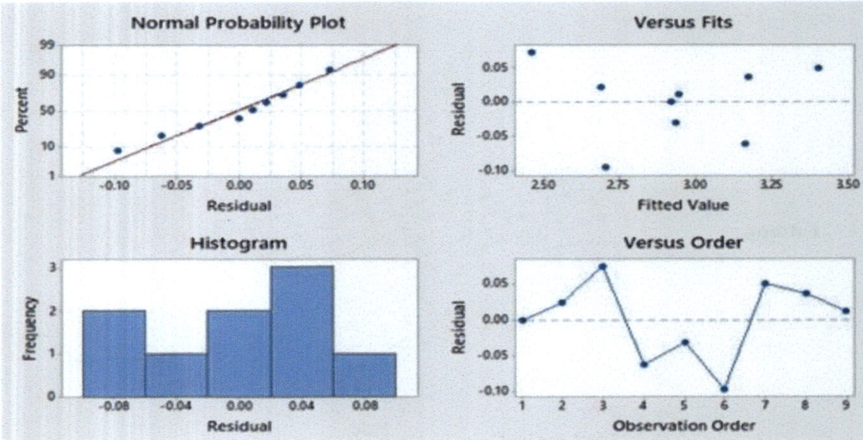

Fig. 28.7 Observation of multiple regression

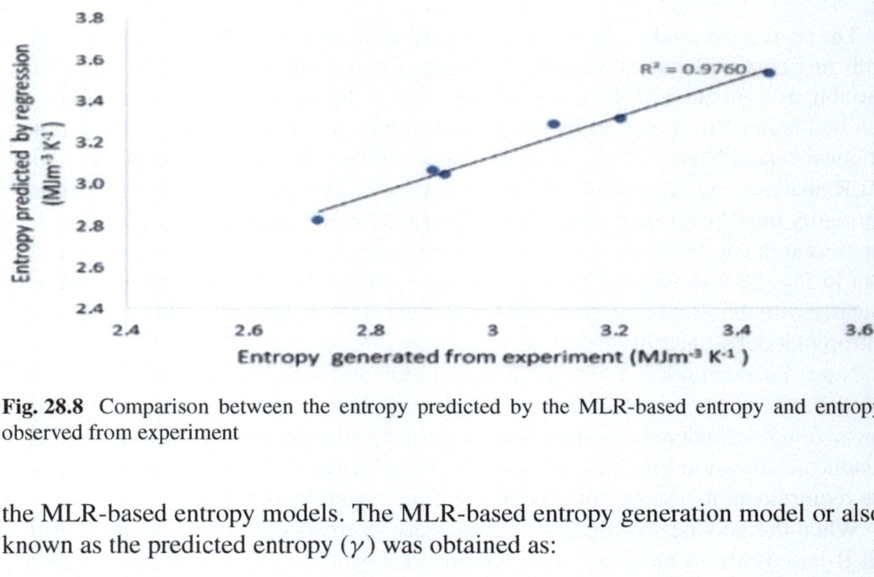

Fig. 28.8 Comparison between the entropy predicted by the MLR-based entropy and entropy observed from experiment

the MLR-based entropy models. The MLR-based entropy generation model or also known as the predicted entropy (γ) was obtained as:

$$\gamma = 5.827 - 0.001148P + 0.8044 \tag{28.7}$$

Once the assumption on MLR-based entropy model was clarified to be acceptable, the models were compared to the experimental values done with another new load, 3000 N. Table 28.1 shows the percentage of the difference between the experimental and predicted data for new load conditions. The difference is less than 10%, and the determined entropy generation well forecasts the experimental data under new load

Table 28.1 The percentage of difference entropy generation with respect to the experimental data

Stress ratio	Experimental entropy	Predicted entropy	% of differences
0.1	2.536	2.608	2.8
0.4	2.607	2.849	9.3
0.7	2.956	3.090	4.6

conditions. This discusses the majority of the predicted entropy generation were near to comparable experimental value.

28.5 Conclusions

This research revealed the principle of entropy generation in magnesium alloy for fatigue crack growth test. The FCG employed temperature level development to demonstrate the magnesium alloy entropy generation. Entropy generation was deployed as an effective way of linking the crack growth behavior of a material with the temperature change during the fatigue process. An approach to develop the MLR relationship between the entropy generation, applied load, and stress ratio was elaborated in this paper. By performing CT tests, various stress ratios of 0.1, 0.4, and 0.7 were used on the specimen with applied load 2600 N and 2800 N. From the test, the highest generated entropy was 2.922 MJ m^{-3} K^{-1} when 2600 N load with a stress ratio of 0.7 was applied to the specimen. As an outcome, the predicted regression model based on the applied load and stress ratio was found to coincide with the outcomes of the experiment, just within 4.6% entropy value difference from the actual experiment and the regression model were also in good agreement. The results were encouraging, where the percentage difference between the MLR-based entropy models was less than 10%, indicating an acceptable alternative method in estimating materials' life expectancy.

Acknowledgements The authors graciously acknowledge the financial support provided by Universiti Kebangsaan Malaysia (FRGS/1/2019/TK03/UKM/02/1) and Universiti Pertahanan Nasional (FRGS/1/2018/TK03/UPNM/03/1).

References

1. Teng Z, Wu H, Boller C, Starke P (2020) Thermography in high cycle fatigue short-term evaluation procedures applied to a medium carbon steel. Fatigue Fract Eng Mater Struct
2. Li X, Zhang W (2019) Probabilistic fatigue evaluation of floating wind turbine using combination of surrogate model and copula model. In: AIAA Scitech 2019 forum
3. Wang XG, Ran HR, Jiang C, Fang QH (2018) An energy dissipation-based fatigue crack growth model. Int J Fatigue 114

4. Kong YS, Abdullah S, Schramm D, Omar MZ, Haris SM (2019) Development of multiple linear regression-based models for fatigue life evaluation of automotive coil springs. Mech Syst Signal Process 118:675–695
5. Yan ZF, Zhang HX, Wang WX, He XL, Liu XQ, Wu GH (2014) Temperature evolution mechanism of AZ31B magnesium alloy during high-cycle fatigue process. Theor Appl Fract Mech 70:30–38
6. Naderi M, Amiri M, and Khonsari MM (2010) On the thermodynamic entropy of fatigue fracture. Proc Roy Soc A: Math, Phys Eng Sci 2009; 466(2114): 423–438

Chapter 29
Constructal Design Applied to Biaxial Elasto-Plastic Buckling Analysis of Square Steel Plates with Cutout

Thiago da Silveira, João Pedro Sarasol Neufeld, Luiz Alberto Oliveira Rocha, Elizaldo Domingues dos Santos, and Liércio André Isoldi

Abstract Thin plates are widely used in structural engineering as components of decks and bottom of ships, offshore structures, aircrafts, and bridges. If these plates are subject to compressive loads, the buckling phenomenon can occur. The presence of cutouts on plates causes a redistribution of the membrane stresses, altering its buckling behavior. In this context, simply supported square steel plates with a centered elliptical hole and subjected to biaxial compression were geometrically evaluated. To do so, the Constructal Design (CD) was applied to propose the geometric configurations (search space); the Finite Element Method (FEM) was used to obtain the ultimate buckling stress of these plates; and the Exhaustive Search (ES) technique allowed to define the optimized plate geometries. The perforation assumed five different sizes, and for each size several variations of its elliptical shape were investigated. As performance parameter, the maximization of the ultimate buckling stress was considered. From the results, it was observed that the optimized geometric configurations are in agreement with the Principle of Optimal Distribution of Imperfections, showing the applicability of CD in evaluation of structural integrity of these plates.

Keywords Constructal design method · Biaxial elasto-plastic buckling · Structural integrity

T. da Silveira (✉) · J. P. S. Neufeld
Federal University of Pampa (Unipampa), 810, Tiarajú Av., Alegrete, RS, Brazil
e-mail: thiagods@unipampa.edu.br

L. A. O. Rocha
Federal University of Rio Grande do Sul, 425, Sarmento Leite St., Porto Alegre, RS, Brazil

T. da Silveira · E. D. dos Santos · L. A. Isoldi
Federal University of Rio Grande (FURG), km 8, Itália Av., Rio Grande, RS, Brazil

G. Lesiuk et al. (eds.), *Structural Integrity and Fatigue Failure Analysis*,
Structural Integrity 25, https://doi.org/10.1007/978-3-030-91847-7_29

29.1 Introduction

Thin-walled structures render a great majority of engineering structures as they provide a natural way of balancing between the structural load carrying capacity and structural mass [1, 2]. Dealing with them involves a wide range of engineering tasks such as development of efficient Finite Element Method (FEM) tools for modeling their behavior [3–5], testing [6], design with respect to various objectives [7], application of novel structural materials like composite laminates [8, 9], etc.

Thin-walled structures are characterized by significantly larger in-plane dimensions if compared to the third dimension denoted as thickness, t. In the case of a flat or a rather shallow shape, they are often referred to as plates.

These structural components are widely used in civil, naval, aerospace, and automotive engineering applications. In addition, for several practical situations the existence of an opening on the plate is required (for inspection, passing pipes, or weight reduction), causing a stress redistribution and modifying its mechanical behavior [7].

An important observation to be done in the structural design field is the consequence of cutouts existence on a fatigue life. Based on experimental and numerical analysis [10], affirm that size and position are the most effective parameters in the crack growth path, which, depending of these features, the fatigue life can be increased or decreased. For instance, in [11] the insertion of circular holes in wire sheaves increased in 123.25% the fatigue life if compared with a no-hole reference.

Another way that a structure can lose its structural integrity is related to the instability phenomenon called buckling. Therefore, if a plate is under compressive loads, a buckling failure can occur. However, a thin plate does not enter in collapse due the occurrence of the elastic buckling but actually it can support loads significantly higher than the critical load without deforming excessively [12]. The load that defines the plate failure, regarding elasto-plastic behavior, is called post-critical or ultimate load (P_u).

It is important to mention that to define the ultimate load of perforated plates is not a simple task; being the computational modeling via FEM an effective approach.

In this context, the main goal of the present work is to optimize the geometry of a centered elliptical cutout on a simply supported square steel plate subjected to biaxial elasto-plastic buckling. To do so, computational modeling (by FEM), Constructal Design method (CD), and Exhaustive Search technique (ES) are jointly employed.

The CD, based on the Constructal theory, is a method that uses degrees of freedom, restrictions, and performance indicators, to define different possible geometric configurations for an engineering system, i.e., to define the search space for a geometric evaluation [13]. Thereby, the performance of all proposed geometries is numerically obtained via FEM and compared among each other, characterizing an optimization by ES technique [14].

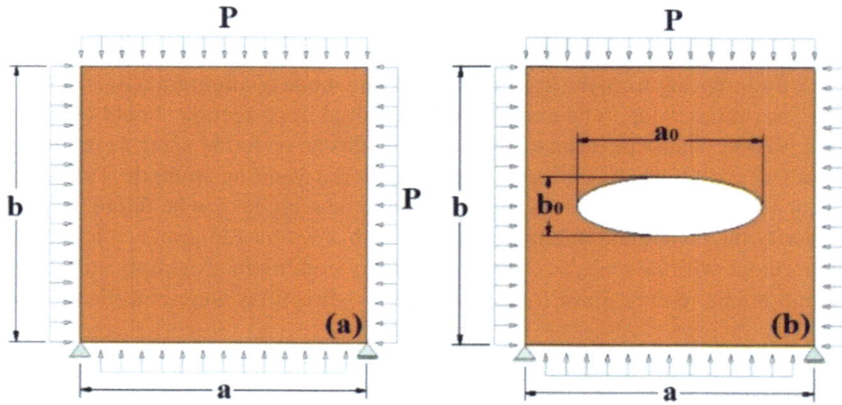

Fig. 29.1 Reference plate **a** and plate with cutout **b** under biaxial compressive loading

29.2 Buckling and Post-Buckling of Plates

Steel plates are important components for structural designs. In many cases, they are subjected to biaxial compressive loading, which make them prone to buckling. In this case, the plates buckle with a single half-wave on x and y—direction. For slender plates, the buckling is elastic; while for sturdy plates, an elasto-plastic buckling occurs [15]. Therefore, the buckling behavior depends of its dimensions: length (a), width (b) (see Fig. 29.1), and thickness (t).

Moreover, plates can support loadings higher than its critical buckling load (elastic), being its ultimate load due the elasto-plastic buckling. This capacity to support load, over the critical buckling, is associated with the formation of a membrane force that stabilizes the displacement through the transverse tension [16]. If the plate has a perforation, the hole presence redistributes the membrane stresses, causing changes on buckling characteristics and, probably, significant reduction in its ultimate load [15].

29.3 Computational Model

When analyzing plates with cutouts subjected to compressive loads, the exact analysis of such components is complex, particularly when the shape of holes and their arrangement are unusual. For that, [17] used the FEM to develop design equations to determine the ultimate load carrying capacity of axially and biaxially compressed square plates with centrally located, circular or square, cutouts.

In the present work, a numerical approach was also used by means the software ANSYS®, which is based on FEM. It was adopted the SHELL281 finite element, which is suitable for analyzing thin to moderately thick shell structures, having eight nodes with six degrees of freedom at each node: translations in the *x*, *y*, and *z*-axes, and rotations about the *x*, *y*, and *z*-axes [18].

Therefore, starting from the first elastic buckling mode configuration, it is defined that the maximum value to be assumed for the initial imperfection should obey the relation $b/2000$. Then, the ultimate load of the plate can be found using as reference the load $Py = \sigma y.t$, where σy represents the material yielding strength [19]. More details about the elasto-plastic buckling computational model can be found in [14].

To verify the computational model, a plate with $a = b = 125$ mm, $t = 6.25$ mm, circular cutout with $a_0 = b_0 = 25$ mm, boundary conditions as simply supported and equal biaxial compressive loads on *x* and *y*—directions were considered (see Fig. 29.1b).

The material used in the investigation was AH-36 steel with $\sigma y = 355$ MPa, $E =$ 210 GPa, and $\nu = 0.30$. From the equation proposed by [17], an ultimate stress of σu $= 257.13$ MPa was reached, while $\sigma u = 276.42$ MPa was numerically obtained. This value represents a difference of 6.98% between numerical and analytical solutions, being acceptable according to [17]. Once verified the computational model, for the proposed problem, it was adopted a converged SHELL281 mesh of 1298 square finite elements of 50 mm side and refined at line around the cutout.

29.4 Constructal Design (CD)

The CD method is based on Constructal Law. This law corresponds to the basis of the Constructal Theory and stats that "for a finite-size flow system to persist in time (to live), its configuration must evolve in such a way that provides greater and greater access to the currents that flow through it." The natural tendency is to evolve freely into flow configurations that offer greater access to what flows, not more flow [20].

Regarding the geometrical evaluation by means CD, it is necessary to define at least one performance parameter (to be improved), the degrees of freedom (variables), and the geometrical restrictions (fixed parameters). The degrees of freedom are free to vary, but it should respect the imposed constraints [14].

Here, as performance parameter the ultimate buckling stress of perforated plates must be maximized; as degree of freedom the ratio b_0/a_0 was considered (see Fig. 29.1b); and as restriction the volume fraction ϕ (defined as the ratio between perforation volume and plate volume) was adopted, assuming five different values kept constant.

It is worth to mention that CD is broadly used for heat transfer and/or fluid mechanics engineering systems, however, its application in structural engineering has not yet been widespread (only few works can be found in literature, e.g., [13, 14]).

29.5 Results and Discussions

Starting from a simply supported steel AH-36 square plate (with no hole), called reference plate (see Fig. 29.1a), with $a = b = 1414.214$ mm, $t = 12$ mm, subjected to equal compressive loading on x and y—direction, perforated plates were proposed having a centered elliptical cutout having $\phi = 0.025$, 0.05, 0.10, 0.15, and 0.20 (see Fig. 29.1b). For each ϕ value, several ratios b_0/a_0 were adopted.

In order to evaluate the mechanical behavior of these plates with different elliptical hole sizes and shapes, a Normalized Ultimate Stress (NUS) was used. The NUS is the ratio between the ultimate stress of holed plate and the ultimate stress of reference plate. From that, NUS for all studied perforated plates are showed in Fig. 29.2.

From Fig. 29.2, one can infer that all ϕ values presented the same mechanical behavior: as b_0/a_0 increases an improvement in the NUS occurs, reaching the maximized NUS; from there the increase of b_0/a_0 promotes a reduction of NUS. As expected for a biaxial compressed square plate, values of $b_0/a_0 = 1$ conducts to the superior performance. However, other b_0/a_0 values also can maximize the NUS. This range of optimized geometries is greater as the ϕ value decreases.

Figure 29.2 also indicates that for all ϕ values the maximized NUS is lower than the reference plate, being this trend previously noticed in [15]. The best performances were obtained for $\phi = 0.025$ and 0.05 where it was achieved NUS = 0.97 which in practice can be considered with the same NUS of reference plate.

In order to understand how the CD defines the optimized geometries, the von Mises stress distribution of some b_0/a_0 ratios of $\phi = 0.20$ were depicted in Fig. 29.3.

It is possible to notice in Fig. 29.3 that the optimized geometries (Figs. 29.3c and d) have more regions submitted to the steel limit stress (represented by red

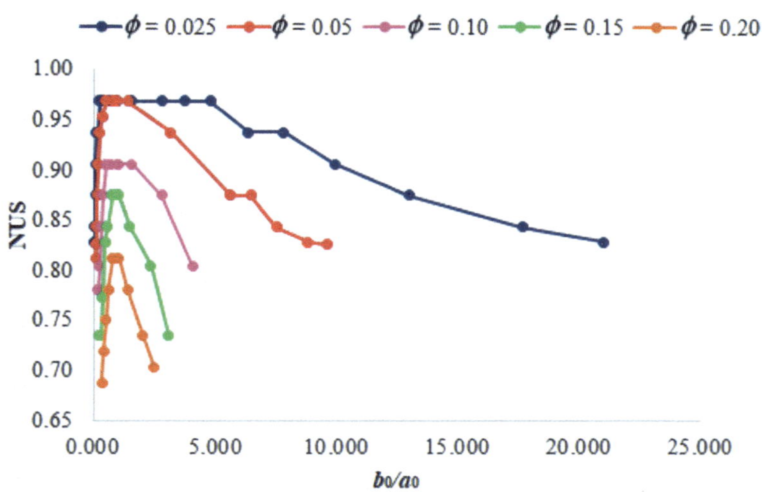

Fig. 29.2 NUS for each volume fraction of the perforated plates

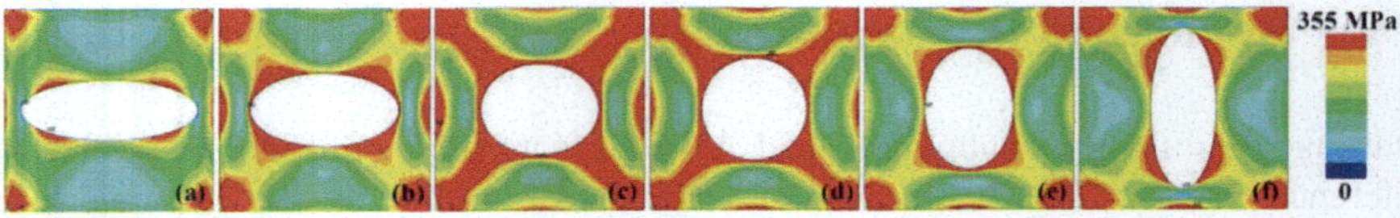

Fig. 29.3 Distribution of von Mises stress for the plates with $\phi = 0.20$ and having b_0/a_0 of: **a** 0.35, **b** 0.51, **c** 0.80, **d** 1.00, **e** 1.42, and **f** 2.52

color); while the other configurations have significantly less regions in this stress state. This fact can be explained by the Constructal Principle of Optimal Distribution of Imperfections [20]. Here, the imperfections are the concentrations of maximum stresses (red color in Fig. 29.3) and the optimized geometries are those that promote its better distribution. It is worth to mention that for the other analyzed ϕ ratios an analogous trend was observed. In addition, it is important to highlight that similar behavior it was already observed in previous works where perforated plates [14] and stiffened plates [21] submitted to elasto-plastic uniaxial buckling were analyzed.

29.6 Conclusion

The results showed the relevance of geometric evaluation on structural engineering. Besides, it was proved the effectiveness of CD method, since the expected best geometries were defined based on the Constructal Principle of Optimal Distribution of Imperfections; being still possible to define other configurations that also conduct to the buckling superior performance of square steel plates with centered elliptical hole.

In addition, it was possible to define the limit stress of the studied perforated plates from which these structures lose its structural integrity regarding the biaxial elasto-plastic buckling phenomenon.

In future works, FEM, CD, and ES can be used to analyze the structural integrity and the life in fatigue for rectangular steel plates with cutouts under biaxial compressive loading, in order to find the optimum geometry that will provide the longest life.

Acknowledgements The authors thank CNPq for the financial support.

References

1. Rama G, Marinkovic D, Zehn M (2018) High performance 3-node shell element for linear and geometrically nonlinear analysis of composite laminates. Compos B Eng 151:118–126
2. Pavlovic A, Fragassa C, Minak G (2017) Buckling analysis of telescopic boom: theoretical and numerical verification of sliding pads. Tehnicki Vjesnik 24(3):729–735

3. Valvano S, Carrera E (2017) Multilayered plate elements with node-dependent kinematics for the analysis of composite and sandwich structures. Facta Univ Ser Mech Eng 15(1):1–30
4. Szilard R (2004) Theories and applications of plate analysis: classical, numerical and engineering methods. JohnWiley & Sons, Hoboken, NJ, USA
5. Marinkovic D, Rama G, Zehn M (2019) Abaqus implementation of a corotational piezoelectric 3-node shell element with drilling degree of freedom. Facta Univ Ser Mech Eng 17(2):269–283
6. Fragassa C, Minak G, Pavlovic A (2020) Measuring deformations in a telescopic boom under static and dynamic load conditions. Facta Univ Ser Mech Eng 18(2):315–328
7. Shojaee T, Mohammadi B, Madoliat R, Salimi-Majd D (2019) Development of a finite strip method for efficient prediction of buckling and post-buckling in composite laminates containing a cutout with/without stiffener. Compos Struct 210:538–552
8. Li Z, Zheng J, Chen Y (2019) Nonlinear buckling of thin-walled FGM arch encased in rigid confinement subjected to external pressure. Eng Struct 186:86–95
9. Saravanakumar K, Lakshminarayanan BS, Arumugam V, Santulli C, Pavlovic A, Fragassa C (2019) Quasi-static indentation behavior of GFRP with milled glass fiber filler monitored by acoustic emission. Facta Univ Ser Mech Eng 17(3):425–443
10. Saber A, Shariati M, Nejad RM (2020) Experimental and numerical investigation of effect of size, position and geometry of some cutouts on fatigue life and crack growth path on AISI1045 steel plate. Theor Appl Fract Mech 107:102506
11. Løland ML, Siriwardane SC, Holen D, Stakkeland F (2016) Optimal design of wire sheave used in offshore drilling operations. J Offshore Mech Arct Eng 138(1):014501
12. Chajes A (1974) Principles of structural theory. Prentice-Hall, Englewood Cliffs, USA
13. Troina G, Cunha M, Pinto V, Rocha L, dos Santos E, Fragassa C, Isoldi L (2020) Computational modeling and constructal design theory applied to the geometric optimization of thin steel plates with stiffeners subjected to uniform transverse load. Metals 10(2):220
14. Helbig D, Silva CCCD, Real MDV, dos Santos ED, Isoldi LA, Rocha LAO (2016) Study about buckling phenomenon in perforated thin steel plates employing computational modeling and constructal design method. Latin Am J Solids Struct 13(10):1912–1936
15. El-Sawy KM, Martini MI (2010) Stability of biaxially loaded square plates with single central holes. Ship Offshore Struct 5(4):283–293
16. Åkesson B (2007) Plate buckling in bridges and other structures. Taylor & Francis, London, England
17. Shanmugam NE, Thevendran V, Tan YH (1999) Design formula for axially compressed perforated plates. Thin-Walled Struct 34(1):1–20
18. Ansys (2009) Element reference manual (version 12.1). Technology Drive, Canonsburg
19. El-Sawy KM, Nazmy AS, Martini MI (2004) Elasto-plastic buckling of perforated plates under uniaxial compression. Thin-Walled Struct 42(8):1083–1101
20. Bejan A, Lorente S (2008) Design with constructal theory. Hoboken, Wiley, NJ, USA
21. Lima JP, Cunha ML, dos Santos ED, Rocha LAO, Real MV, Isoldi LA (2020) Constructal design for the ultimate buckling stress improvement of stiffened plates submitted to uniaxial compressive load. Eng Struct 203:10

Chapter 30
Numerical Study of FPSO Platform Brackets for Different Geometric Configurations Subjected to Environmental Loads

Kauê Louro Martins, Vinícius Torres Pinto, Cristiano Fragassa, Mauro de Vasconcellos Real, Luiz Alberto Oliveira Rocha, Liércio André Isoldi, and Elizaldo Domingues dos Santos

Abstract In the present work, it is developed a computational methodology for prediction of mechanical behavior of brackets used for support of fairleads of mooring systems subjected to the maximum environmental load. With the developed methodology, different geometric configurations are studied for brackets using thinner plates than that commonly employed in existent brackets of P-66 Petrobras oil production platform. Theoretical recommendations about geometrical configurations that reduce the von Mises maximum stress in brackets' plates are obtained using Constructal Design. For the solution of distribution of stresses along with the solid domain of the brackets, it is used a code based on the Finite Element Method, more precisely the ANSYS Mechanical APDL. The brackets studied here are composed of AH36 steel plates. Three different configurations for the base of superior brackets are compared: rectangular, rectangular/triangular, and rectangular/trapezoidal. Results indicated the importance of geometrical investigation in this kind of problem. The rectangular bracket led to mechanical performance, nearly 35%, and 38% superior to that achieved with rectangular/trapezoidal and rectangular/triangular bases, respectively.

Keywords Mooring systems · Brackets · Computational modeling · Geometrical investigation · Mechanical behavior

---

K. L. Martins · V. T. Pinto · M. de Vasconcellos Real · L. A. Isoldi
School of Engineering, Federal University of Rio Grande (FURG), km 8, Itália Av., Rio Grande, RS, Brazil

C. Fragassa
Department of Industrial Engineering, University of Bologna, Viale Risorgimento 2, 40136 Bologna, Italy

L. A. O. Rocha · E. D. dos Santos (✉)
Department of Mechanical Engineering, Federal University of Rio Grande do Sul (UFRGS), Sarmento leite St., 425, Porto Alegre, RS 90040-001, Brazil
e-mail: elizaldosantos@furg.br

G. Lesiuk et al. (eds.), *Structural Integrity and Fatigue Failure Analysis*, Structural Integrity 25, https://doi.org/10.1007/978-3-030-91847-7_30

30.1 Introduction

Mooring systems are responsible for maintaining the position of oil production platforms under environmental loads. One of the mooring system components is the brackets, which are robust structures composed of thick steel plates responsible for fixing the fairlead, or other equipment, in the hull of the vessel. Figure 30.1a illustrates a global view of the mooring system and Fig. 30.1b depicts a local view of the fairlead with the two brackets (superior and inferior) [1].

The study of mooring systems has been more devoted to the investigation of anchorage lines. For instance, Finucane [2] elaborated on a technical report related to the rupture of four anchorage lines of an FPSO platform situated in the North Sea. The author concluded that it was necessary to develop a new project for the mooring system and improve the dynamic placement tests. Afterward, Qiao et al. [3] numerically investigated the effect of corrosion in anchorage lines and ties composed of steel and polyester. The primary purpose was to comprehend the influence of corrosion over the number of cycles supported by the lines under fatigue. Results indicated that the corrosion effects are more critical in the region composed of ties than in the cables. Azcona et al. [4] developed a computational code to predict a catenary anchorage line's behavior subjected to harmonic loads with different periods. Results showed the importance of knowing the stress distribution and displacement of the lines, mainly when subjected to high-frequency loads. Shen et al. [5] developed a numerical study to predict the anchorage line of a wind turbine subjected to cyclic load considering the region's interactions where the line is disposed on the seabed. Recently, some studies improved the comprehension of the interaction between the mooring system and boat displacement in the ocean [6–8]. Despite these important investigations, little has been reported on other equipment of the mooring system, such as fairleads and brackets.

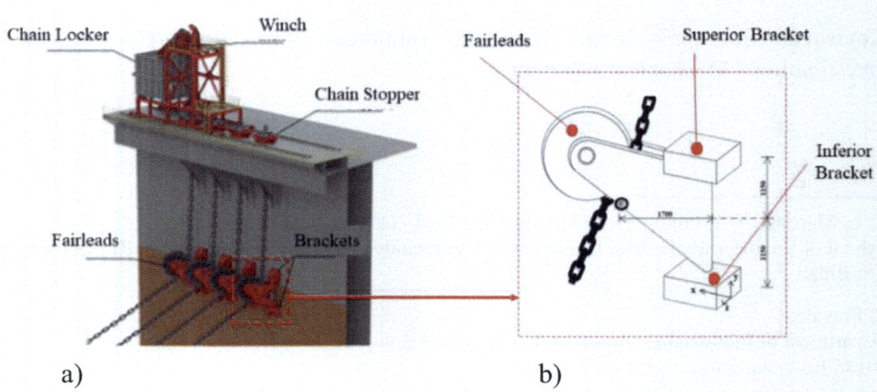

Fig. 30.1 Mooring system of an oil production platform: **a** mooring system mounted in the platform hull, **b** local view of fairlead, and brackets components

In the present work, it developed the computational modeling of brackets used for support of fairleads subjected to the maximum environmental load. One important subject in this problem is the achievement of design solutions for the employed plates, which would reduce the costs for mounting and maintenance of the structure in the vessel [1]. Therefore, it is also performed a study case reaching the theoretical recommendations about geometrical configurations that reduce the von Mises maximum stress in the plates of brackets. The studied geometries are based on the existent brackets of the P-66 Petrobras platform. More precisely, three different configurations for the base of superior brackets are compared in light of Constructal Design [9–12].

30.2 Definition of Environmental Loads and Reaction Forces

The methodology is defined in various steps; Step 1 consists on the estimation of the environment loads (wind, sea currents, and waves) over an FPSO platform with a spread mooring type composed of 24 lines using data available in the Campos basin (Brazil) by Det Norsk Veritas (DNV) standards [13].

It is considered a platform hull with height and length of $H = 40$ m and $L = 280$ m, respectively. From those, $A_c = 2240\ \text{m}^2$ faces the sea current while $A_w = 4480\ \text{m}^2$ faces the wind flow. Moreover, it is considered an ocean depth of $d = 1355$ m; velocities of $V = 1.6$ m/s and 35 m/s for water current and wind, respectively; and densities of $\rho_{\text{water}} = 1040\ \text{kg/m}^3$ and $\rho_{\text{air}} = 1.3\ \text{kg/m}^3$ [13].

In the present work, the oscillatory loads generated by winds and waves are disregarded since it is the first development step of the model. However, they can be incorporated into the modeling in future studies, since DNV standards supply the needed data for calculating these loads. DNV [13] also proposes different angles of action of the waves, currents, and wind loads. Despite this, only one angle is considered (perpendicular to the hull loads) to obtain a higher safety factor on the analyzed devices.

Table 30.1 shows the environmental loads calculated with the DNV standard used for predicting the ultimate limit state (ULS) and accidental limit state (ALS). For the ULS, the environmental load is shared between 24 mooring structures

Table 30.1 Environmental loads calculated for an FPSO (magnitudes in MN)

Environmental load type	Total load	Per line (24 lines)	Per line (20 lines)
Wind	6.69	0.28	0.33
Current	5.96	0.25	0.30
Waves	11.43	0.48	0.57
Total	24.08	1.01	1.20

distributed along the hull while the ALS, the distribution is done for only 20 structures (considering the failure of 4 components).

Step 2 of the methodology is predicting forces in the mooring lines caused by the environmental loads and the weight of the wet line. A free body diagram in the anchorage line is performed, and the calculation of reaction forces in the centroid of the fairlead is obtained. An inextensible line is taken into account in the calculation of reaction forces to make a more conservative evaluation. Later, Step 3, the reaction forces are transferred to the pin that connects the two brackets and, consequently, to the brackets. This methodology was presented in Martins et al. [14].

30.3 Computational Model for Prediction of Stress in Brackets

Step 4 consists of the solution of the computational domain, which represents the bracket under reaction forces. Figure 30.2 illustrates one sketch of the fairlead-brackets which composes the mooring system, Fig. 30.2a, the computational domain of one of the brackets, Fig. 30.2b, and a scheme of application of reaction forces applied in a distributed way in different quadrants of the hole of bracket, Fig. 30.2c. In the surface welded to the hull, it is imposed fixed support, while the other surfaces are free to displace.

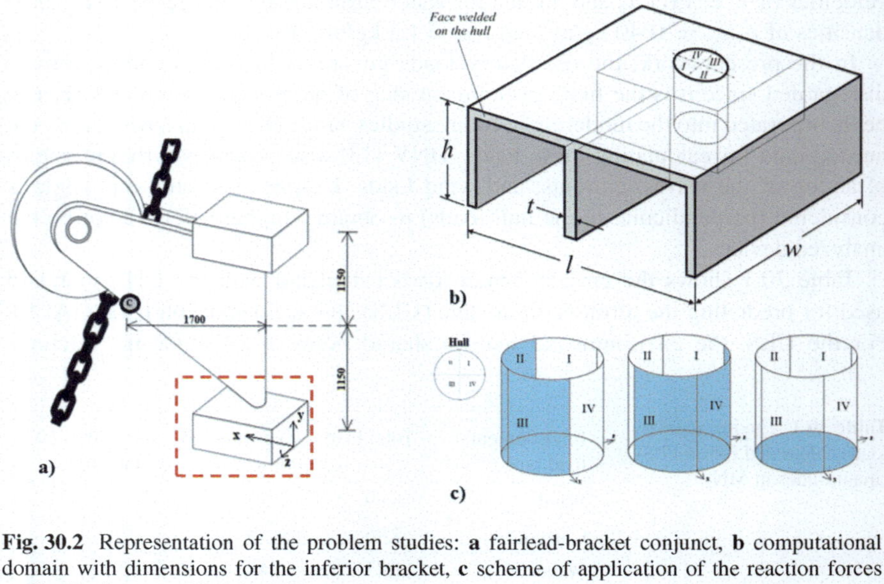

Fig. 30.2 Representation of the problem studies: **a** fairlead-bracket conjunct, **b** computational domain with dimensions for the inferior bracket, **c** scheme of application of the reaction forces applied in the centroid of the fairlead

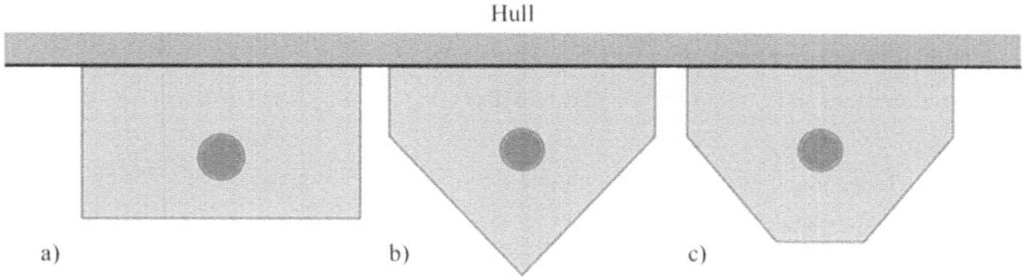

Fig. 30.3 Representation of different configurations of brackets studied here: **a** rectangular, **b** rectangular/triangular, **c** rectangular/trapezoidal

The brackets studied here are composed of AH36 steel plates, with a yielding limit of 355 MPa, modulus of elasticity $E = 210$ GPa and a Poisson's ratio $\nu = 0.30$. It is considered 10 mm thick plates (four times lower than that commonly found in real applications) based on the bracket. The hole, where the pin that connects the two brackets is placed, has a radius of 0.15 m. Moreover, it is 0.65 m distant from the hull and centralized in the x-direction of the plate. The height of the vertical plates of the bracket is also constant at $h = 0.6$ m.

Finite Element Method (FEM) is currently the most powerful method in the field of structural analysis [15] that enables a wide range of structural analyses [16–18]. In the present work, the numerical approach is performed with a code based on the FEM and developed ANSYS Mechanical APDL [19, 20]. It adopted the SOLID186 finite element, composed of 20 nodes with 3 degrees of freedom per node, being the translation on the x, y, and z axes [15]. The tetrahedral shape of the SOLID186 was adopted, allowing a more accurate discretization in the region around the hole that receives the support pin of the fairlead.

Step 5 consists of the comparison between the von Mises stresses for three different geometries of the plate basis: rectangular, rectangular/triangular, and rectangular/trapezoidal. These configurations are illustrated in Fig. 30.3a, b and c. All configurations are subjected to the same area constraint of building, i.e., in light of Constructal Design [9–12].

30.4 Results and Discussion

Table 30.2 shows the maximum von Mises stresses under ULS and ALS conditions and the three different geometric configurations studied in the present work. The rectangular structure led to the mechanical performance of nearly 35% and 38% superior to that achieved with rectangular/trapezoidal and rectangular/triangular bases for the ULS condition. For the ALS condition, the worst configurations reached to maximum stress higher than the yielding limit. Results evidenced that the investigation of geometry can be an essential strategy for designing the brackets found in mooring systems. It is also important to mention that the present results were achieved using plates four

Table 30.2 Results of ultimate limit state (ULS) and accidental limit state (ALS) for three different configurations of superior brackets

Configuration/Limit state	ULS (MPa)	ALS (MPa)
Rectangular	248.51	279.55
Rectangular/Triangular	338.04	380.09
Rectangular/Trapezoidal	333.78	382.43

times thinner than that commonly used in real brackets, which is an indication that the present modeling is promising to obtain general recommendations to minimize the costs of building of mooring systems regarding its stress limit.

30.5 Conclusion Remarks

In the present work, it was developed a computational model for prediction of mechanical behavior in brackets found in mooring systems. Results obtained indicated that this methodology can be promising for the rationalization of the design of this kind of equipment. Moreover, it was performed a comparison between three different geometric configurations for the brackets. The bracket with a rectangular basis led to a 35% and 38% superior performance than the rectangular/triangular and rectangular/trapezoidal configurations, showing the importance of design in this problem.

Future studies are recommended for analysis of other mechanical conditions considering dynamic loads in the problem and other geometrical configurations.

Acknowledgements K. L. Martins thank CAPES for Master Dissertation Scholarship (Finance Code 001). C. Fragassa thanks the MAECI as part of the "Two Seats for a Solar Car" international project. M.V. Real, L.A.O. Rocha, L.A. Isoldi and E.D. dos Santos thank CNPq for research grant (Processes: 304211/2018-4, 307791/2019-0; 306012/2017-0, 306024/2017-9).

References

1. Gordon RB, Brown MG, Allen EM (2014) Mooring integrity management: a state-of-the-art review. In: Offshore technology conference, Texas, pp 1–12
2. Finucane M (2012) Details of gryphon incident of 4th February 2011 and lessons learned. Technical report, Maersk Oil
3. Qiao D, Yan J, Ou J (2014) Fatigue analysis of deepwater hybrid mooring line under corrosion effect. Pol Marit Res 21(3):68–76
4. Azcona J, Munduate X, González L, Nygaard TA (2017) Experimental validation of a dynamic mooring lines code with tension and motion measurements of a submerged chain. Ocean Eng 129:415–427
5. Shen K, Guo Z, Wang L (2019) Prediction of the whole mooring chain reaction to cyclic motion of a fairlead. Bull Eng Geol Environ 78(4):2197–2213

6. Hollyhead CJ, Townsend NC, Blake JIR (2017) Experimental investigations into the current-induced motion of a lifeboat at a single point mooring. Ocean Eng 146:192–201
7. Pradana MR, Qian X, Ahmed A (2019) Efficient discrete element simulation of managed ice actions on moored floating platforms. Ocean Eng 190:106483
8. Touzon I, Nava V, Gao Z, Mendikoa I, Petuya V (2020) Small scale experimental validation of a numerical model of the HarshLab2.0 floating platform coupled with a non-linear lumped mass catenary mooring system. Ocean Eng 200:107036
9. Bejan A, Lorente S (2008) Design with constructal theory. Wiley, Hoboken, U.S.A
10. Bejan A (2020) Freedom and evolution: hierarchy in nature, society and science. Springer, Switzerland
11. Troina GS, Cunha ML, Pinto VT, Rocha LAO, Dos Santos ED, Fragassa C, Isoldi LA (2020) Computational modeling and constructal design theory applied to the geometric optimization of thin steel plates with stiffeners subjected to uniform transverse load. Metals 10:220–249
12. Lima JPS, Cunha ML, Dos Santos ED, Rocha LAO, Real MV, Isoldi LA (2020) Constructal design for the ultimate buckling stress improvement of stiffened plates submitted to uniaxial compressive load. Eng Struct 203:109883.109883-16
13. DNV, Det Norske Veritas: OS-301. Offshore standard, position mooring (2015)
14. Martins KL, Pinto VT, Rocha LAO, Dos Santos ED, Isoldi LA (2019) Numerical evaluation of the mechanical behavior of an FPSO mooring system fairleads foundations due to maximum environmental loads. Res Eng Struct Mater 5:355–366
15. Marinkovic D, Rama G, Zehn M (2019) Abaqus implementation of a corotational piezoelectric 3-node shell element with drilling degree of freedom. Facta Univ Ser Mech Eng 17(2):269–283
16. Pérez R, Hernández L, Quesada A, Pino J, Zayas E (2020) Finite element modeling of temperature fields on the cutting edge in the dry high-speed turning of AISI 1045 steel. Facta Univ Ser Mech Eng 18(2):205–217
17. Tigh Kuchak AJ, Marinkovic D, Zehn M (2020) Finite element model updating—case study of a rail damper. Struct Eng Mech 73(1):27–35
18. Fallahi N, Viglietti A, Carrera E, Pagani A, Zappino E (2020) Effect of fiber orientation path on the buckling, free vibration, and static analyses of variable angle tow panels. Facta Univ Ser Mech Eng 18(2):165–188
19. ANSYS, Inc: ANSYS mechanical APDL theory reference. Release 15.0 (2018).
20. Zienkiewicz OC, Taylor RL, Zhu JZ (2005) The finite element method: its basis and fundamentals. Elsevier

Chapter 31
Experimental Identification of Viscoelastic Properties of Plates Made of Quiet Aluminum

Pasquale Grosso, Alessandro De Felice, and Silvio Sorrentino

Abstract The present study is aimed at the identification of equivalent viscoelastic models for layered thin-walled structures, obtained from vibration measurement only. Accurate modeling of modal properties is fundamental for describing metal fatigue, caused by forced vibration on structural components. A new approach is proposed, based on a definition of an equivalent modal damping ratio applied to the circle-fit technique, to overcome the difficulties related to the identification of modal parameters when adopting non-conventional viscoelastic models. When the structural internal dissipative effects are dominant, this procedure identifies the parameters of an equivalent Young's modulus in the frequency domain. The proposed procedure is applied to the analysis of plates made by quiet aluminum, adopting the linear fractional Kelvin-Voigt viscoelastic model and assessing the accuracy of the identified parameters by comparison of numerically simulated with experimentally measured frequency response functions.

Keywords Damping · Viscoelastic models · Fractional derivative models · Modal parameters · Experimental vibration analysis · Quiet aluminum

31.1 Introduction

Accurate evaluation of the damping properties of layered thin-walled structures is a relevant problem for several industrial applications in the field of vibration control and noise reduction [1, 2]. But, direct identification from vibration data is generally

P. Grosso (✉) · A. De Felice · S. Sorrentino
Department of Engineering Enzo Ferrari, University of Modena and Reggio Emilia, Via Pietro Vivarelli 10, 41125 Modena, Italy
e-mail: pasquale.grosso@unimore.it

A. De Felice
e-mail: alessandro.defelice@unimore.it

S. Sorrentino
e-mail: silvio.sorrentino@unimore.it

G. Lesiuk et al. (eds.), *Structural Integrity and Fatigue Failure Analysis*,
Structural Integrity 25, https://doi.org/10.1007/978-3-030-91847-7_31

a difficult task [3], especially when more refined models than the viscous damping or hysteretic ones are necessary for getting a sufficient accuracy.

The problem of modeling vibrations of layered damped plates has been extensively investigated adopting conventional damping models, but only recently, an approach was developed to identify storage modulus and loss factor with frequency-dependent characteristics [4].

In the present study, a different indirect approach is proposed, based on the concept of equivalent modal damping ratio (ζ_n). It is well known under which assumptions ζ_n can be defined, by considering a linear viscous dissipative model (integer-order Kelvin-Voigt model) or a hysteretic model [5]. However, this theoretical parameter shows a dependency on the related natural frequency that in most cases fails in fitting the experimental data on relatively broad frequency ranges. Better results can be achieved by means of non-integer-order differential models [6].

To overcome the difficulty of finding analytical expressions of ζ_n in case of non-conventional dissipative models, a method of general validity is developed, introducing the concept of equivalent modal damping ratio applied to the circle-fit technique [7, 8]. This identification method is based on the assumption that the Nyquist plot of a frequency response function (FRF) for any mode n can be approximated by a circumference, which is still acceptable when considering non-conventional viscoelastic models [9]. An identification algorithm is then proposed, valid when the structural internal dissipative effects are dominant with respect to those due to air [10], adopting the fractional Kelvin-Voigt model [6] and applying it to the analysis of a vibrating plate made of quiet aluminum (QA). Among applications, accurate modeling of modal properties would improve the results of vibration-based methods for estimating fatigue in metal or metal-composite structures like QA panels.

31.2 Identification Method

According to the circle-fit technique [8] applied to the integer-order Kelvin-Voigt model (in the following, simply referred to as Kelvin model), almost circular experimental Nyquist plots of the mobility allow the estimate of the related modal damping ratios ζ_n and natural angular frequencies ω_n, which are given by:

$$\zeta_n = \frac{1}{2\omega_n}\left[\frac{\Omega_{2n}^2 - \Omega_{1n}^2}{\Omega_{2n}\tan(\gamma_{2n}) + \Omega_{1n}\tan(\gamma_{1n})}\right] \tag{31.1}$$

where the symbols refer to Fig. 1a and the natural angular frequency ω_n is identified as the frequency of maximum sweep rate of the circle (say Ω_{0n}; where the assumption $\omega_n = \Omega_{0n}$ is valid for the Kelvin model, not in general). The result of the identification procedure is a set of experimental modal parameters (ω_n, ζ_n).

However, if the aim of the identification procedure is getting an equivalent Young's modulus in the frequency domain, then the classical dissipative models (Kelvin and

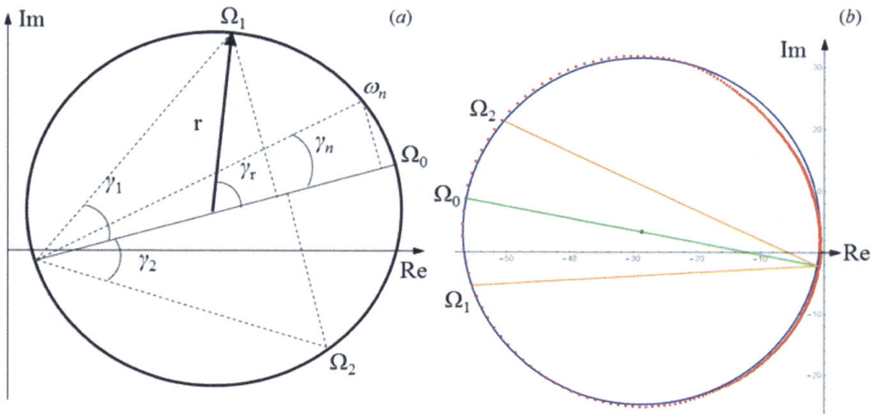

Fig. 31.1 Nyquist plot of mobility, general scheme for mode n (**a**). Experimental Nyquist plots of mobility, mode at 320 Hz (**b**)

also hysteretic) may not be accurate enough. As a consequence, more refined models are often needed. The fractional Kelvin-Voigt model [8] is herein adopted (simply referred to as fractional Kelvin model):

$$E(\omega) = E_0 R(\omega) = E_0\left[1 + (i\tau\omega)^\alpha\right], \quad E(s) = E_0 R(s) = E_0\left[1 + (\tau s)^\alpha\right] \quad (31.2)$$

where α is a non-integer (or fractional) derivative order (values between 0 and 1).

When considering a complex Young's modulus $E(\omega) = E_0 R(\omega)$, the modal mobility can be written as a function of the real and imaginary parts of $R(\omega)$, yielding the following possible definition for an equivalent modal viscous damping coefficient:

$$c_n^* = \frac{A_n}{2r} = \frac{k_n \mathrm{Im}[R(\Omega_{0n})]}{\Omega_{0n}} \quad (31.3)$$

where k_n is the modal stiffness, A_n is a dimensionless modal coefficient (assumed as real, without loss of generality), r is the radius of the modal Nyquist plot (assumed having circular shape), and Ω_{0n} identifies the maximum sweep rate of the modal circle.

Then, a definition for an equivalent modal viscous damping ratio can be given as:

$$\xi_n^* = \frac{1}{2}\frac{c_n^*}{\sqrt{k_n m_n}} = \frac{1}{2}\omega_n \frac{\mathrm{Im}[R(\Omega_{0n})]}{\Omega_{0n}} \quad (31.4)$$

which, compared to Eq. (31.1), yields a direct link between the parameters in the constitutive function R and the experimental data:

$$\mathrm{Im}[E_0 R(\Omega_{0n})] = \frac{\Omega_{0n}}{\lambda_n^2}\left[\frac{\Omega_{2n}^2 - \Omega_{1n}^2}{\Omega_{2n}\tan(\gamma_{2n}) + \Omega_{1n}\tan(\gamma_{1n})}\right] = p_n \quad (31.5)$$

where λ_n is a modal parameter (which can be separately estimated considering a model for the undamped structure). As a possible solution, therefore, an iterative procedure is developed, aimed at identifying the values of the constitutive parameters that minimize an error function between the experimental and the numerically simulated FRFs.

31.3 Experimental Results

The proposed method is applied adopting the fractional Kelvin-Voigt model for identifying the equivalent viscoelastic properties of a QA plate, without any extrapolation beyond the limits of the frequency interval of available data. A uniform square plate was tested with respect to flexural-free vibration (height = width = 300 mm; thickness of aluminum layers = 0.9875 mm; thickness of intermediate layer = 0.025 mm; total mass = 0.52 kg; equivalent density = 2889 kg/m^3). The specimen was restrained to a frame by means of flexible couplings, in order to approximate the free-free boundary conditions. It was excited by means of an instrumented ICP hammer in correspondence of the mid-point, and acceleration responses were evaluated by means of a miniaturized ICP piezoelectric accelerometer placed in the same excitation point, on the opposite face of the plate. An experimental Nyquist plot of modal mobility is displayed in Fig. 1b (mode at 320 Hz).

The modal parameters were computed according to the Rayleigh–Ritz method. The equivalent Poisson's ratios were estimated according to [11], after testing the sensitivity of the natural angular frequencies of the undamped plate. It resulted that the equivalent Poisson's ratio of QA does not vary significantly with respect to that of aluminum ($\nu = 0.33$, with maximum percentage errors in the natural angular frequencies of less than 0.1%). So, also, its dependency from frequency was neglected.

The identified equivalent constitutive parameters are reported in Table 31.1, in which $E_{0\mathrm{Al}} = 7.1 \times 10^{10}$ [N m^{-2}] is the standard value for Young's modulus of aluminum. A reduction of equivalent static Young's modulus was expected, and the low values of the fractional exponents mean that the overall dissipative behavior is closer to the hysteretic behavior than to the viscous one.

The experimental estimates of ζ_n are displayed vs. the natural frequencies ω_n in Fig. 31.2a. On the experimental data (dots) superimpose the curve $\zeta_n(\omega_n)$, plotted with the identified equivalent constitutive parameters in the Fractional Kelvin model.

Table 31.1 Identified equivalent constitutive parameters in the fractional Kelvin model

$E_0/E_{0\mathrm{Al}} = 0.825$	$\alpha = 0.219$	$\tau = 4.829 \times 10^{-8}$ s	($\rho = 2889$ kg/m^3, $\nu = 0.33$)

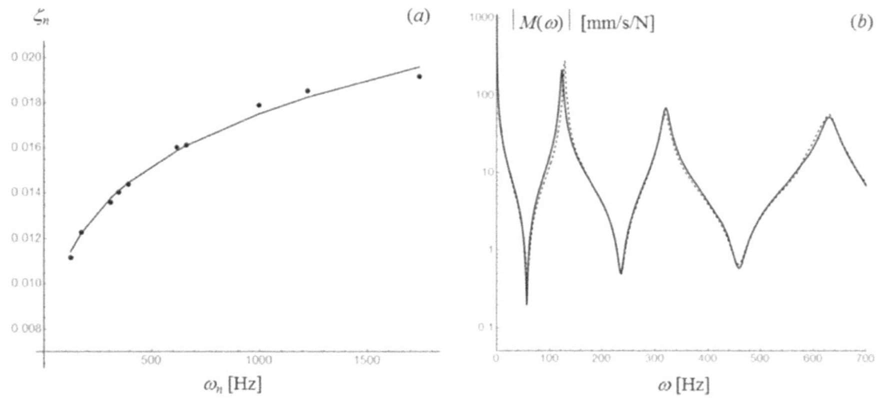

Fig. 31.2 Modal damping ratios ζ_n versus natural frequencies (**a**): experimental identified data (dots) and fractional Kelvin model interpolation (continuous curve). Mobility modulus versus frequency (**b**): experimental data (dotted) and fractional Kelvin model numerical simulation (continuous)

The accuracy of the identified parameters in the fractional Kelvin model is assessed by comparison of numerical with experimental FRFs (modulus of mobility), as shown in Fig. 31.2b (where the peaks are related to Modes 3, 5, and 11 of the free plate).

The same experimental FRFs were then compared with those obtained adopting an equivalent complex Young's modulus descending from either the Kelvin or the hysteretic models. In this case, the viscoelastic parameters were identified considering a single mode (actually Mode 3). Both models are clearly unable to fit the experimental FRFs, as shown in Fig. 31.3a (where '*v*' identifies the FRFs computed with the Kelvin model, '*h*' those computed with the hysteretic model, and '*f*' those already displayed in Fig. 2b).

Better results were obtained by using whole sets of identified modal parameters, either (ω_n, ζ_n) for viscous damping or (ω_n, η_n) for hysteretic damping. These sets of modal parameters were used for building a viscous damping matrix and a hysteretic damping matrix, respectively. The FRFs, thus, obtained are displayed in Fig. 31.3b. Also in this case, the hysteretic damping model gives better results than the viscous damping one. However, notice that the viscous and hysteretic models are not accurate out of the resonances.

31.4 Conclusions

The possibility of applying the circle-fit technique to the identification of non-conventional viscoelastic models from vibration data has been investigated on a plate made of quiet aluminum, discussing the limitations due to the adoption of the viscous damping and hysteretic models. A novel procedure has been proposed

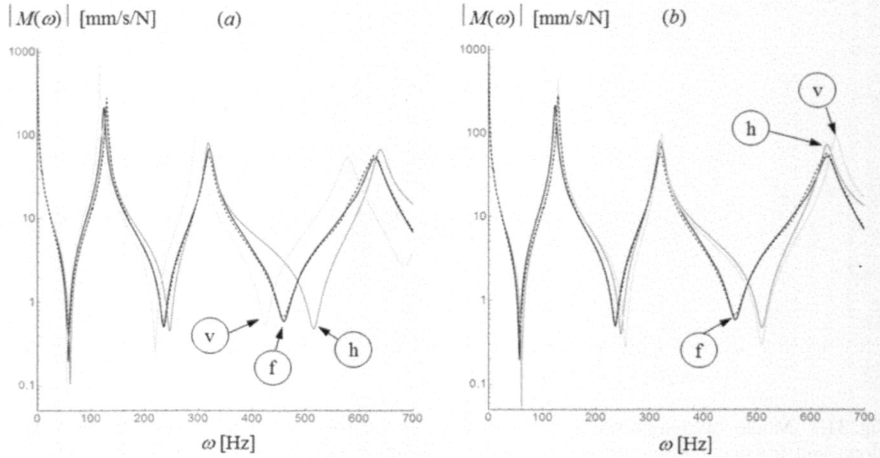

Fig. 31.3 Mobility modulus versus frequency. Experimental (dotted) and numerical simulations. Kelvin and hysteretic models (**a**) compared with identification based on modal viscous damping and modal hysteretic damping (**b**)

to overcome the difficulties related to the identification of modal parameters when adopting non-conventional viscoelastic models (in this case the linear fractional Kelvin-Voigt model), based on the concept of equivalent modal damping ratio and aimed at identifying an equivalent Young's modulus when the structural internal dissipative effects are dominant. The proposed identification procedure may also be suitable for application in the finite element method, for studying dissipative effects in thin-walled structures of general shape.

References

1. Oveisi A, Nestorovic T (2016) Mu-synthesis based active robust vibration control of an MRI inlet. Facta Univ Ser Mech Eng 14(1):37–53
2. Gabbert U, Duvigneau F, Ringwelski S (2017) Noise control of vehicle drive systems. Facta Univ Ser Mech Eng 15(2):183–200
3. Tigh Kuchak AJ, Marinkovic D, Zehn M (2020) Finite element model updating—case study of a rail damper. Struct Eng Mech 73(1):27–35
4. Sun W, Wang Z, Yan X, Zhu M (2018) Inverse identification of the frequency-dependent mechanical parameters of viscoelastic materials based on the measured FRFs. Mech Syst Signal Process 98(1):816–833
5. Gaul L (1999) The influence of damping on waves and vibrations. Mech Syst Signal Process 13:1–30
6. Mainardi F (1997) Fractional calculus: some basic problems in continuum and statistical mechanics. In: Fractals and fractional calculus in continuum mechanics, Springer, Wien
7. Kennedy CC, Pancu CDP (1947) Use of vectors in vibration measurement and analysis. J Aeronaut Sci 37:603–625

8. Ewins DJ (2000) Modal testing: theory, practice and application, 2nd edn. Research Studies Press, Baldock
9. Catania G, Sorrentino S (2009) Experimental evaluation of the damping properties of beams and thin-walled structures made of polymeric materials. In: 27th international proceedings on IMAC, Orlando
10. Wesolowski M, Barkanov E (2016) Air damping influence on dynamic parameters of laminated composite plates. Measurement 85:239–248
11. Ege K, Roozen NB, Leclere Q, Rinaldi RG (2018) Assessment of the apparent bending stiffness and damping of multilayered plates; modeling and experiment. J Sound Vib 426:129–149

Chapter 32
Machine Learning Approach Based on the Finite Element Method to Predict Reaction Forces

Michał Smolnicki

Abstract Machine learning is an important tool used in many science branches. In this paper, neural network is created to predict reaction forces in the plate under tension. The automatized Python script is used to generate model, solve model, obtain results and process data in SIMULIA Abaqus environment. The script was then used with over 15,000 combinations of 6 independent input parameters. Reaction forces in three directions were obtained in this process. Created in this way, dataset is normalized and used as the training dataset for a neural network. Errors obtained for simple structure in this approach are small and comparable to the error of the finite element method itself. This approach demands preparing a large dataset, but it can be worth it if the application lets analysis in real time, i.e., construction monitoring or medicine.

Keywords Neural network · Finite element method · Machine learning

32.1 Introduction

Machine learning is a branch of informatics based on the approach that computers can not only follow deterministic algorithms but also widen their knowledge based on defined rules, etc. Neural networks are a machine learning tool, which is based on the neural system of humans. It lets computers find complicated, abstract relations between sets of data (input, output) or classify objects defined by input parameters into some categories—predefined or not. Currently, machine learning is widely used outside informatics—in chemistry, pharmaceuticals, economy and much more [1]. However, there is a lack of papers that are focused on neural networks trained on a dataset based on the finite element method or even other connection between mechanics and machine learning. Existing papers are focused on simple examples like harmonic oscillators [2], simple shell arc structure [3]. Another set of papers is not using the finite element method as a source of data for neural networks, but

M. Smolnicki (✉)
Wrocław University of Science and Technology, Wrocław, Poland
e-mail: michal.smolnicki@pwr.edu.pl

G. Lesiuk et al. (eds.), *Structural Integrity and Fatigue Failure Analysis*, Structural Integrity 25, https://doi.org/10.1007/978-3-030-91847-7_32

rather uses some machine learning to define a constitutive model of material [4] or analyze damage of material stiffness [5]. The main problem which must be faced when using the described approach is to generate enough data point for the learning of a neural network. However, if one can do that, a neural network can be much faster than the standard finite element method, which can be crucial, for example, in medicine or live monitoring of constructions [6]. In this paper, author presents application of machine learning to predict reaction forces in plate under tension. The finite element method along with Python script is used to generate data which will be used to train neural network. Accuracy of such neural network will be revealed. This paper is part of larger project to analyze more complicated cases and is focused on assuring that the implementation of algorithms is proper.

32.2 Methods

32.2.1 Problem Description

As mentioned in the introduction, the main purpose of the presented research is to test the implementation of neural network script and model generation script as well as checking data proceeding procedure. So, the chosen problem is fairly simple. Abaqus was chosen as finite element method software because it has embedded Python that allows automatization of model generation. Plate modeled as a shell is used as a construction. Geometric parameters of the plate such as width (w), length (l), thickness (t) as well as boundary conditions such as displacement in *x*-, *y*- and *z*-direction were used as variable parameters. All parameters were defined as millimeters (however, Abaqus has not defined any units at all—output data units are dependent on input data units) (Fig. 32.1).

Creation of model in Abaqus is required from a user to define the material model. In this research, the variability of material will not be analyzed. Thus, to not over-complicate the problem, isotropic (same properties in all three axes) material was chosen. Such material model requires to define two from main material properties (Young Modulus, Poisson coefficient and Kirchoff Modulus). The exact values do not influence the rest of the research (besides exact result values). However, steel

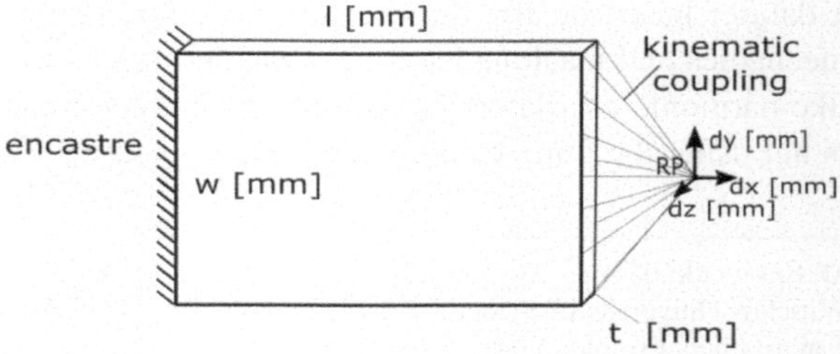

Fig. 32.1 Definition of input parameters of a model

Table 32.1 Input parameters definition and ranges

Quantity [unit]	Min. value	Max. value	Cardinality [−]
w	10	90	5
l	10	90	5
t	1	9	5
dx	1	9	5
dy	1	9	5
dz	1	9	5

was used as reference material ($E = 210{,}000$ MPa and $\nu = 0.3$). For each described above input parameters, the range of values was chosen. These ranges are presented in Table 32.1.

Only six input parameters with a cardinality of value sets equal to 5 give over 15 thousand combinations. It should be enough to train a simple neural network.

32.2.2 Data Generation Process

The large number of models that must be processed causes that some automation must be implemented. In SIMULIA Abaqus environment, it is possible by using Python script. Such a script can be used to generate model, send it to the solver and then process the data after results are obtained. The first step of model creation is to define the geometry. The dimensions are based on the input data for the current model as width, length and thickness vary between models. The next step is defining the material model which is done using isotropic material as mentioned before. Mesh generation—so the discretization process is fairly simple in the case of the described model. The effect of such a process on an exemplary model is presented in Fig. 32.2. In simulations, standard shell elements were used (4-nodes shell elements suitable

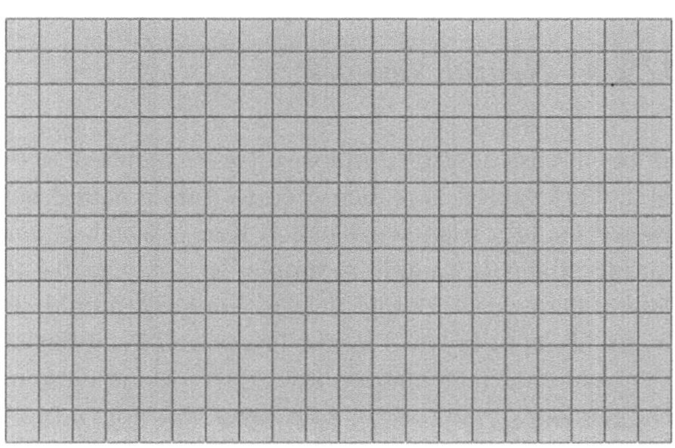

Fig. 32.2 Discrete model of the shell plate used in research

for thin and thick shells) with hourglass effect control and reduced integration—in Abaqus denoted as S4R elements.

The next stage of defining a numerical model is to define boundary conditions. In the case of analyzed shell plate, BCs are defined as shown in Fig. 32.1. So, one side is fully restrained (6 degree of freedoms—DOFs—are taken), and on the other side, displacement is forced using kinematic coupling to the reference point. The displacement is defined in 3DOFs (*x*, *y*, *z*), and three other DOFs are taken away. Such definition was used because it enables simple calculation of reaction under these boundary conditions. The whole reaction will be concentrated in reference points. Proper field and history output are defined in propose to obtain resultant data. Described steps lead to the ready numerical model which can be analyzed using Abaqus solver. For each combination of 6 input data, it is required to generate a new model and then send it to the solver. After that, data can be obtained.

Creation of model was automatized by using Python script. In Fig. 32.3, the flowchart of such script is presented. SIMULIA Abaqus shares Abaqus scripting interface which allows using all capabilities normally accessible in Abaqus CAE (graphic user interface mode) and even more. There is also possible to access the .odb files (resultant files) and explore these result bases.

Example of assembly obtained in the process is presented in Fig. 32.4. For this case, input parameters were as follows: width—80 mm, length 50 mm, thickness 5 mm, displacement in *x*-direction—5 mm, displacement in *y*-direction—3 mm, displacement in *z*-direction—7 mm.

The script read values of reaction force in three directions (RF1, RF2 and RF3) form the reference point (where the displacement was defined). The effect of the script is a text file with defined input and output parameters. The structure of such a file is presented in Fig. 32.5.

The form of data can affect results achieved by using a neural network. This the reason why input data should be normalized; i.e., all input parameters should be from the range from 0 to 1. It can be achieved by using scale factors. In Table 32.2, statistical data before normalization is shown, and In Table 32.3 statistical data after normalization is presented.

32.2.3 Neural Network

In the process described in the previous subchapter, over 15,000 records were gathered. This dataset was then used to train a neural network. A neural network was trained using a framework called Keras, which is available as a library to Python. An investigated example is simple, so it is possible to suspect that the relationship between input and output data is similarly simple. Thus, a neural network with 5 layers (1 input layer, 3 dense layers and 1 output layer) and sequential dataflow (it means that information flows only in one direction between layers) was used. The number of neurons in each layer was not so big. In the input layer, there are 6

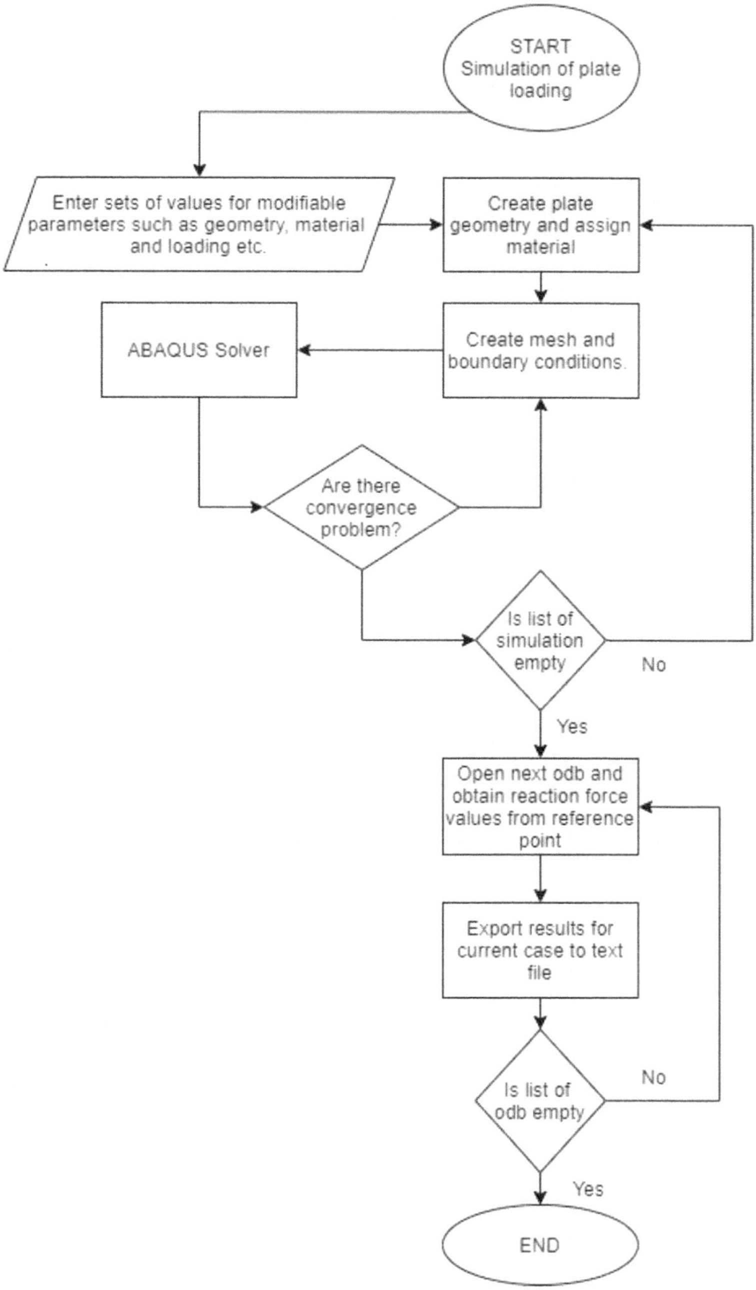

Fig. 32.3 Flowchart of the automated script to generate and solve a model for each combination of input data from the defined range

Fig. 32.4 Assembly view of one of the models created during the research

```
LP.    w     l     t     dx    dy    dz    RF1    RF2    RF3
...
199    40    80    5     0     8     8     0      237.2  243.1
...
```

Fig. 32.5 Structure of result data file

Table 32.2 Statistics of input parameters before normalization

	w	*l*	*t*	d*x*	d*y*	d*z*
Count	15,624	15,624	15,624	15,624	15,624	15,624
Mean	50	50	5	5	5	5
Std	31.62	31.62	31.62	3.162	3.162	3.162
Min	10	10	1	1	1	1
Max	90	90	9	9	9	9

Table 32.3 Statistics of input parameters after normalization

	w	*l*	*t*	d*x*	d*y*	d*z*
Count	15,624	15,624	15,624	15,624	15,624	15,624
Mean	0.5	0.5	0.5	0.5	0.5	0.5
Std	0.27	0.27	0.27	0.27025	0.27025	0.27
Min	0	0	0	0	0	0
Max	1	1	1	1	1	1

neurons (number of independent input parameters), and dense layers have 200, 50 and 10 neurons. The last layer has 1 neuron—for one of the reaction force directions. The number of neurons should be small enough, so the number of trainable parameters is not much bigger than the dataset size. In another case, there is a risk

```
Model: "sequential"
_________________________________________________________________
Layer (type)                 Output Shape              Param #
=================================================================
dense (Dense)                (None, 250)               1750
_________________________________________________________________
dense_1 (Dense)              (None, 50)                12550
_________________________________________________________________
dense_2 (Dense)              (None, 20)                1020
_________________________________________________________________
dense_3 (Dense)              (None, 1)                 21
=================================================================
Total params: 15,341
Trainable params: 15,341
Non-trainable params: 0
```

Fig. 32.6 Structure of the neural network used in research

of overfitting, which could decrease the accuracy of the new data. In Fig. 32.6, the structure of the used neural network is presented.

Dataset was split into three parts for cross-validation. The biggest part (60% of the size) was the train set, then the second part (20% of the size) was used as a 'teach' set, and finally, last 20% of the dataset was used to obtain the final accuracy. It is worth underlining that this part of the dataset was never 'seen' by neural network neither as train data nor as data used to generate loss function. The whole training dataset was passed to the neural network 20 times (number of epochs) in smaller subsets called 'mini-batches.' This approach lets us avoid other overfitting potential causes.

32.3 Results

Each single finite element analysis results in *.odb file with full information about displacement, stresses, etc. In this paper to further analysis, only reaction forces were used, but the rest of the data can be used to improve the technique. In Fig. 32.7, field of reduced stresses (hypothesis of Huber-Mises-Hencky) is presented for exemplary simulation. The input parameters were as follows: width—80 mm, length 50 mm, thickness 5 mm, displacement in x-direction—5 mm, displacement in y-direction—3 mm, displacement in z-direction—7 mm.

Neural network during the process of learning with the supervisor is getting information about current accuracy. To estimate this accuracy, loss function is widely used—it shows the difference between predicted values and the real ones. Loss function during the learning process should converge to near 0 values. In Fig. 32.8, the loss function in the function of epochs number for train and test data was shown.

Fig. 32.7 Stress field (HMH) in megapascals—example of simulation result

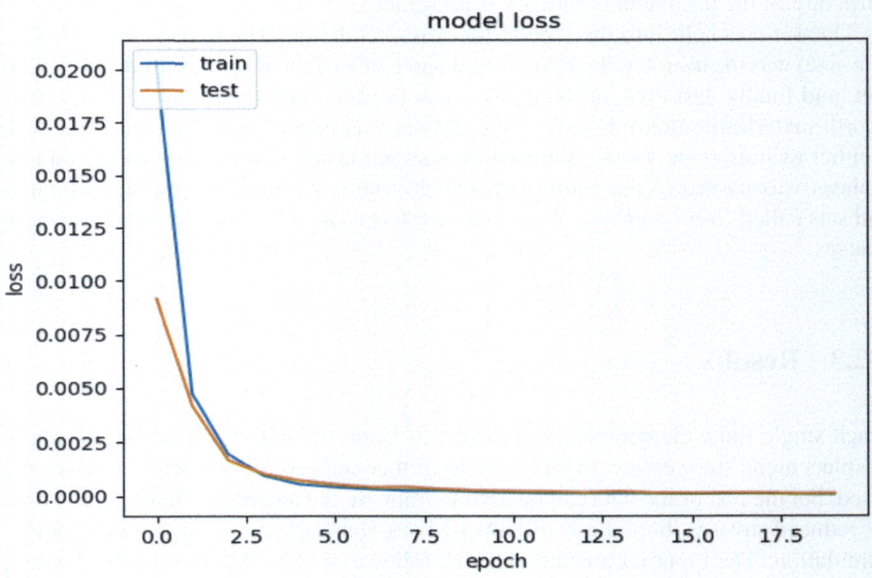

Fig. 32.8 Loss function of the training process of the neural network

The effect of the training process is the resultant set of weights. Using these weights, it is possible to calculate a result for any input data (from the assumed domain). The loss function is needed to supervise the learning process. Final results are described by measures like rooted mean squared error (RMSE) or mean squared error (MAE). Values of these errors are less than commonly widely adopted error of the finite element method.

32.4 Conclusions

Conducted research proves that this is possible to use SIMULIA Abaqus along with a Python scripting environment to automate the creation of a simple numerical model for a large combination of variable parameters. This can lead to the creation of a dataset that can be used by the neural network in process of learning. Input parameters to the script are the same as first-layer neuron values in a neural network. Output parameters of automating process are the same as the reference value used in the learning process. Obtained data must be normalized—that process increases the accuracy of the neural network. Described approach was used to deal with a simple mechanical problem—determination of reaction force in the plate under tension in three directions. The analogical approach could be implemented to deal with more complicated problems. However, even for this fairly simple problem, a large dataset was needed to deal with the problem of overfitting. This can lead to the huge time needed to generate such a dataset for complex problems. On the other hand, in some cases such as live monitoring of construction, more important is to decrease the time needed to detect problems. In such cases, this approach can be successful even if training of the neural network is time- and cost-consuming.

Acknowledgements Calculations have been carried out using resources provided by Wroclaw Centre for Networking and Supercomputing (http://wcss.pl), grant no. 531.

References

1. Abiodun OI, Jantan A, Omolara AE et al (2018) State-of-the-art in artificial neural network applications: a survey. Heliyon 4:e00938
2. Kononenko O, Kononenko I (2018) Machine learning and finite element method for physical systems modeling
3. Ćojbašić Ž, Nikolić V, Petrović E et al (2014) A real time neural network based finite element analysis of shell structure
4. Javadi AA, Tan TP, Zhang M (2003) Neural network for constitutive modelling in finite element analysis. In: Computer assisted mechanics and engineering sciences, pp 523–529

5. Gulikers THE (2018) An integrated machine learning and finite element analysis framework, applied to composite substructures including damage
6. Liang L, Liu M, Martin C, Sun W (2018) A deep learning approach to estimate stress distribution: a fast and accurate surrogate of finite-element analysis. J R Soc Interface 15:20170844. https://doi.org/10.1098/rsif.2017.0844

Chapter 33
Electric Moto3 Welded Chassis-Structural Analysis and Measurements

Michał Skraburski and Paweł Zielonka

Abstract Contraction weld stress creates distortions that are never desired especially in a vehicle such as a racing motorcycle. Every dimension deviance may have an impact on drivers' safety, assembly accuracy, or performance. The following article presents a motorcycle frame and swingarm (constructed by students of Wroclaw University of Science and Technology) after they were welded. Those items were 3D scanned and compared with their 3D models. Figures present the most significant distortion, which is later discussed. Additionally, some welds were inspected and analyzed using an X-ray scanner. The conclusion reveals that it is only possible to minimalize deviances by actions such as positional welds and properly constructed welding stance.

Keywords Weld structures · Electric motorcycle · X-ray scan

33.1 Introduction

Motostudent is an international contest for students whose objective is to construct a fully functioning Moto3 (with engine capacity maximum 250 cc) category motorcycle. There are two categories: petrol and electric. Each team is obliged to prepare a detailed concept of their vehicle and provide documentation of every part that had to be carefully chosen or manufactured. An event takes place at Motorland Aragón track in Spain where dynamic and static tests are conducted on every prototype. The team that qualifies to the next level takes part in a race in their category, which consists of 5 laps (around 40 km).

Light Electric Motorcycle (LEM) project is a series of motorcycles designed by Wroclaw University of Science and Technology students involved in Team under the name of LEM Wroclaw. 2020 project LEM Photon (Fig. 33.1) is the first Moto 3 category motorcycle fully designed and constructed by the team. Every prototype

M. Skraburski · P. Zielonka (✉)
Faculty of Mechanical Engineering, Wroclaw University of Science and Technology, Smoluchowskiego 25 St, 50-370 Wroclaw, Poland
e-mail: pawel.zielonka@pwr.edu.pl

G. Lesiuk et al. (eds.), *Structural Integrity and Fatigue Failure Analysis*, Structural Integrity 25, https://doi.org/10.1007/978-3-030-91847-7_33

Fig. 33.1 LEM Photon. Fully electric Moto3 category motorcycle

created by the organization since 2013 is an electric motorcycle which indeed is the future in fast-developing areas [1, 2].

LEM Photons' main parameters are:

- 42 kW (peak) power—a permanent-magnet synchronous motor (PMSM) Motor.
- 197 km/h maximum velocity.
- 6.5 kWh battery.
- 4 s 0–100 km/h.

Motorsport is a valid, developing department of automotive since companies are on a never-ending lookout for better technologies to improve vehicles' performance. Students following their steps want to test themselves in real-life tasks such as constructing a race motorcycle. Chassis is one of the essential parts of the motorcycle without which every other component may work perfectly yet still give no satisfying results [3]. It ensures the safety of a driver and enables reaching the best performance [4]. A lot of solutions are tested such as carbon chassis, cast (at least partially), or welded ones [5]. Every type has its pros and cons, but welded constructions are the cheapest among previously listed. That is an important factor when we talk about a prototype and not a serial production.

Students from LEM Wroclaw organization designed welding stances in order to prevent contraction stress following dimension deviations [6]. However, it is not possible to remove them completely. A team has successfully minimalized the stress which will be depicted later. After the frame and swingarm of the motorcycles were welded and soaked, they were 3D scanned in order to check deviations. Conducted 3D scan has an accuracy of 0.1 mm. To comprehensively analyze all possible causes of distortions, an X-ray scan has been conducted to check the correctness of the welds. Additionally, temperature and weather conditions have been taken into account.

This article consists of sufficient information for any team willing to properly plan their welding stances and welds. Young engineers are full of ideas that are often atypical, which does not imply that they are bad. Some of the results of the further depicted analysis may help fulfill their project.

33.2 Material and Methods

33.2.1 Materials

The swingarm is made of 6082 T6 aluminum alloy and the frame is made of Docol® 8 25CrMo4 steel. Particular mechanical properties give Table 33.1. Moreover, the chemical composition of this Docol® 8 steel is presented in Table 33.2.

Chosen aluminum shows including good corrosion resistance, high strength, good formability, and low density. T6 heat treatment is used to obtain the desired mechanical properties [7]. Its' main properties are presented in Table 33.3. Moreover, the chemical composition of this 6082 T6 aluminum is presented in Table 33.4.

Table 33.1 Main properties of 25CrMo4 steel according to data provided by the manufacturer

Property	Value
Yield stress ($R_e/R_{0.2}$) [MPa]	690
Ultimate tensile stress, R_m [MPa]	800
A_5 [%]	8

Table 33.2 Chemical composition of 25CrMo4 according to data provided by the manufacturer

C (max %)	Si (max %)	Mn (max %)	P (max %)	S (max %)	Al (max %)	Cr (max %)	Cr + Mo (max %)
0.15	0.30	1.70	0.020	0.010	0.015	0.105	0.155

Table 33.3 Main properties of 6082 aluminum according to data provided by the manufacturer

Property	Value
Yield stress ($R_e/R_{0.2}$) [MPa]	260
Ultimate tensile stress, R_m [MPa]	310
A_5 [%]	8

Table 33.4 Chemical composition of 6082 aluminum according to data provided by the manufacturer

Si (max %)	Fe (max %)	Cu (max %)	Mn (max %)	Mg (max %)	Cr (max %)	Zn (max %)	Ti (max %)	Others (min %)
1.30	0.50	0.10	1.0	1.20	0.25	0.20	0.10	0.15

These materials are characterized by great durability and decent weldability [8–10]. Both were chosen to sustain previously calculated loads. Another advantage of these materials is that they are commonly used, which is why it is not problematic to find particular profiles.

33.2.2 Welding Stances

Welding stances both for the frame (Fig. 33.2) and swingarm (Fig. 33.3) have been designed in a way to easily dismount them and take out welded parts. Frame stance had been fully created by connecting parts (which were previously laser cut) using bolts and swingarm stance has been mostly welded. Stances were made out of S355 steel of thickness ranging from 4 to 10 mm.

Figures 33.4 and 33.5 depict a complete model of swingarm and frame created using SolidWorks software.

Fig. 33.2 Half of the frame in a welding stance

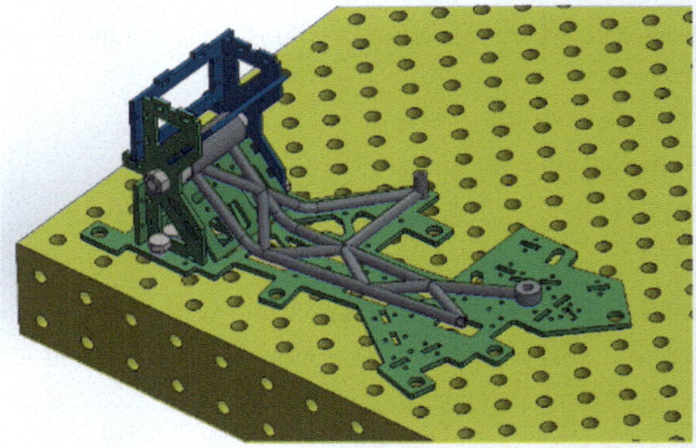

Fig. 33.3 Swingarm in a welding stance

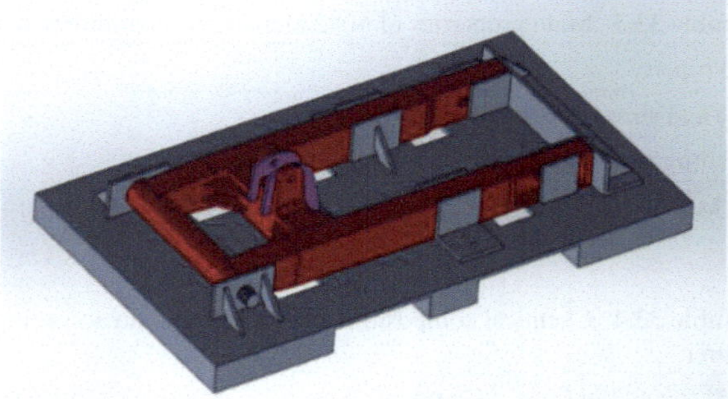

Fig. 33.4 Swingarm assembled with rear wheel

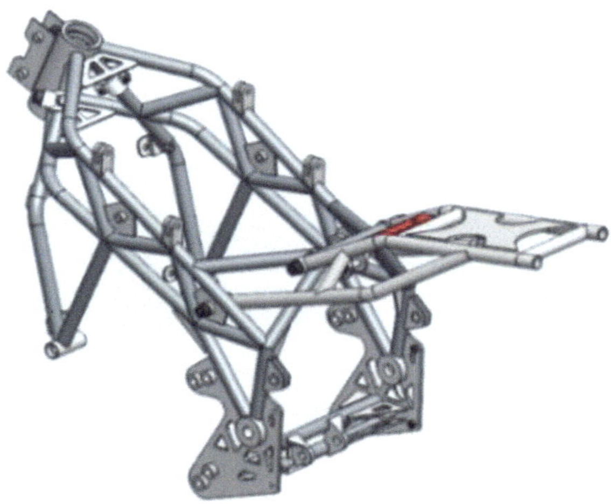

Fig. 33.5 Frame assembled with subframe

33.2.3 3D Scan

Motorcycle chassis has been scanned using ATOS Triple Scan, S029. Acquired data has been compared with the 3D model (Figs. 33.6 and 33.7). The accuracy of the scan is 0.1 mm.

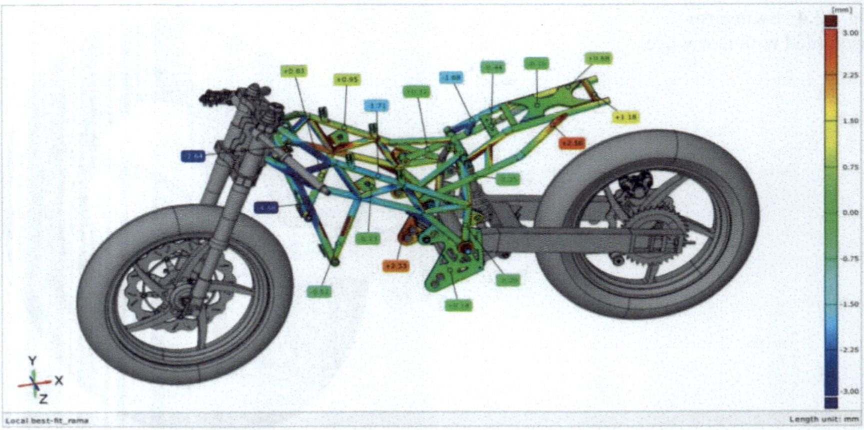

Fig. 33.6 Frame surface comparison

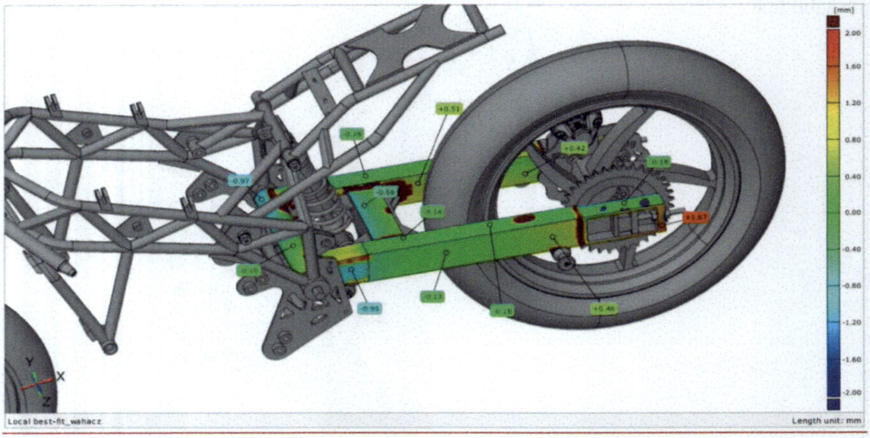

Fig. 33.7 Swingarm surface comparison

33.2.4 X-Ray Scan

Swingarm welds have been created properly. There is significant root penetration, no impurities, and only small insignificant holes. On the other hand, the subframe has a lot of holes and there is not a full root penetration which is naturally inexpedient. The subframe has been later analyzed and passed all durability tests. In order to conduct the analysis, 2 spots have been chosen. One on a subframe (Fig. 33.8) and one on a swingarm (Fig. 33.9), which roentgen analysis are shown consecutively on (Figs. 33.10 and 33.11).

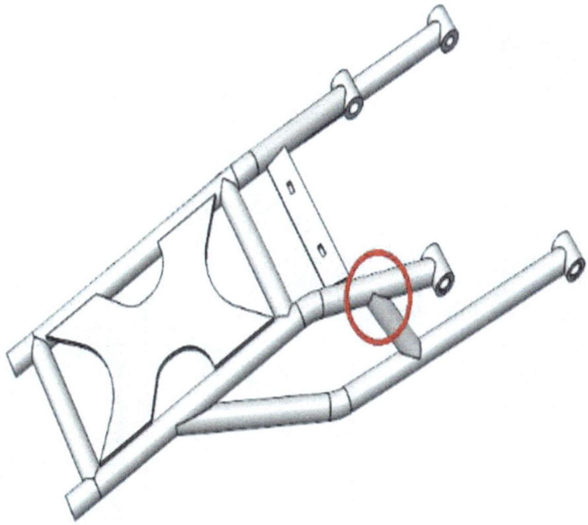

Fig. 33.8 Analyzed part of the subframe

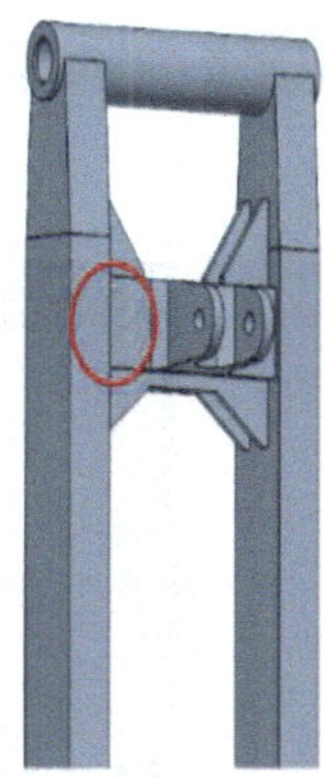

Fig. 33.9 Analyzed part of the swingarm

Both elements have been welded using the Tungsten Inert Gas (TIG) method for better control over welding [11]. Filler metals are consecutively EA 5356 and ER 80SD2. The action took place in a closed room where the temperature was 10 °C. The welder had 5 years of experience in this method.

To analyze the correctness of the welds it was possible to choose either X-ray or UT method [12, 13]. X-ray scan was a much better choice since the thickness of materials was between 2 and 10 mm and all of the parts were previously powder painted.

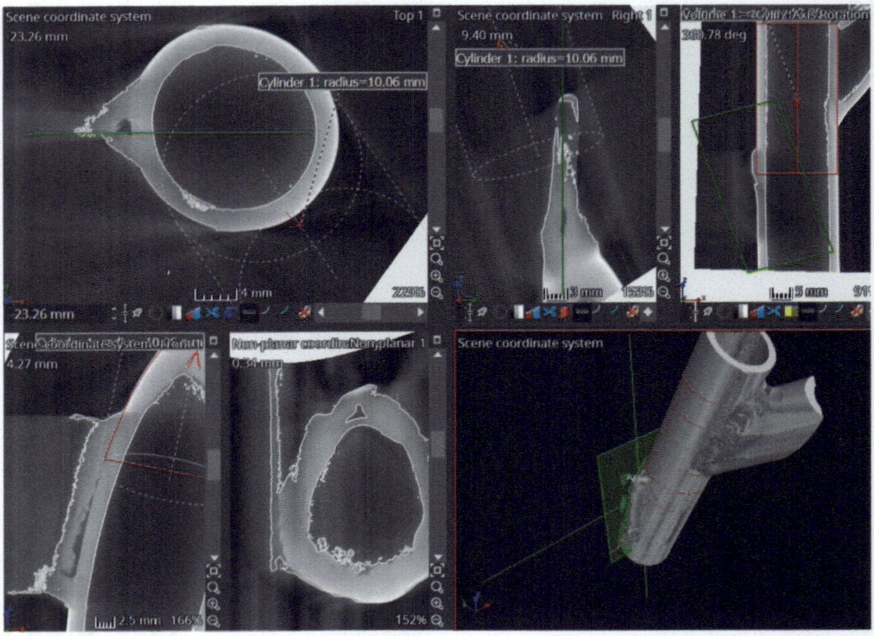

Fig. 33.10 Subframe X-ray weld analysis

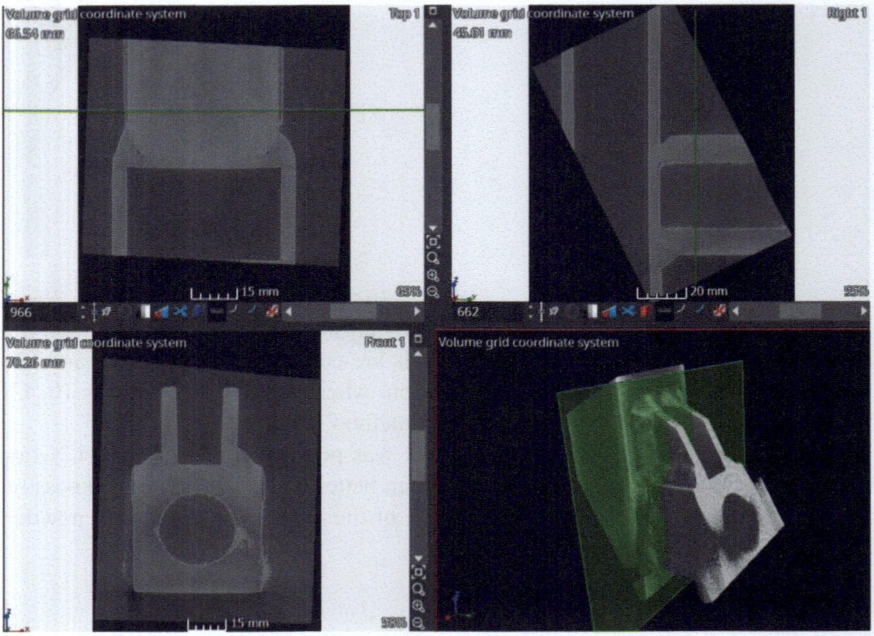

Fig. 33.11 Swingarm X-ray weld analysis

33.3 Results and Discussion

What might be observed is a fact that areas, where welds were placed correctly are less buckled. In addition, it applies also to places where it was much harder for the nozzle to get to. It is highly possible that it is simply connected with the welding accuracy.

Another dependence that can be observed is that most deviations are symmetrical compared to a motorcycle plane created on a rear wheel.

Swingarm which was welded in a stance that was fully welded has buckled less than the frame. That implies that correctly prepared stances, which are welded are better than those connected with screws. Incorrections that would usually occur in a welding stance should already create dimension' deviations, yet they did not. 0.3 mm deviation is significant but can be easily corrected. The swingarm is 1.5 mm shorter in reality probably due to imprecision in preparing profiles.

33.4 Conclusion

Some deviations are going to impact the motorcycle's performance and easiness of assembling all parts. What is most important is that the chassis is safe for the driver and all distortions can be easily corrected with pads or by grinding surfaces. It is highly important to place welds correctly not only because of deviations but also for safety [14].

Welding in cold temperatures requires the welder to set the amperage level up much higher in order to create a proper weld puddle. It is directly connected with root penetration and heat-affected-zone, which impacts the structure of the material near the weld. It might be more brittle which is inadvisable for constructions that are subjected to random nonlinear loads [15]. It is always better (if possible) to weld a room temperature.

Every team or company that is planning to create a prototype motorcycle chassis should carefully design it in order to enable a robot or welder to easily place a nozzle at a proper angle to ensure proper root penetration. It is highly important to create positional welds and to weld part of the profiles so that contractional stress can reduce itself.

References

1. Satiennam T, Satiennam W, Tankasem P, et al (2014) A study of potential electric motorcycle use to support a low carbon society: case of a developing Asian city. Adv Mater Res 541–545
2. Tsakalidis A, Thiel C (2018) Electric vehicles in Europe from 2010 to 2017: is full-scale commercialisation beginning. Jt Res Cent
3. Foale T (2006) Motorcycle handling and chassis design: the art and science. Tony Foale

4. Thakur SS, Roy EA, Dhakad SK. (2020, February). Design of Electric Motorcycle. In: 2020 IEEE International Students' Conference on Electrical, Electronics and Computer Science (SCEECS) 1–4, IEEE
5. O'Dea N (2011) Motorcycle swingarm redesigned in carbon composite. Reinf Plast 55(6):38–41
6. Michaleris P, DeBiccari A (1997) Prediction of welding distortion. Weld J Weld Res Suppl 76(4):172s
7. Cornacchia G, Cecchel S (2020) Study and characterization of EN AW 6181/6082-T6 and EN AC 42100–T6 aluminum alloy welding of structural applications: metal inert gas (MIG), cold metal transfer (CMT), and fiber laser-MIG hybrid comparison. Metals (Basel) 10:441. https://doi.org/10.3390/met10040441
8. Mrówka-Nowotnik G, Sieniawski J (2005) Influence of heat treatment on the microstructure and mechanical properties of 6005 and 6082 aluminium alloys. J Mater Process Technol 367–372
9. Tomita Y, A KO-MT, 1985 undefined Mechanical properties of 0.40 pct C-Ni-Cr-Mo high strength steel having a mixed structure of martensite and bainite. Springer
10. Szczucka-Lasota B, Węgrzyn T, Łazarz B et al (2020) MAG welding of structures used in means of transport and made of steel DOCOL 1200M. Biul Inst Spaw 64(4):65–71. https://doi.org/10.17729/ebis.2020.4/6
11. Ericsson M, Sandström R (2003) Influence of welding speed on the fatigue of friction stir welds, and comparison with MIG and TIG. Int J Fatigue 25:1379–1387. https://doi.org/10.1016/S0142-1123(03)00059-8
12. Raj B, Jayakumar T, Thavasimuthu M (2002) Practical non-destructive testing. Woodhead Publ
13. Blitz J, Simpson G (1995) Ultrasonic methods of non-destructive testing (Vol. 2). Springer Sci Bus Media
14. Guo S, Shah L, Ranjan R et al (2019) Effect of quality control parameter variations on the fatigue performance of aluminum friction stir welded joints. Int J Fatigue 118:150–161. https://doi.org/10.1016/j.ijfatigue.2018.09.004
15. Galvão I, Oliveira J, Loureiro A, Rodrigues D (2011) Formation and distribution of brittle structures in friction stir welding of aluminium and copper: influence of process parameters. Sci Technol Weld Join 16:681–689. https://doi.org/10.1179/1362171811Y.0000000057

Chapter 34
Analysis of the Impact of Vibrations on a Micro-Hydraulic Valve by Sequence Graph Method

Michał Stosiak, Krzysztof Towarnicki, Marian A. Partyka, and Adam Deptuła

Abstract The paper defines areas where a working machine or device may be a source of external forces on the control elements of hydraulic valves. The influence of external mechanical vibrations carries a number of unfavorable and undesirable effects on the operation of hydraulic valves. During the experimental tests, the valve was affected by external mechanical vibrations of known amplitude and frequency. The test stand of the hydraulic measuring system is described. Exemplary amplitude-frequency characteristics for selected parameters are presented. The work describes the application of the sequential graph method to analyze the impact of vibrations on a micro-hydraulic valve. The procedures for determining the sequence of the importance of parameters in relation to the graphs used were defined. The analysis of iterative systems, i.e., those that contain a certain set of interval chains with the same basic cell (interval) but with a different rank of importance, is an important tool for the diagnostic analysis of systems built on functions. This will eventually allow multi-valued decision trees to be used for analysis in the intervals concerned.

Keywords Graph method · Mechanical vibrations · Micro-hydraulic valve

M. Stosiak · K. Towarnicki (✉)
Department of Technical Systems Operation and Maintenance, Faculty of Mechanical Engineering, Wroclaw University of Science and Technology, Wroclaw, Poland
e-mail: krzysztof.towarnicki@pwr.edu.pl

M. Stosiak
e-mail: michal.stosiak@pwr.edu.pl

M. A. Partyka
Department of High Voltage Technology, Faculty of Electrical Engineering Automatic Control and Informatics, Institute of Electrical Power Engineering and Renewable Energy, Opole University of Technology, Opole, Poland
e-mail: m.partyka@po.edu.pl

A. Deptuła
Department of Management and Production Engineering, Faculty of Production Engineering and Logistics, Opole University of Technology, Opole, Poland
e-mail: a.deptula@po.edu.pl

G. Lesiuk et al. (eds.), *Structural Integrity and Fatigue Failure Analysis*,
Structural Integrity 25, https://doi.org/10.1007/978-3-030-91847-7_34

34.1 Introduction

The operation of machinery and equipment that contains micro-hydraulic systems is a source of mechanical vibration over a wide spectrum of frequencies [1]. The various environmental sources of vibration can generally be divided into determined, random, external and internal. Internal vibration is generated by the running machines that are seated on floor slabs and foundations. This type of vibration can also include vibrations caused by mechanical and plumbing systems, although they are mostly acoustic. Extraneous vibrations are transmitted through the ground, and their sources can be: (a) traffic, the frequency of which depends on the speed of a moving vehicle (car, truck), while for a traveling streetcar, the frequency of ground vibration caused by that type of forcing usually ranges from 3 to 40 Hz [2], (b) blow-ups (e.g., in obtaining stone material from working faces in quarries), explosions, implosions, (c) machines operating in the vicinity of buildings, factories, (d) railway traffic [3].

The forces causing vibrations are often due to reactions from other elements of a machine or device. In addition to the forces varying periodically, the vibration can originate from forces the values of which are constant in time, but their directions and points of application do change (e.g., for centrifugal forces). In machines and devices, vibration often originates with kinematic forcing (e.g., while a mobile machine is traveling over uneven ground).

In hydraulic systems, mechanical vibration is mainly caused by pressure pulsation due to two reasons. First one is the kinematics of the pump displacement components [4] and the second is the excitation by the vibrating components of a micro-hydraulic system. This concerns, in particular, the excitation of micro-valve control elements (e.g., micro-distributor slides or lift micro-valve crowns). Because of the pumping pressure tending to increase in gear pumps [5, 6], the pressure pulsation generated by displacement pumps is expected to increase. The pressure pulsation frequency corresponds to performance pulsation frequency [7].

34.2 Test Bench

The tests to determine the impact of external mechanical vibration on a hydraulic micro-valve are presented in Fig. 34.1a.

The measurement station consists of a three-phase electric motor steered by a control cabinet inverter, which drives an external-meshing gear micro-pump. This type of pump was used because it has a relatively low purchase price, is resistant to pollution, generates high pressures and is often used in micro-hydraulic systems. The pressure pulsations depend strictly on the number of pump's displacement components and driveshaft speed. In a gear pump, it is mainly the number of teeth that affects the pressure pulsation. A heat exchanger was used to stabilize the hydraulic system temperature. Measurements were performed for a constant vibration amplitude with a forcing frequency between 200 and 900 Hz. A series of measurements

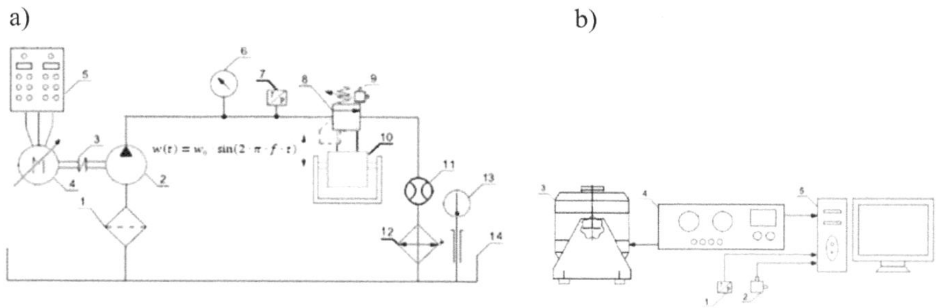

Fig. 34.1 **a** Diagram of the hydraulic system for testing the effect of external mechanical vibration on hydraulic micro-valves: 1—oil filter, 2—gear pump: WPH PZ3A1G, 3—flexible coupling, 4—three-phased electric motor, 5—control cabinet, 6—pressure gauge: Parker service junior SCJN-400–01, 7—piezoelectric pressure sensor: PCB piezotronics 105C23, 8—micro-valve being tested, 9—piezoelectric vibration inductor: TiraVib, 11—flow meter: Parker KSCVF-002–10-07X, 12—oil coller with a controller, 13—thermometer: Elmetron PT-217, 14–oil tank, **b** measuring system: 1—piezoelectric pressure sensor: PCB piezotronics 105C23, 2—piezoelectric acceleration sensor: PCB piezotronics 340A16, 3—electrodynamic inductor of mechanical vibration: TiraVib, 4—measuring amplifier: BA 1000TiraVib, 5—PC

were carried out using Azolla 22AF oil in the hydraulic system, $\rho = 865$ [kg/m^3], $\upsilon(40\ °\text{C}) = 22.5$ [mm^2/s].

The test bench includes a system to measure the vibration and its impact on the operation of micro-hydraulic components, Fig. 34.1b. The piezoelectric pressure sensors (1) and acceleration sensors (2) provided a signal which was then sent to a computer with a built-in measuring card, where analog data were converted to digital data further acquired. Measuring card communication was enabled by a dedicated PUMA software for remote starting, ending and recording measurements. During measurements, an overflow micro-valve weighing 700 g was used. Its design was based on an overflow valve used in conventional hydraulics. It is a classic miniaturized design with a sharp-edged socket [8] and a cone-shaped closing element, the so-called crown. The flare angle is 30°. The spring is pre-tensioned using an adjusting screw so that the valve's opening pressure can be changed. To attach the valve to the vibrator, a special clamp was bolted to the valve.

34.3 Selected Results of Experimental Tests

The selected measurement results of valve flow rates of 1, 0.8 and 0.6 [dm^3/min] for the overflow valve opening pressure of 10 [MPa]. Figure 34.2 shows the results for an external forcing frequency of 210 Hz and a flow rate of 1 dm^3/min, Fig. 34.3 shows results for a frequency of 210 Hz and flow rates of 0.8 dm^3/min.

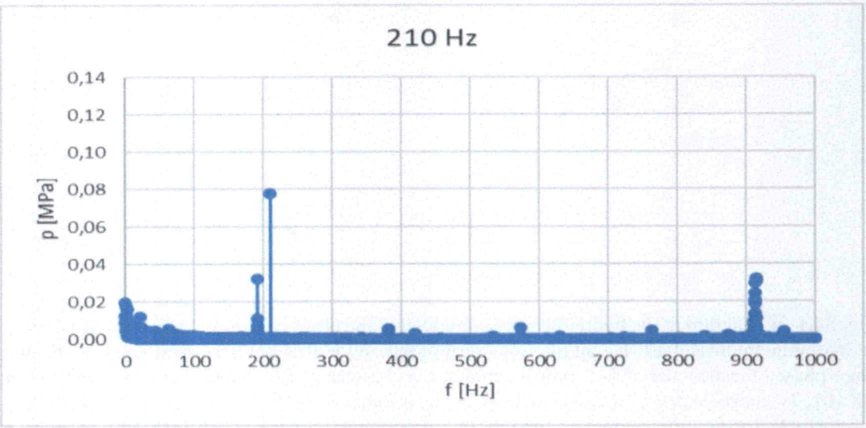

Fig. 34.2 Amplitude-and-frequency spectrum for a flow rate of 1 [dm^3/min], with a mean pressure of 10 [MPa] and external forcing frequency of 210 [Hz]

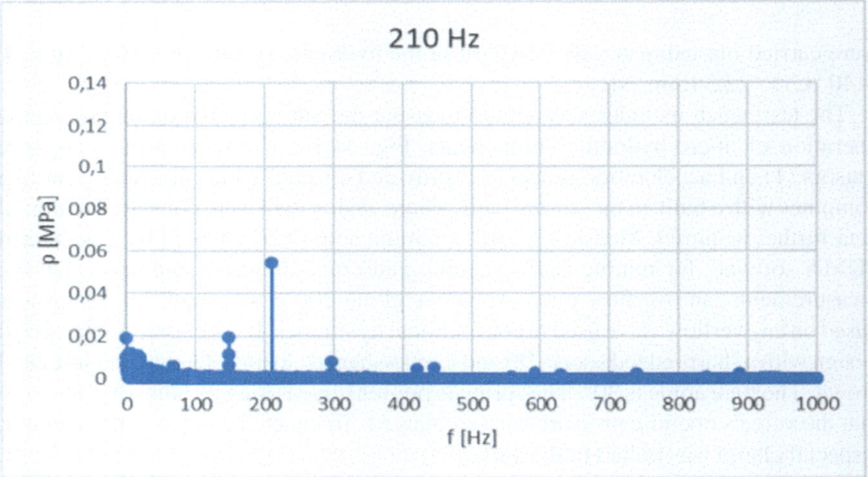

Fig. 34.3 Amplitude-and-frequency spectrum for a flow rate of 0.8 [dm^3/min], with a mean pressure of 10 [MPa] and external forcing frequency of 210 [Hz]

34.4 Using the Sequence Method to Include the Evaluation Measures of Decision Rules

The decision rules can be induced by two approaches, predicted classification of new objects and their description. It's important that classification-oriented induction uses training examples to build a set of decision rules to be used to classify new objects.

The description-oriented induction is aimed to analyze data to find sets of decision rules featuring regularities (sometimes anomalies or exceptions), that are interesting, useful and understandable to different types of users [9, 10]. The descriptive patterns are therefore oriented toward the interpretation of knowledge extracted from data [11].

Traditional machine learning algorithms are developed with a classification approach. The second approach to rule induction helps gain knowledge while increasing the demand for patterns to be searched for. However, in the context of the exploration of databases for knowledge, description becomes more important than prediction. Associative rules represent significant interdependencies between attribute values, but the concept of decision classes is not considered, and links apply only to conditional attributes.

34.4.1 Problems in Applying Decision Rules and Machine Learning Theory in the New Object Classification Approach

For the analyzed algorithm, experiments are recommended to compare it with other algorithms running them on test data. It is a standard to use data collected using multiple algorithms.

The papers [12, 13] present the methodology and examples of such comparative studies. The results of this and other studies indicate that there is no universal algorithm for all data. The determinants are the nature of the data and strategies used in the algorithm (so-called induction orientation). The paper [12], in describing the results of comparative experiments, states that "the performance (of the rules) was virtually indistinguishable from decision trees, but offered a better mental fit for a human". Similarly, [14] emphasizes that the rule-based methods, unlike the black-box techniques (e.g., discriminatory functions, neural networks) offer a clear representation of knowledge to facilitate understanding and explanation of decision problems (method groups differ in a property referred to as data fit vs. mental fit).

With respect to methods using the theory of rough sets, paper [15] indicates that it is possible to achieve the classification accuracy comparable to that of standard symbolic classifiers derived from machine learning.

Previous co-authored studies on the analysis of vibration effects on a micro-hydraulic valve have involved the use of inductive decision trees (with machine learning and neural networks) [16] and interval classification [17]. The problem of large data sets is often solved by reduction. Both the number of objects and attributes can be reduced. In the first case, a common solution is to randomly select a sample of appropriate size from the original table and train a classifier based on the sample [18, 19]. There are three ways to create a set:

- a single sample,
- incremental creation of the training set,
- randomly splitting data into subsets.

A well-defined problem of supervised training should include conditional attributes that are mutually independent. Research on dependent attribute detection and elimination is conducted in both machine learning [20], and rough set theory.

34.4.2 The Possible Use of Sequential Graphs

In papers [16, 17], there is the analysis for the interval of external forcing frequency (200–900 [Hz]) with a step of 10 [Hz], for:

- three values of the rate of flow through the valve: 0.6, 0.8, 1 [dm^3/min],
- value of the opening pressure of the overflow valve $p = 10$ [MPa],
- spring stiffness $c = 7.49$ [N/mm].

A total of 213 plots were generated (71 for each flow rate). For each plot, 4 most important intervals (frames) were determined, and ordered hierarchically: the most important (coefficient *IV*), important (coefficient *III*), less important (coefficient *II*), and least important (coefficient *I*) interval range for the entire range of the frequency spectrum [16].

The plots (Fig. 34.4) show examples of areas with specific intervals for flow rates: 0.8 (Fig. 34.4a) and 1 [dm^3/min] (Fig. 34.4b).

The paper [16] presents logical encodings of a fixed flow rate $Q = 0.8$ [dm^3/min] and frequency variation ranges < 200–900 > and arithmetic encodings for bar graphs from highest to lowest $\gg$ 4*, 3*, 2*, 1* $\ll$ specifying the range of importance ranks of intervals.

The graphs method can be a nonparametric approach to supervised training for classification and regression. The objective is to create a model that predicts the importance of the target variable value by generating decision rules inferred on the

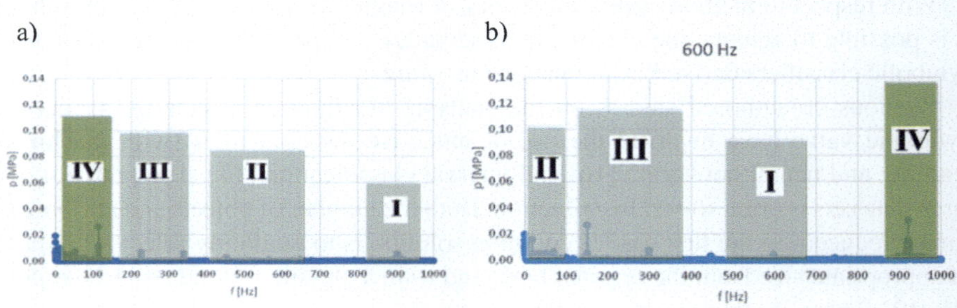

Fig. 34.4 The most important framework intervals for the amplitude-and-frequency spectrum for: **a** a flow rate 0.8 [dm^3/min] and external forcing frequency of 650 [Hz], **b** a flow rate 1 [dm^3/min] and external forcing frequency of 600 [Hz]

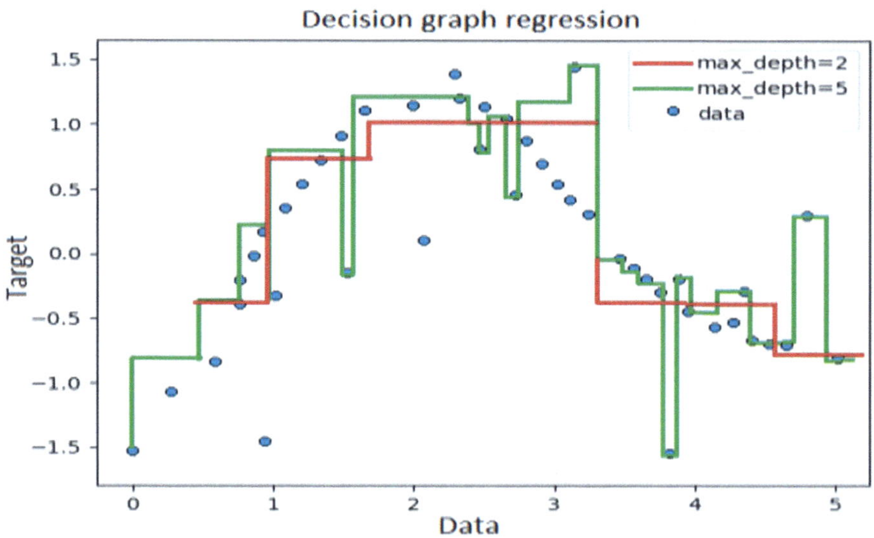

Fig. 34.5 Sample interval ranges with the possible sequence of computation

basis of dependency characteristics. Then, the tree from the graph can be seen as a continuous, fragmentary approximation (Fig. 34.5).

For example, in the example above (Fig. 34.5), the decision trees are trained by the data to approximate the plot of the amplitude-frequency spectrum using a set of "if–then-else" decision rules. The deeper the tree, the more complex the decision rules and the model fit.

34.5 Analysis with Sequence Graphs

A graph is called an ordered pair $G = (V, E)$, where V is a finite set of elements called vertices of the graph, and E is the set of pairs $\left(v_i, v_j\right)\left(v_i, v_j \in V\right)$ called the edges of the graph. To fully define a graph, one must also specify the relationship P that holds for the elements of the set of vertices $V(G)$ and edges $E(G)$. Then the graph can be called the three objects ordered in this way:

$$G = (V, E, P) \tag{34.1}$$

where:

- V the set of the vertices of the graph,
- E the set of the edges (branches of the graph),
- P a ternary relationship $P \subset V \times E \times V$ that meets the following conditions.

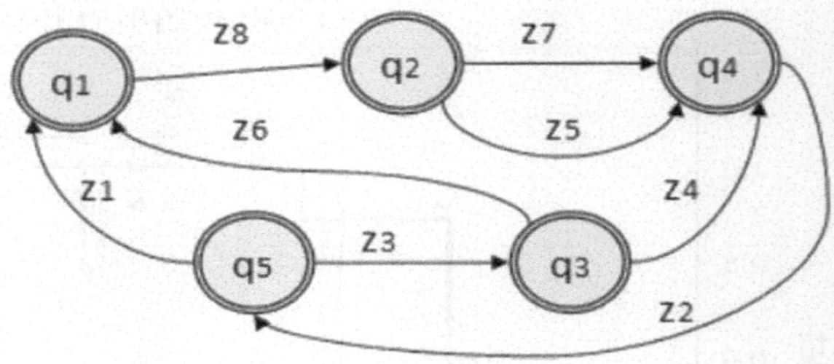

Fig. 34.6 Parametrized dependence graph

(a) For each edge *e*, there exists such a pair of vertices $x, y \in V$ that $< x, e, y >\in P$;

(b) If, for branch *e*, there exist $< x, e, y >\in P$ and $< w, e, z >\in P$, then $x = w$ and $y = z$ or $x = z$ and $y = z$.

For parametrized dependence graphs, there is a notation introduced by Deptuła [21, 22]: $G = (Q, Z)$, where Z is a set of pairs $\left(z_i, z_j\right)\left(z_i, z_j \in Z\right)$. Figure 34.6 shows an example of a parametrized dependence graph. In Fig. 34.6, the directed dependence graph consists of this set of vertices Q:

$$Q = \{q_1, q_2, q_3, q_4, q_5\} \tag{34.2}$$

and the set of edges Z:

$$Z = \{z_1, z_2, z_3, z_4, z_5, z_6, z_7, z_8\} \tag{34.3}$$

that is, an ordered pair of vertices.

A path in a graph $G = (Q, Z)$ is called a sequence of edges $\left(z_{i_1}, z_{i_2}\right), \left(z_{i_2}, z_{i_3}\right), \ldots, \left(z_{i_{k-1}}, z_{i_k}\right)$, where for each $j \in \{2, 3, \ldots, k\}\left(z_{i_{j-1}}, z_{i_j}\right) \in R$ and vertices $q_{i_1}, q_{i_2}, \ldots, q_{i_k}$ are not equal.

34.5.1 *The Possible Use of Graphs to Analyze Sequentially the Function Properties*

Let us assume that each cell of a certain interval contains combinational logic and that all cells are connected unilaterally, i.e., function propagation is analyzed in one direction.

Let graph $G = (E, U)$, $U \subseteq E \times E$ be an oriented graph and X *be* a set of intervals. Then, G is said to be an ordered pair graph (Fig. 34.7) if:

$$\begin{aligned} E &= \{e/e = a/b = (a, b) \in X \times X, a \neq b\} \\ U &= \left\{u/\forall z\left[u = \left(e, \hat{x}(e, z)\right)\right]\right\} \end{aligned} \tag{34.4}$$

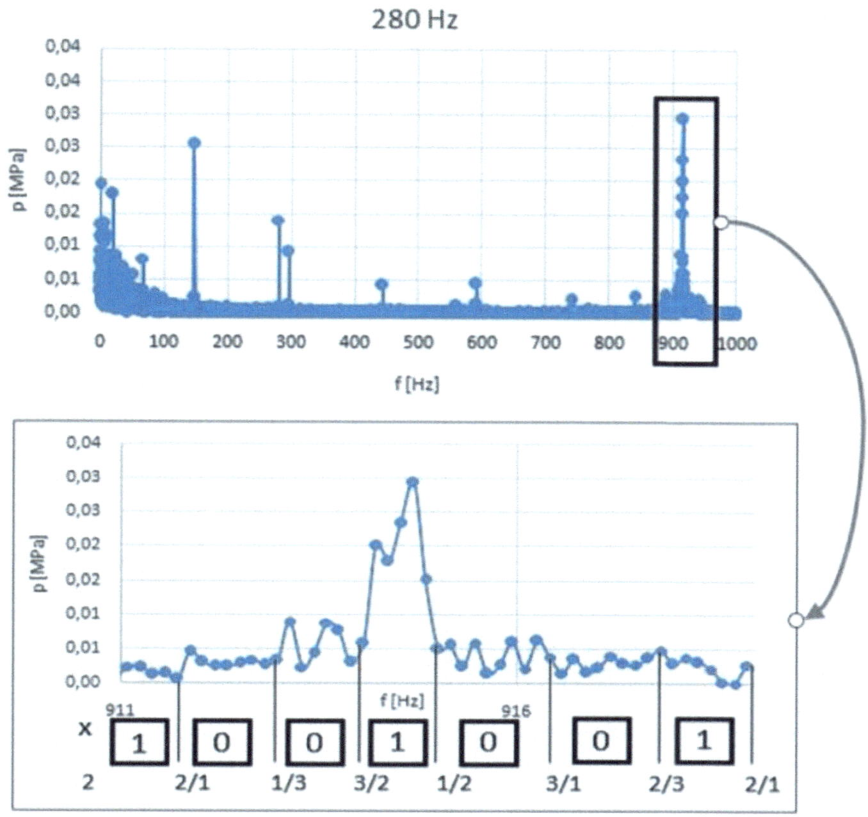

Fig. 34.7 An example of using a sequential description of a sample frame interval for a selected range

Figure 34.7 shows an example of applying the sequential description of a sample frame interval for an amplitude-and-frequency spectrum at a flow rate of 0.6 [dm^3/min] and an external forcing frequency of 280 [Hz]. Figure 34.8 shows the analysis of a sequential system to create a graph and a transition table.

There is a finite comparing sequence such that it is assigned to a certain elementary contour (2/1, …, 2/*y*, 2/1) in the graph, where $u = (2/y, 2/1)$ is a chord with a weight (in this case $z = 1$, but other weights will be used). In our case, an elementary path 00,100 will be obtained by shifting the sequence progressively by one interval to the left, and 6 tests will be obtained. For each *i*, *j* that $i \neq j$, the input from cell *i* is independent of the input from cell *j*. The behavior of any interval of the function can be described using an array (e.g., by logical encoding). The input *x*'s are called states and each input is assigned a column. The paper only preliminarily describes an example frame interval for the amplitude-frequency spectrum at a given flow rate and a given external forcing. Using induced decision trees, a comprehensive

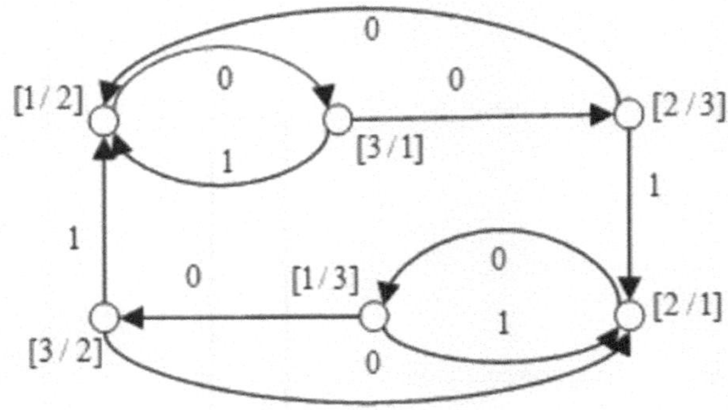

	0	1
1	3	2
2	1	2
3	2	1

Fig. 34.8 The analysis of a sequential system through a graph and a transition table

analysis should be performed for all the most important intervals selected in earlier co-authored papers.

34.6 Summary and Conclusions

The paper provides a rough description of the application of sequential graph method for the analysis of vibration on a micro-hydraulic valve. This type of analysis should receive special attention because of the demands placed on the components and systems and since the disturbance forces are sometimes of the order of the control forces. In experimental testing, the micro-hydraulic valve was subjected to external mechanical vibration with a specified amplitude and frequency.

Considerations show that procedures for ranking the parameter importance can also be useful in applying graphs. Systems that contain a set of interval chains with the same basic cell (an interval), but with a different rank of importance are called iterative systems. They are an important tool in diagnostic analysis of systems based on functions. Another approach to reducing large decision tables may be to use sequential graph theory, where the initial decision table is decomposed into a set of "sub-tables" of acceptable size. The decomposition employs the (directed) measurement relationships to create vectors of pairs (attribute, value) confirmed by a sufficiently large number of examples. The graph-based inference methods can be used to generate decision rules from tables.

This will eventually allow multi-valued decision trees to be used for analysis in the intervals concerned. The properties of logical decision trees have combinatorial significance for systematically reviewing the sub-intervals of variation of a fixed set of independent variables, and subsequently determining the importance rank of those variables.

References

1. Stosiak M (2015) Identification of vibration impact and methods of their reduction in selected hydraulic valves. Publishing house PWR, Wrocław
2. Engel Z, Zawieska M (2010) Noise and vibrations in work processes: sources, assessment, threats. CIOP-PIB, Warszawa
3. Grajnert J (1997) Vibration isolation in machines and vehicles. Publishing house PWR, Wrocław
4. Kollek W (2011) Basics of design, modelling operation of microhydraulic components and systems. Publishing house PWR, Wrocław
5. Rundo M (2017) Theoretical flow rate in crescent pumps. Simul Model Pract Theory 71:1–14
6. Lambeck RP (1983) Hydraulic pumps and motors. Marcel Dekker INC, New York
7. Kudźma Z, Kułakowski K, Stosiak M (2014) Selected problems in the operation of microhydraulic systems. Drives Control 16(4):78–83
8. Kudźma Z, Stosiak M (2016) Micro-max valve. Patent PL221214B1
9. Fayyad UM, Piatetsky-Shapiro G, Smyth P, Uthurusamy R (1996) Advances in knowledge discovery and data mining. AAAI/MIT Press, Cambridge Mass
10. Joshi BM, Chanddorkar MC (2014) Two-motor single-inverter field-oriented induction machine drive dynamic performance. Sadhana 39:391–407
11. Mannila H (1996) Data mining: machine learning, statistics and databases. In: Proceedings of the eight international conference on scientific and statistical database management June 18–20, Stockholm, pp 1–8
12. Mitche D, Spiegielhalter CJ, Taylor C (red) (1994) Machine learning, neural and statistical classification. Elis Horwood
13. Kourou K, Exarchos TP, Exarchos KP, Karamouzis MV, Fotiadis DI (2015) Machine learning applications in cancer prognosis and prediction. Comput Struct Biotechnol J 13:8–17
14. Błaszczyński J, Słowiński R, Szeląg M (2011) Sequential covering rule induction algorithm for variable consistency rough set approaches. Inf Sci 181(5):987–1002
15. D'eer L, Cornelis C, Yao Y (2016) A semantically sound approach to Pawlak rough sets and covering-based rough sets. Int J Approximate Reasoning 78:62–72
16. Stosiak M, Towarnicki K, Partyka MA, Deptuła A (2020) Analysis of the impact of vibrations on the microhydraulic valve, including neural networks and induction trees. Design and operation of working machines PO, Opole
17. Stosiak M, Towarnicki K, Partyka MA, Deptuła A (2020) Analysis of the impact of vibrations on the microhydraulic pressure relief valve taking into account the interval classification of induction trees. In: Stryczek J, Warzyńska U (eds) Advances in hydraulic and pneumatic drives and control 2020. NSHP 2020. Lecture notes in mechanical engineering. Springer, Cham
18. Cios KJ, Pedrycz W, Świniarski RW (1999) Data mining methods for knowledge discovery. Dordrecht, Kluwer
19. Bellazzi R, Ferrazzi F, Sacchi L (2011) Predictive data mining in clinical medicine: a focus on selected methods and applications. WIREs Data min Knowles Discover 1:416–430
20. Kononenko I (1994) Estimating attributes: analysis and extensions of RELIEF. In: Proceedings European conference on machine learning, pp 171–182
21. Deptuła A (2014) Application of multi-valued weighting logical functions in the analysis of a degree of importance of construction parameters on the example of hydraulic valves. Int J Appl Mech Eng 19(3):539–548
22. Deptuła A, Partyka MA (2018) Application of complex game-tree structures for the Hsu graph in the analysis of automatic transmission gearboxes. J Mach Eng 18(4):96–113

Index

G. Lesiuk et al. (eds.), *Structural Integrity and Fatigue Failure Analysis*, Structural Integrity 25, https://doi.org/10.1007/978-3-030-91847-7

The manufacturer's authorised representative in the EU is Springer Nature Customer Service Centre GmbH, Europaplatz 3, 69115 Heidelberg, Germany. If you have any concerns regarding our products, please contact ProductSafety@springernature.com

Printed and bound by CPI Group (UK) Ltd, Croydon, CR0 4YY
30/09/2026
02225498-0016